Impasse

IMPASSE

Climate Change and the Limits of Progress

ROY SCRANTON

STANFORD UNIVERSITY PRESS
Stanford, California

Stanford University Press
Stanford, California

©2025 by Roy Scranton. All rights reserved.

No part of this book may be reproduced or transmitted in any form or by any means, electronic or mechanical, including photocopying and recording, or in any information storage or retrieval system, without the prior written permission of Stanford University Press.

Library of Congress Cataloging-in-Publication Data
Names: Scranton, Roy, author.
Title: Impasse : climate change and the limits of progress / Roy Scranton.
Description: Stanford, California : Stanford University Press, [2025] | Includes bibliographical references and index.
Identifiers: LCCN 2024043480 (print) | LCCN 2024043481 (ebook) | ISBN 9781503640030 (cloth) | ISBN 9781503643161 (ebook)
Subjects: LCSH: Climatic changes—Social aspects. | Climate change adaptation—Social aspects. | Progress.
Classification: LCC QC903 .S524 2025 (print) | LCC QC903 (ebook) | DDC 363.7—dc23/eng/20241120
LC record available at https://lccn.loc.gov/2024043480
LC ebook record available at https://lccn.loc.gov/2024043481

Cover design: Derek Thornton / Notch Design
Cover art: Manabu Ikeda, *Foretoken* (detail), 2008, pen, acrylic ink on paper, mounted on board, 190×340cm, Collection of Sustainable Investor, Co., Ltd., Tokyo, Photography by Shu Nakagwa, © IKEDA Manabu, Courtesy of Mizuma Art Gallery

The authorized representative in the EU for product safety and compliance is: Mare Nostrum Group B.V. | Mauritskade 21D | 1091 GC Amsterdam | The Netherlands | Email address: gpsr@mare-nostrum.co.uk | KVK chamber of commerce number: 96249943

For Rosalind

> Where then are the gods you made for yourselves?
> Let them come if they can save you . . .
> —JEREMIAH 2:28

CONTENTS

PREFACE	Beginning with the End	xi
	Introduction	1

Part One: The Broken Thread

ONE	The Limits of Progress	13
TWO	The Failure of Climate Politics	48
THREE	The Age of Acceleration	87
FOUR	The End of the World	111

Part Two: The Leap

FIVE	Get Happy	141
SIX	A Melancholy Hue	160
SEVEN	OK, Doomer	189

Afterword: The Children of Ruin	213

Acknowledgments 221

Notes 225

Selected References 311

Index 339

Preface
BEGINNING WITH THE END

Our Subaru bounced and rumbled over the rugged dirt road to Chaco Canyon. My partner, my daughter, and I were on our way from upstate New York to Los Angeles, where Sara had been offered a postdoctoral fellowship at USC and I planned to spend the year on sabbatical. Rosalind, our toddler, had just turned two. A couple of nights in the New Mexico desert offered one last adventure before a season of rest and research.

We set up camp then drove through the park, struck by the austere beauty of the thousand-year-old pueblos, luminous in the late afternoon sun and incongruous as alien wreckage. It hardly seemed plausible that a whole civilization had once flourished in this arid and unforgiving place. As we walked around Kin Kletso, one of the smaller Great Houses, my sense of wonder grew into an objectless unease that seemed as if to resonate from the land itself. Chaco's burst of growth and rapid collapse, blighted with dark hints of cannibalism, presented an alluring and disquieting enigma.[1] There was something ominous in the Navajo (Diné) story of how the Chacoan Great Houses had been built by men enslaved to the god Noqoìlpi, "the Gambler," and how the Navajo called them Anasazi, meaning "enemy ancestors."[2] I assumed my unease was a projection, mere exoticism maybe, or vestigial stress shaken loose in transit, and I focused on building a fire, making dinner, and putting our daughter to bed.

Day passed into night. Our sun set and other suns brightened against the gathering dusk, at first only a dim few keeping Venus company, then a bright scatter, then as the canyon spun into night an infinite spill of nuclear fire burning across the icy void of space. Only the absolute dark could reveal so much light, and I thought back to other skies in other places where I'd been granted similar visions, understanding once again why the Romans called our galaxy the Via Lactea and the Greeks called it γαλακτικός κύκλος, the "milky circle." As was often the case in such circumstances, I couldn't help but think of Kant's famous lines from the *Critique of Practical Reason*, carved on his tombstone in German and Russian: "Two things fill the mind with ever new and increasing admiration and reverence, the more often and more steadily one reflects on them: the starry heavens above me and the moral law within me."[3]

For Kant, our ability to apprehend and recognize the sublimity of the heavens and the rightness of moral truth was evidence of our free reason, that spark lit within us by God connecting merely mortal humanity to the divine. A more modern, materialist view might see awe as a biological phenomenon evolved to promote social cohesion, as argued, for example, by psychologist Dacher Keltner. "Why did awe become part of our species' emotional repertoire during seven million years of hominid evolution?" Keltner asks. "A preliminary answer is that awe binds us to social collectives and enables us to act in more collaborative ways that enable strong groups, thus improving our odds for survival."[4] Both perspectives express a similarly benign view, and both link awe to human social networks, even if in one case the human is a sacred participant in the rational order that makes the universe holy, and in the other we're basically widgets. Most important, both perspectives take our ability to make sense of our *inability* to make sense of reality as paradoxically confirming the power of human reason. Another view, less optimistic, might interpret the awe we feel in apprehending the sublime as an expression of humility at our cognitive limits, as we teeter between ignorance and illusion, poised between the vastness of an infinite, incomprehensible cosmos and the ephemeral human constructions we throw up against it, like the Great Houses of Chaco Canyon.

The next morning we took a guided tour of Chetro Ketl, one of the several houses making up what archaeologists call "Downtown Chaco." It's an imposing structure, something like a prehistoric shopping mall, with a straight, 450-foot-long back wall running against the cliffs and another outer wall curving south toward the valley in a D shape, altogether enclosing more than three acres. Everything is a ruddy ocher. Pine beams cut from mountain forests fifty miles away and carried across the desert on foot support the building's three floors, which likely rose to a fourth, now collapsed. It's estimated that thousands of logs, "tons upon tons" of mortar, and more than fifty million pieces of stone were used in building this one Great House—"a prodigious task for the rather small population of Chetro Ketl," observes Edgar Hewett, who mapped and excavated the site early in the twentieth century.

> This, it must be remembered, was repeated proportionately on each of the large communities of the Chaco Canyon, and an unknown number of small villages. It was no unwilling work under the lash of priestly or kingly taskmasters; the American Indians were never so ruled. It was the spontaneous impulse of a virile people, comparable to the heaping up of great mounds far in excess of actual needs, by insect communities. Other examples might be pointed out of the excessive activities of the human species as the building of the earth mounds of the Mississippi Valley, the Egyptian pyramids, the Great Wall of China and the European cathedrals of the Middle Ages.[5]

Comparing Chetro Ketl to other monuments seems apt, though scholars disagree about whether or not the Puebloans who built Chaco worked "under the lash." How hierarchical Chacoan culture may have been is a point of dispute, and although evidence points to significant inequality, hoarding, and ritual cannibalistic violence, the evidence is equivocal enough to support multiple interpretations. Archaeologist Stephen Lekson calls the Great Houses "palaces," Chaco a "city," and imagines a militaristic ruling class led by a "king."[6] Others are more circumspect.[7] One thing most agree on, however, is that the remarkable architectural efflorescence at Chaco represents something more than just a population boom.[8]

The picture of Chaco that has emerged from the painstaking work of archaeologists and anthropologists is not that of a bustling desert metropolis like modern-day Phoenix, but rather "the administrative center of a . . . system of hierarchical social and economic integration" managing ritualized seasonal trade among a host of communities across the

26,000-square-mile San Juan Basin.⁹ This trade empire, sometimes called "Chaco World," is revealed not only through the building materials at Chaco and trade goods found there from as far away as the Pacific Northwest and Mexico (such as cacao and macaws) but also through regionally shared architecture, "shared decorative designs on pottery, wood, stone and textiles," and a far-reaching network of roads and signal towers.¹⁰ The mark of Chaco's wide reach is visible on Chetro Ketl's very face: the main structure is fronted by a colonnade, a stylistic flourish "virtually nonexistent in the Puebloan Southwest," likely inspired by Toltec architecture in Central Mexico.¹¹ The ruins of Chetro Ketl are a testament to Chaco's grandeur, a monument to human pride, and—like all ruins—a warning.

I've always been drawn to ruins, perhaps in part because they didn't really exist where I grew up. The Chinook, Tillamook, Salish, and Kalapuya peoples indigenous to the Pacific Northwest had wintered in dirt-banked, semi-excavated cedar houses, long since obliterated, and summered in even more temporary abodes, and the white settlers who took the Oregon Trail out west had been building there only about two hundred years. Yet an early Cub Scout trip to Camp Rilea, a National Guard training camp built on the Oregon coast in 1927, gave me the chance not only to see an evocative wooden replica of Fort Clatsop, Lewis and Clark's outpost overlooking the mouth of the Columbia River, but also to run wild amidst the "maze of tunnels, cave-like chambers, and thick concrete walls" of Fort Stevens, a massive system of artillery emplacements built for coastal defense during the Spanish-American War.¹²

There was something unreal in the cyclopean scale of the abandoned fortifications: the vast concrete structures blistering the hills overlooking the ocean seemed to belong to Atlantean gods. And when I later stumbled across philosopher Paul Virilio's obsessive photographs of German military forts on the French coast, I felt a shock of déjà vu: "The geometry is no longer affirmative, but eroded, worn. The angle is no longer right, but depressed, resisting apprehension; the mass is no longer anchored in the ground, but centered on itself, independent. . . . This architecture floats on the surface of an earth which has lost its materiality."¹³ Virilio's description of abandoned Nazi fortifications evoked a disturbing metaphysical disjunction that I recognized from the empty, graffitied tunnels of Fort Stevens and felt wherever I saw human hubris make its mark upon the Earth: in Saddam

Hussein's Ba'athist monuments rising up over Baghdad and the ancient Ziggurat of Dur-Kurigalzu, in Berlin's Kaiser-Wilhelm-Gedächtniskirche and the Soviet War Memorial at Treptower Park, and yet again in Chetro Ketl.

"All men take a secret delight in beholding ruins," writes Chateaubriand. "This sentiment arises from the frailty of our nature, and a secret conformity between these destroyed monuments and the caducity of our own existence."[14] In her survey of the role ruins play in Western culture, *The Ruins Lesson*, poet and literary critic Susan Stewart writes:

> From their earliest recorded notice, ruined structures are evocative not only of ghosts but also of our relations to living persons: to our distinctions between appearance and interiority, to the visibility of our surfaces and invisibility of our organs, to losing and saving face in relation to reputation, to imagining the breakdown of social relations. . . . Ruins, like trees, live beyond us and raise the haunting thought that what we have made may outlive us, even as a species.[15]

Ruins are both monuments and memento mori, emblems of our arrogance and warnings against our pride, concrete remains of our human glory and decaying reminders of our human monstrosity, which in the end are one and the same.

After Chetro Ketl, lunch, and a nap for Rosalind, we drove out to the largest and best-known Great House, a breathtaking structure of almost seven hundred rooms and more than thirty kivas called Pueblo Bonito. Bonito was the center of the Chacoan World, though it likely housed only a few people, perhaps a hundred, very probably the elites of the Chacoan ruling class, or maybe priests.[16] It was, in the words of archaeologist Jill Neitzel, "the most powerful place in its world . . . the center of the center—the most important structure in the most important settlement in the most powerful society ever to develop in the prehispanic northern Southwest."[17]

Meanwhile, dark clouds had gathered at the horizon. We realized we'd left the rainfly off our tent, leaving our gear exposed, so Sara offered to go back and take care of it while Rosalind and I started along the trail that circled around the back of the pueblo. We went up around a tumble of boulders, losing sight of the pueblo's walls as clouds enveloped the sun. The day darkened like somebody had flipped a switch. I wondered if we'd made a mistake.

The trail curved west then south, ascending to a vista overlooking the pueblo and the valley beyond. The sky was the color of lead. A purple curtain of rain swept across the land. I read the interpretative sign to Rosalind:

> Imagine how it might have been if you had been standing here then, amidst those bustling crowds, within the towering, gleaming walls of Pueblo Bonito. Imagine all the sights, sounds, and smells you might have enjoyed—the brilliantly colored ornaments and clothing, the colorful macaw feathers, shells, and turquoise—the pounding drum beats, flutes, conch-shell trumpets, and hypnotic chanting of singers echoing off the canyon walls—and the fragrances of foods and fires wafting near and far.[18]

I imagined the Puebloan elites in their finery, the commoners kneeling below, all of them looking up with alarm and expectation at the darkening sky, then urged Rosalind down the steps and along the trail, which curved north again, then again west, offering tantalizing peeks at the pueblo before descending along the structure's high, artfully masoned back wall. We came to the rear entrance just as the skies opened, rain falling not in drops but in a wall of water. Suddenly drenched, I picked Rosalind up and ran to the only shelter I could see, a narrow alcove under an ancient awning, where we stood clutching each other and watching the water pour down.

Pueblo Bonito, like the other Great Houses at Chaco, was built to take ingenious advantage of local climate and landscape features such as canyon breezes, ambient heat coming off the cliffs, and runoff from summer rain. Sophisticated irrigation systems including dams, canals, and gridded fields made the most of sparse precipitation. Chaco was like a great intricate watch, a complex system exploiting and modifying the ecological patterns of an entire region to sustain human flourishing in an otherwise marginal zone, connecting thousands of people across hundreds of miles in networks of food production, commerce, narrative, symbolic organization, and cosmic correspondences.[19]

Why did Chaco develop when it did, and what happened? It may have had something to do with rain. The region averages less than nine inches of precipitation a year, most of it falling in sudden summer bursts like that one that had caught Rosalind and me, but the canyon's geography acts as a natural basin: "A series of side canyons, or *rincones*, channel runoff to the main canyon floor," which lies between the ecologically rich Chacra Mesa to the east and Escavada Wash to the west, where the water table is near the surface.[20] Chaco thus serves as a natural oasis and shows evidence of human habitation going back ten thousand years.[21] Significant settlements had been established there by 500 CE, recognizably Puebloan structures and settled agriculture appeared around 700 CE, and the first Great Houses were built between 850 and 950 CE.

Over the four centuries between 800 CE and 1200 CE, Chaco emerged as the trading and religious center of the San Juan Basin, a cosmopolitan "melting pot" connecting peoples of diverse identities and origins, with most of the major construction in the valley happening between 1000 and 1150 CE.[22] Research from tree-ring analysis shows that this coincided with a period of generally above-average rainfall; Chaco's decline coincided with an extended, decades-long drought.[23] While archaeologists tend to shy away from strict environmental determinism, the general consensus on Chaco's mysterious rise and fall is that it was largely driven by climate change: a slight increase in precipitation providing the conditions for sudden, unprecedented growth, which a slight decrease in rain then parched.

Yet as anthropologist Joseph Tainter points out in his book *The Collapse of Complex Societies*, "The Chacoans had . . . previously survived droughts in the mid tenth, the early eleventh, and the late eleventh centuries without collapsing."[24] Tainter, a leading expert on societal collapse, is generally critical of explanations that rely on environmental stressors, for the simple reason that dealing with such stress is just what complex societies do. "Complex societies are problem-solving organizations," he argues, which evolve and adapt in response to environmental conditions.[25] Thus explaining the collapse of any particular society requires looking not at external causes but at the society itself.[26]

How is it that a complex society like Chaco could survive one drought but not another? Why do some societies collapse and others not? Are there consistent reasons why different societies fall apart?[27] Numerous archaeologists, anthropologists, and historians have tried to answer such questions, and Tainter's answer is among the best, as elegant as it is illuminating. In short, he argues, the problem is complexity itself. As the energy costs of social complexity rise, Tainter theorizes, the returns on that complexity decline, to the point where simplification becomes more economical than additional complexity. When societies reach such an impasse, they face three possibilities: the discovery of new energy sources; slow, managed simplification; and sudden, unmanaged simplification, or what Tainter calls "collapse." The idea of collapse is often conflated with disappearance, eradication, or apocalyptic destruction, but this is not what most archaeologists who work on the subject mean by the word. "Collapse does not imply a permanent end," writes archaeologist Donna Glowacki. "It means there was a breakdown or a critical loss of the effectiveness of some aspect of society and *things can no longer be the way they were*."[28] As Ronald Faulseit puts it, "Collapse is probably best understood as the fragmentation or disarticulation of a particular political apparatus."[29]

Thus it seems best to understand collapse primarily as a sociopolitical phenomenon, not a cultural one, not least because while particular political formations can be identified as discrete entities with beginnings and ends, cultures are looser, more porous and rhizomatic.[30] Making this distinction helps clear up confusions like the claim that such-and-such a society didn't collapse because its people or culture survived, often invoked with regard to indigenous peoples of the Americas. Such conflations emphasize narratives of ethnocultural continuity over historically significant disjunctions. The problem here can be seen by looking at Rome: Latin peoples live in their ancestral home to this day, elements of ancient Roman culture continue to survive and flourish, and aspects of twentieth-century Italian political culture leaned heavily on claims of long-term continuity with the Caesars, but such observations hardly help us understand how, why, or whether the Western Roman Empire collapsed in the fifth century CE. The same point applies to the Classic Maya or the Ancestral Puebloans.

Essential to Tainter's theory of collapse is understanding human societies as living systems that, like all forms of life, survive by managing energy extracted from their environment. The ways we organize energy flows are just what give our society its shape and character: from human labor powered by caloric intake to complex global phenomena like the petroleum industry, the material affordances of energy production and distribution structure human social life.[31] "Energy flow and sociopolitical organizations are opposite sides of an equation," writes Tainter. "Neither can exist, in a human group, without the other, nor can either undergo substantial change without altering both the opposite member and the balance of the equation. Energy flow and sociopolitical organization must evolve in harmony."[32]

In Tainter's model of social organization, complexity increases as solutions are developed in response to specific problems: specialists and resources are allocated to address new challenges, leading to greater division of labor, increased central management, increased hierarchy, more robust dependency networks, more resources directed to support problem-solving organizations, and greater energy costs for society as a whole. "As society evolves toward greater complexity," he writes, "the support costs levied on each individual will also rise, so that the population as a whole must allocate increasing portions of its energy budget to maintaining organizational institutions."[33] As a society develops new structures and specialized roles to address specific problems, it has to pay not only the one-time cost of development but continued and often increasing maintenance costs, as well as additional costs of integrating these new elements with existing ones, plus unforeseen costs that emerge from inefficiencies in the system, unexpected challenges, and political conflict. As time goes on, costs tend to rise. "At some point in the evolution of a society," writes Tainter, "continued investment in complexity as a problem-solving strategy yields a declining marginal return."[34] By the time a society reaches this point, it's spending so much energy maintaining itself that it lacks the capacity to address the inevitable stressors in response to which societies develop complexity in the first place, such as drought, disease, war, internal conflict, and climate change.

"What is interesting in this view," writes archaeologist Guy Middleton in *Understanding Collapse*, "is that collapse is itself an adaptation."[35] Societies coping with declining marginal returns on complexity may "choose" to collapse and may begin to dis-integrate through political disaffiliation well before disaster strikes, with people detaching from centralized bureaucracies, disengaging from public fora, and disidentifying with unifying narratives and ideology.[36] As is the case with any complex system, positive feedbacks accelerate the transition to a new equilibrium state: as a complex society struggling to maintain itself and respond to external stressors fragments internally, it will face increasing demands on its ability to manage resources with decreasing effectiveness, thus aggravating strain across the whole of society and speeding simplification.

In the case of Chaco, argues Tainter, "it is possible that the collapse of Chacoan society was partially a consequence of its own success."[37] The complex regional civilization that emerged in the San Juan Basin more than a thousand years ago was capable of integrating outlying communities, distributing resources, and coping with fluctuations in climate and productivity. "If our present understanding of the Chacoan system is correct," Tainter writes, "it would seem that the population of the San Juan Basin obtained a valuable return on its investment in complexity by lowering the administrative costs, and increasing the effectiveness, of an energy averaging system. Beyond its initial establishment, however, further expansion of this system may not have been so advantageous."[38]

Indeed, the costs of maintaining this far-flung system were demonstrably high: the massive structures at Chetro Ketl and Pueblo Bonito, the great kivas, the intricate irrigation, the elaborate road and signal tower networks, ritual violence, political violence, and the symbolic structures aligning the ruling elite at Chaco with the passage of the sun and moon and stars all sustained a system whose energy costs began to exceed what it could extract from its environment. "By increasing subsistence security and reducing the natural checks on population," Tainter writes, "the Chacoan system allowed regional population to rise to a level that would otherwise have never been attained."[39] Declining returns on complexity, population stress, costly labor investment in increasingly marginal agricultural land, and internal political conflict made the Chaco Empire vulnerable; a fifty-year drought in the twelfth century was the coup de grâce. "Given the trends in marginal productivity that the Chacoans faced," Tainter writes, "this complex society would have ultimately collapsed with or without the final drought."[40] From this systemic and ecological perspective, the picturesque ruins in which Rosalind and I huddled out of the rain weren't lugubrious reminders of the caducity of our existence, but evidence of social complexity responding to environmental stress—and of the rapid simplification that followed.

We shivered in the lee of a centuries-old wall, soaked to the skin, wondering how long the storm would last. Then I saw something so strange I couldn't make sense of it: a faceless purple figure gliding erratically toward us, weaving in and out of the broken walls of the pueblo. At first I thought it must be a kachina, come to exact revenge for some trespass, and it took me a long, vertiginous moment to realize that the figure was in fact Sara, her jacket pulled over her head. I picked Rosalind up and ran to Sara through the pueblo, and we all raced back together to the car, where we blasted the heater until the rain passed.

Later, after dinner, while Sara put Rosalind to bed, the gloomy unease I'd felt the first night washed over me again. To the east, the dimming sky was clear, but to the west, the setting sun fell burning behind a thick, coal-colored, evil-looking mass of clouds, and I was gripped by an irrational yet inescapable fear that something terrible was going to happen. I couldn't say what or why or how, but the intuition felt as real as anything else I might have claimed to know about the world.

I talked about it with Sara, trying to convince myself that it was just a weird day, weird weather, or free-floating anxiety, but I couldn't shake the nagging sense of doom. The feeling stuck with me all the way to Los Angeles, where it faded, as such feelings do, behind the traffic, phone calls, dirty dishes, and laundry that make up the rhythms of day-to-day life. We had a bungalow on a hill in Highland Park that looked out over the palms toward Mt. Washington. We drove our daughter to preschool in Silver Lake. We saw friends in Santa Monica, Koreatown, Los Feliz, and Eagle Rock. We went to the Los Angeles Zoo and the Long Beach Aquarium and concerts and museums. Sara worked on her book, while I read Hans Jonas, Carolyn Merchant, and Murray Bookchin.

The doom of Chaco came back to me in February, as reports of a novel coronavirus from Wuhan, China, began to dominate the news. The story kept changing: The virus was under control, but new cases were appearing in Washington, Wisconsin, California, and Illinois. The head of the CDC, Robert Redfield, said "It's important to note that this virus is not spreading within American communities at this time," and that same day, in an editorial for the *New England Journal of Medicine*, Redfield, H. Clifford Lane, and Anthony Fauci compared the virus to the flu, but the next day saw the first reported American death.[41] National health officials discouraged people from buying or wearing masks, and on March 2, the surgeon general tweeted, "Seriously people, STOP BUYING MASKS! They are NOT effective in preventing general public from catching #Coronavirus."[42] But by March 6, eighteen people had died; the governors of Washington and California had declared statewide emergencies; and President Trump had signed a bill committing $8.3 billion to fight the virus. A few days later, Trump said the virus was "very much under control," while CNN declared the outbreak a pandemic. Los Angeles locked down bars, theaters, and gyms, then restaurants and schools. Governor Newsom issued a statewide quarantine. President Trump declared the state a disaster zone.

A year later, at least 2.6 million people around the world had died from COVID-19, more than 529,000 of them in the US—the most reported deaths of any single country—and those numbers almost certainly underrepresent actual deaths from the disease and related complications.[43] The entire global human community, a Rube Goldberg machine held together with fiber-optic cable, fossil fuels, shipping containers, and narrative, had been stressed beyond the capacities of its leaders and institutions to cope. All our wisdom, all our tech, all our science and knowledge and power barely kept the virus from upending the whole contraption. The damage COVID did may never be healed. The weaknesses it revealed in the system cannot be unseen.

I used to think that even if it was too late to stop planetary and societal transformation because of climate change, we could at least hope to manage the transition intentionally. After witnessing COVID and conducting the research for this book, I'm much more pessimistic. The problem we face with climate change isn't just unsustainable carbon waste from excessive energy use, and it isn't just the planetary metabolic disequilibrium our excessive energy use has caused.[44] The deeper problem is complexity itself: our global industrial civilization has grown unsustainably expensive as well as vulnerable to environmental stress, which climate change and related ecological crises will provide in spades. And although numerous smart people have labored earnestly to imagine systemic transformation, international agreements, carbon taxes, carbon scrubbers, energy transition, sea walls, cold fusion, communist revolution, degrowth, donut economics, geoengineering, and countless other technology and policy hacks, what we face is not just a "wicked problem," and not even what Kelly Levin et al. call a "super wicked problem," but an intractable predicament—a tragic failure—an "interminable catastrophe."[45] Every fix faces insurmountable challenges, beginning with inertia and vested opposition, quickly compounded by resource limits, energy costs, political dysfunction, and competing priorities, while the long-term nature of key problems requires commitments from future actors impossible to compel. Even worse, if Tainter's theory of the diminishing returns on complexity is right, then the increasing complexity required to address our situation will only accelerate rapid, large-scale simplification.

Meanwhile, the narratives and conceptual frameworks we use to make sense of reality grow increasingly incoherent, having proved manifestly incapable of dealing with the challenges we face. The world as we know it seems to be coming to an end. We have no idea what the future holds beyond a daunting extrapolation from our current trajectory, nor do we have any idea what kind of world our children or grandchildren may inherit. Yet despite our ignorance, impotence, and vulnerability, we are nevertheless obliged to get up in the morning and go to work. We have to explain things to our children. We have to explain things to ourselves. The narrative thread connecting past and future has been broken, but somehow we must get on with our lives. This book is an attempt to think through this impasse—and make a leap into the future.

Impasse

INTRODUCTION

According to best estimates, global atmospheric warming will soon cross the 2°C redline scientists have warned may be the tipping point into abrupt transformation and will likely hit 4°C above pre-industrial temperatures by 2100, perhaps as early as the 2060s.[1] There is no way humans can continue living as we do now under such conditions, and adaptation will likely prove difficult.[2] Abrupt, widespread ecological collapse may occur within decades.[3] Yet instead of unified, binding international efforts toward energy transition, carbon mitigation, and adaptation, we see instead persistent fossil fuel dependence (including record-breaking use of coal), increasing carbon emissions, divisive politics, and armed conflict.[4] Strikingly, more than half of all historic cumulative CO_2 emissions occurred *after* the United Nations Intergovernmental Panel on Climate Change (UN–IPCC) delivered its first monitory report in 1990.[5] And planetary warming is only one of the crises we face—including biodiversity loss, widespread deforestation, pervasive toxic pollution, unsustainable human development, disruptive technologies, and war—amidst indications that the global economy may be careening out of control.[6]

We face a cascade of interconnected social and environmental challenges that pose grave threats to modern civilization, life as we know it, and long-term human survival.[7] CO_2 spikes similar to the one we are currently causing are associated with most of the major extinction events in Earth history.[8] There is substantial historical evidence that climate variation has had significant impacts on human society, including contributing to political destabilization and societal collapse.[9] Meanwhile, insurmountable obstacles block the way to rapid decarbonization and green energy transition, while limits on cheap energy production from fossil fuels grow increasingly likely to prove transformative.[10] Energy expert Vaclav Smil writes: "Designing hypothetical roadmaps outlining complete elimination of fossil carbon from the global energy supply by 2050 is nothing but an exercise in wishful thinking that ignores fundamental physical realities."[11]

What's more, the predicament we face today is so immense, complex, and abstract that it not only exceeds our ability to respond but outstrips our ability to make it make sense.[12] We see as if through a broken kaleidoscope, apprehending fragments of a pattern through lenses that misrepresent and warp reality. Overreliance on highly questionable models creates a false sense of security.[13] Optimistic narratives about the future distort our comprehension of the present. Cognitive and institutional biases inhibit effective and timely action.[14] And we are blinded by our faith in progress: underlying the numerous challenges we face, we have exceeded the planetary boundary conditions for human flourishing, yet overcoming material conditions is the very meaning and promise of progress.[15] Modern global civilization is predicated on the belief that we can improve, control, and surpass nature, and any challenge to this faith is dismissed out of hand or attacked as socially regressive, unenlightened, Malthusian, nihilistic doomerism—in a word, heresy.

Thus despite an observable trajectory of ecological overshoot, clearly signaled by overwhelming evidence, as well as numerous other good reasons for being pessimistic about our future, an indefatigable chorus of optimists continues to insist we can solve the most complex, intractable problem humanity has ever faced, while also sustaining global economic growth, advancing social justice, and preserving liberal democracy. At the same time, most research on global warming and adaptation remains focused on either telling us what we already know or, worse yet, exploring increasingly unlikely outcomes, rather than addressing the devastating consequences we're likely to see.[16] Astonishingly, the IPCC has focused *more* on lower-end estimates of temperature increase as those estimates have grown less probable.[17] From UN Conference of the Parties meetings to corporate greenwashing campaigns, from Silicon Valley effective altruists to left-wing political activists, from Wall Street to the White House, one finds self-congratulating optimists ignoring probable outcomes in favor of gambling our collective human future on vacuous abstractions like "net-zero emissions," "green jobs," and "winning the climate war."[18] Such optimism in the face of climate change is not only unsupportable but irresponsible, and the paradigmatic civilizational narrative upon which it depends—the Myth of Progress—can no longer guide effective action in a changing world.

This book, *Impasse: Climate Change and the Limits of Progress*, argues that we face an incomprehensible concatenation of environmental and social problems, that optimism in the face of this impasse is unwarranted and irresponsible, and that given the stakes, our best response to this situation is a form of virtue ethics I call ethical pessimism. My project describes the paradoxical challenges our predicament poses to our understanding and attempts to answer the central question that emerges: How to live ethically in a world of catastrophe? As numerous thinkers have argued, this is not a question that can be adequately answered by appeal to prevailing frameworks, since such frameworks fail to address the scale and urgency of our impasse and are often implicated in some commitment to progress. Nor can empty gestures toward hope or "the youth" satisfy. There exists one tradition of modern thought, however, which although marginal and often disparaged, seems plausibly capable of sustaining a commitment to human futurity and ethical action in the face of our impasse: the tradition of philosophical pessimism.

Such a claim may seem counterintuitive. Pessimism is antithetical to progressivist faith as it rejects any confidence in the long-term improvement of the human condition. Critics of pessimism often conflate it with quietism, nihilism, defeatism, and misanthropy. They often assert that pessimism leads inevitably to despair, paralysis, and even worse fates. Yet such criticisms are mistaken. On the contrary, as I will show, pessimism is a robust philosophical response to modern conceptions of time and suffering, as well as a resilient and compassionate way to apprehend reality and make ethical choices. The cluster of challenges we face includes not only climate change, ecological overshoot, and unsustainable civilizational complexity but also the end of progress as an effective and coherent narrative. Philosophical pessimism offers a particularly useful response to our impasse precisely because it emerged historically alongside and in reaction to Enlightenment progressivism.

My method is interdisciplinary, synthetic, and perspectivist, focused on the larger problem while consistently and intentionally shifting conceptual, disciplinary, and epistemological lenses. I am trained as a literary scholar and my prior scholarly book, *Total Mobilization: World War II and American Literature*, sits squarely within the field of literary studies, but while *Impasse* is concerned with narrative, it is not a work of literary criticism. I also have some training in philosophy, as well as abiding interests in anthropology and intellectual history, and have written many works of journalism and public-facing essays for large-audience platforms. This book reflects my motley intellectual trajectory and dispositional resistance to specialization, as well as my commitments to methodological perspectivism, epistemic diversity, and public scholarship. By "methodological perspectivism," I mean a practical commitment to the following propositions, if only as heuristics:

- Reality exists independently of human consciousness.
- Reality is "inhabited by different sorts of subjects, or persons, human and non-human, which apprehend reality from distinct points of view."[19]
- No single point of view offers privileged access to reality.
- Human language and culture (including modern science) are embodied, embedded, extended, and socially distributed forms of cognition.
- A more accurate understanding of reality may be reached through the comparative multiplication of perspectives, even if that understanding may never be comprehensive.

As much as I draw from multiple disciplines and perspectives, I also hope to be useful to a variety of readers, including not only philosophers, historians, and literary critics but also political scientists, policymakers, journalists, artists, and activists.

This is where *Impasse* makes its intervention: working through a perspectivist, interdisciplinary approach to reckon with the incomprehensibility of our situation and establish robust grounds for ethical action committed to human futurity. The need for such work is pressing, yet the risks are significant. The project's reach cannot help but exceed its grasp. The very nature of the problem resists apprehension. As philosopher Stephen Gardiner has pointed out, climate change on its own is so complex that a "comprehensive survey of the relevant literature would be impossible."[20] In the words of philosopher Dale Jamieson, "There are no experts on climate change, only experts on particular aspects of the problem and generalists who are skilled at integrating diverse material. Anyone who tries to tell the larger story inevitably trespasses on the terrain of specialists."[21] And unfortunately, specialists in one field are often dismissive of the work of specialists in others.

Yet the difficulty of the task in no way frees us from the obligation to attempt it: *Hoc opus, hic labor est.* Thus we philosophize, as Plato wrote, "standing aside under the shelter of a wall in a storm and blast of dust and sleet"—or in the lee of an alcove as the rain pours down.[22] The core argument of this book is that there are good reasons for being pessimistic about our ability to respond to or even comprehend our impasse and that embracing our pessimism can free us from the dangerous optimism of progress, thus opening new forms of commitment to the future.

Impasse is organized into seven chapters, divided into two parts. In "Part One: The Broken Thread," I consider some of the narrative, cognitive, political, and conceptual problems that we face in trying to understand, come to terms with, represent, and respond to our impasse. I discuss the limits of progress, problems in climate change communication, political extremism, conceptual issues, and the idea of the end of the world, concluding that our situation is not comprehensible within the contemporary value system of progressive modernity and may not be comprehensible at all. I argue that our biases, psychological dispositions, political and cultural frameworks, core social values, and dependence on narrative obstruct both understanding and action. We find ourselves in a paradoxical situation, where the best empirical evidence suggests we face a future at once extremely dangerous and wholly unimaginable. In "Part Two: The Leap," I begin to build a framework for ethical decision-making in such an impasse, which I call ethical pessimism. Despite its roots in ancient wisdom, pessimism is best understood as a modern philosophical response to Enlightenment optimism and the Myth of Progress; it is thus particularly relevant today, as we confront their limits. Part One, comprising chapters one through four, may be considered "the problem," and Part Two, chapters five through eight, "the solution"—though as readers will discover, the only real solution is to learn to live with the problem.

In important ways, *Impasse* builds upon and develops my previous work on the philosophy of climate change, in *Learning to Die in the Anthropocene* (2015) and various essays, some of them collected in *We're Doomed. Now What?* (2018). There is a consistent through line in this work, arguing that global elites have failed to adequately address the existential threat climate change poses and that we have good reasons for being pessimistic about their willingness or capacity to do so in time to prevent serious social destabilization. There is also a consistent through line in my emphasis on the need for spiritual and philosophical reflection in response to this impasse, rather than reactive politics, violence, emotive rhetoric, "political will," or any of the other performative gestures one sees so often. *Impasse* offers a more grounded and rigorous articulation of some of the arguments that were necessarily condensed or only sketched out in *Learning to Die in the Anthropocene*, given that book's intentional brevity, and expands upon some of the philosophical ideas undergirding that book's approach.

Impasse is also in some ways a response to various criticisms and objections to the positions I've advanced in *Learning to Die in the Anthropocene* and elsewhere. There are two main charges *Impasse* meets. The first is that my position is "doomist" or that I am a "doomer," which charge implies that I am erroneously overstating the negative implications of climate change, trying to discourage people, or somehow welcome the catastrophic future I warn against. As I hope this book demonstrates, I am concerned with what the evidence shows, including both climate science and the historical record of political and social failure to address climate change. I believe that the high probability of a catastrophic future, suggested by the evidence, means that we face urgent social and philosophical challenges that must be addressed without the comforting fantasies of technological utopianism or progressivist teleology. If that makes me a doomer, so be it, but it seems to me the only responsible position to take at this time. Our situation is dire, and while one may hope for a miracle, it's foolish to plan on one.

Which brings me to the second charge, that my position is fatalist, nihilist, quietist, or a counsel of despair. As I take pains to show, pessimism should be understood as a philosophical stance on the meaning of suffering and time that is defined by its skepticism toward progressivist teleology. The pessimist believes that suffering and failure are ineradicable aspects of human existence and that any ethical approach grounded in the promise of doing away with human suffering is fundamentally flawed. Thus pessimism is distinct from nihilism, cynicism, fatalism, etc., most centrally because the pessimist believes that human suffering matters and that we are obliged to do what we can to mitigate their effects, even if the results may be only temporary. This is not a position of despair but one of intellectual humility, compassion, and resilience.

It is true that my position is not one that easily supports revolutionary or liberatory politics, but that's not because of any principled opposition to political change as such. Rather, the pessimist position recognizes how much time, effort, and sheer luck it takes to change society and sees an unbridgeable gap between the amount of time meaningful social change requires and the amount of time we have left before shifting conditions rip such prospects from our hands. I am a pessimist, that is, because I recognize how much work change takes. I find it ironic to be accused of advocating "giving up" when so much of my adult life has been committed to effortful labor, both toward distant, difficult goals (like writing books) and in service to my community.

Indeed, I regard my work on climate change itself as a kind of service. While I find the writing and research intellectually rewarding and am pleased that people find my work useful, I also find the subject depressing and the discourse wearisome. In the weeks after publishing my 2018 essay, "Raising My Child in a Doomed World," in the *New York Times*, I was subject to a deluge of personal attacks on social media and by email from people across the political spectrum. Some told me I was a terrible father and should have my daughter taken away. Others told me I should kill myself. One of my senior colleagues at Notre Dame, the political theorist Patrick Deneen, tweeted that he was ashamed our university had hired me. I was attacked for being too Marxist, not Marxist enough, too leftist, not leftist enough, and too pessimistic (yet never, to my recollection, not pessimistic enough, though that's probably nearer the truth). All this is more or less standard behavior on social media these days, and my point is not to portray myself as the victim of undeserved aggression but rather to illustrate the character of public conversation on climate change, which is both noxious and dull. Motivated reasoning, ad hominem attacks, tone policing, tribalism, and ideological one-upmanship are endemic to the discourse. Indeed, the frustrating and even bruising experience of discussing climate change in an increasingly dysfunctional public sphere was part of my motivation in writing this book. I wanted to understand why the subject is so hard to talk about, so fraught, and so prone to tendentious framing and bad-faith arguments. The research I conducted did not inspire optimism.

Thus *Impasse* differs from *Learning to Die in the Anthropocene* and is a more pessimistic book. When I wrote *Learning to Die*, I had some hope that the United States' intellectual elites, reading public, and ruling class might begin to recognize the awful impasse we're in and act accordingly. I didn't think my little book would do it alone, but I saw myself as part of a group of people making similar arguments, including at that time Dale Jamieson in *Reason in a Dark Time*, Elizabeth Kolbert in *The Sixth Extinction*, Mark Lynas in *Six Degrees*, Christian Parenti in *Tropic of Chaos*, Naomi Oreskes and Erik Conway in *The Collapse of Western Civilization*, Bill McKibben in *Eaarth*, Naomi Klein in *This Changes Everything*, James Hansen in *Storms of My Grandchildren*, Anna Tsing in *Mushroom at the End of the World*, Dipesh Chakrabarty and Susanne Moser in their various articles, and numerous other people too many to name here.

And indeed, the years following the publication of *Learning to Die in the Anthropocene* saw a wave of important books and articles confronting the topic with the seriousness and gravity it demanded, including David Wallace-Wells's *The Uninhabitable Earth*, Pablo Servigne and Raphaël Stevens's *How Everything Can Collapse*, and Amitav Ghosh's *The Great Derangement*. Those years also saw a genuine shift in the conversation around climate change, some deepening recognition of the stakes, increased interest in renewable energy, and the emergence of new activist movements such as Extinction Rebellion, the Sunrise Movement, and Greta Thunberg's Fridays for the Future. At the same time, those years also saw the election of reactionary, climate-denying presidents in the United States and Brazil, the French *gilets jaunes* protest, Brexit, increasing Hindu nationalism, the Mediterranean refugee crisis, the Rohingya crisis, political and humanitarian crises in Venezuela and Yemen, a wave of climate refugees heading north from Central America, increasing deforestation in the Amazon, increasing global carbon emissions, and increasing political fragmentation. Then COVID hit, providing a case study in mismanagement, elite failure, political cynicism, public hysteria, profiteering, geopolitical division, the politicization of science, and diminishing marginal returns on complexity. And while 2020 gave us the pandemic, it also gave us shattered heat records, bleached coral reefs, near-record lows in Arctic sea ice, the western United States' worst fire season in history, unprecedented wildfires in both the Amazon and the Arctic, the most active Atlantic hurricane season on record, and the collapse of the last fully intact ice shelf in the Canadian Arctic.[23] More alarming, observed increases in atmospheric methane were so large that if they were to continue—which they have—they'd soon overwhelm pledged emissions reductions from the Paris Agreement, if those reductions had been happening—which they haven't.[24]

Meanwhile, global capitalism revealed itself to be an increasingly precarious Ponzi scheme on the verge of total collapse, as documented by Adam Tooze in his books *Crashed* and *Shutdown*. "In 2020," Tooze writes, "at least as far as the financial system is concerned, managerialism once again prevailed, but it was less an exercise in all-powerful technocratic manipulation than a scrambling effort to preserve a dangerous status quo. 'Too big to fail' has become a total systemic imperative."[25] The COVID pandemic not only revealed the fragility of global networks, supply chains, markets, and governance but made it worse. If Joseph Tainter's theory of the decreasing marginal returns on complexity is correct, it seems plausible that global capitalist civilization could see rapid simplification in the next hundred years regardless of whether or not it has to deal with a chaotic transition to a new planetary climate system. But it does have to deal with such a transition, and the outcome doesn't look good.

On a personal note, the pandemic also derailed my research agenda, which had been focused on finding or developing a system of thought that could bridge the metaphysical gap between Nature and Culture constitutive of the modern social imaginary—which is to say, constructing an "ecological humanism" that might combine an ideal of biological diversity for its own sake, a respect for scientific knowledge, and a commitment to human justice and equality. For a variety of reasons, as you will see in this book, I no longer believe that such a project is plausible. If a "new ecological consciousness" is to somehow emerge in human culture, it won't be because some academic worked it out at their desk. Social change is messier and more mysterious than the vanity of intellectuals is willing to credit; in even the best of circumstances we generally have a poor idea of what's actually happening around this teeming, beautiful, bloody world in which we live, and we are not working in the best of circumstances. Capitalism is accelerating itself to death, the post-1945 political order has come apart, the internet is making us all dumber, and then there's climate change. The only things we can really be sure of are that human reason has limits, suffering is inevitable, and tomorrow will be different from today. Thus, as this book argues, any ethical commitment to human futurity must be apophatic and pessimistic in orientation: a commitment to an unknowable future we will never see and that has nothing to do with our notions of progress.

PART ONE

The Broken Thread

One
THE LIMITS OF PROGRESS

In the fall of 1965, as gingkoes turned gold across the campus of Bryn Mawr College, British literary critic Frank Kermode gave a series of talks about the future. The title of the series was *The Long Perspectives*, and Kermode's main concern, broadly framed, was the fictional structures through which we makes sense of time and experience. He was particularly interested in how we relate an individual human life, with its beginning, middle, and end, to grander stories, historical or cosmological in character, about the fate of the world. He was skeptical, though not unsympathetic, toward the all-too-human desire to see some kind of concordance between the two—the desire, that is, to believe that time has a shape like life and that we happen to occupy a privileged place within it: the beginning, perhaps, or the end. Times being what they were, two decades after a global political cataclysm had killed seventy or eighty million people and just a few years after Russia and the United States had nearly started a nuclear war over Cuba, Kermode began with the "ways in which, under varying existential pressures, we have imagined the ends of the world."[1]

Kermode published his talks as a book, *The Sense of an Ending: Studies in the Theory of Fiction*, which was received with "almost unanimous acclaim" when it appeared in 1967 and quickly became a modern classic of literary criticism.[2] It is a sophisticated work, replete with insight, and intimidatingly learned, written with donnish authority and earthy wit, ranging in reference from medieval apocalypticism to French avant-garde novelists, with an abiding concern throughout for the poetry of Wallace Stevens. It's also one of the last great books of Anglo-American literary criticism to appear before the discipline was transformed by French post-structuralist theory, and although the work introduces its own special concepts and terms ("in the middest," for instance, and the temporal distinction between *chronos*, *kairos*, and *aevum*), it's written without reliance on post-structuralist jargon.[3] Nothing is deconstructed or "deeply imbricated with the kind of social and historical individuals we are," nor is there any discussion of how subject positions are interpellated within hegemonic epistemes or marked under erasure by the Symbolic.[4] *The Sense of an Ending* is nevertheless not an especially easy book to make sense of, despite being elegantly written, and its focus is limited primarily to French, British, Irish, and American literature, as well as almost exclusively to white, male authors. But if one is at least passingly familiar with the works that made up those canons in the 1960s, *The Sense of an Ending* remains a pleasurable, illuminating, and provocative read. It also happens to be strikingly relevant, today, as we struggle with our own anxious sense of an ending.

"We have our Terrors, and specific images of them," Kermode writes, which fact in his view did "not distinguish us essentially from other apocalyptists."[5] Yet much has happened since Kermode gave his talk in 1965, when global atmospheric carbon dioxide concentrations passed 320 parts per million for the first time in more than two and a half million years.[6] In the six short decades since, cumulative US carbon dioxide emissions have increased by more than 212%, from more than 134 billion tonnes to more than 421 billion tonnes; China's cumulative emissions have increased by more than 3,000%, from less than 8 billion tonnes to almost 250 billion tonnes; and global land and ocean surface temperatures have increased by more than a full degree Celsius.[7] Because of these trends, among others, we face a situation some call a polycrisis or predicament, but I prefer to call an impasse, in which several connected and cascading crises bring into question the survival of modern civilization.[8] In addition to war and the ongoing threat of nuclear Armageddon, familiar from Kermode's day, we face accelerating climate change, biodiversity collapse, political instability, increasing economic inequality, unaddressed trauma and complications from the COVID pandemic, hormone-disrupting toxic pollution, and disruptive technological innovation.

These are real phenomena, not fairy tales, yet Kermode's point that the specific character of our apocalypse does not distinguish it from other apocalypses holds, at least insofar as the stories through which we comprehend our impasse are not new. We make sense of reality through inherited narrative and conceptual schemata, political commitments, biological affordances, and psychological dispositions that shape our perception of reality in various ways, many of which are poorly adapted to the challenges we face. We can only comprehend our impasse through cognitive, phenomenological, symbolic, and conceptual filters that often misrepresent and obscure our situation.

Paradoxically, that is, the very ways we understand and narrativize reality impair our ability to comprehend it. Our desire to see our situation as tractable leads us to accept easy answers and unfeasible solutions.[9] Our reliance on narrative structure makes it difficult to embrace a radically unknowable future. Widespread faith in progress hinders our capacity to imagine societal collapse. And our bias for optimism impedes our judgment and blinds us to likely outcomes. The challenge we face is not only technical, political, and ethical, but epistemological, philosophical, narrative, and, in a deep sense, critical, since the conceptual and narrative structures that give life meaning today are fundamentally inadequate to our situation. Thus the pressing task of the thinker, as I see it, is neither to square our impasse with one or another preexisting program nor to more thoroughly analyze and clarify the impasse itself nor to discover a new narrative framing that could better motivate "the public" (whatever that is) but rather to come to terms with the inescapably apophatic character of our relation to the future.

THE ETHICAL CHALLENGE OF FICTIONALISM

Kermode's main argument proceeds from the observation that we live, as he puts it, "in the middest." We stumble into life well after the party started and are forced to leave before it concludes. Our perspectives on time, history, the cosmos, and existence are necessarily partial, contingent, and to some degree arbitrary, formed in the shadow of our inevitable death. A devoted reader of Beckett, Sartre, and Wallace Stevens, Kermode understands narrative as an existential enterprise: reality has no fixed or inherent meaning, yet human beings have evolved as self-conscious social animals who find such groundlessness repugnant, nearly incomprehensible, and all but impossible to live with, so we make sense of our being in the world, the world itself, and our ends by producing symbolic structures, or "fictions," that give coherent meaning to suffering, death, and time. Kermode writes: "Men, like poets, rush 'into the midst,' *in media res*, when they are born; they also die *in mediis rebus*, and to make sense of their span they need fictive concords with origins and ends, such as give meaning to lives and poems."[10]

As inevitably as the sun "rises" in the east and "sets" in the west, human consciousness weaves sensory data, feelings, and desire into a composite spatial, temporal, and social tapestry, turning noise into information by organizing it into practices, artifacts, and structures of meaning such as Chetro Ketl, the novel *Beloved*, the periodic table, the American Association for the Advancement of Science, funeral rites, literary studies, climate models, and ABBA's "Waterloo," which hang together in an aggregate chronotopological and ontological gestalt we call "the world."[11] I adopt the term *chronotope* from the Russian Formalist critic Mikhail Bakhtin, who applies it to describe the weave of time and space used in the novel to model the texture of social reality. Bakhtin writes:

> We will give the name chronotope (literally, "time space") to the intrinsic connectedness of temporal and spatial relationships that are artistically expressed in literature. . . . In the literary artistic chronotope, spatial and temporal indicators are fused into one carefully thought-out, concrete whole. Time, as it were, thickens, takes on flesh, becomes artistically visible; likewise, space becomes charged and responsive to the movements of time, plot, and history. This intersection of axes and fusion of indicators characterizes the artistic chronotope.[12]

Whereas Bakhtin is careful to restrict his usage of the term to intentionally literary narratives, my concern is with more ad hoc world-producing cultural and social structures. Making worlds out of words may be unique to Homo sapiens, or maybe not, but it is something for which we have exceptional talents. Evidence suggests that humans have communicated and crafted tools as long as we've existed, having inherited these capacities from our ancestors, and while researchers have observed cultural differentiation in chimpanzees, bonobos, orangutans, and dolphins, the development of abstract thought and syntactical language more than seventy thousand years ago triggered significant and apparently singular advances in humans' ability to create collective systems of meaning and memory.[13]

The problem, as Kermode saw it in 1965—with the Cold War in full swing, nuclear annihilation looming, the memory of World War II bright in his mind, US combat troops deploying to Vietnam, Watts on fire, and unarmed civil rights activists being savagely beaten and tear-gassed as they marched from Selma to Montgomery—was that our fictions don't always match reality and are sometimes woefully mistaken. This was why he focused his lectures on "the End," taking a close look at various attempts to imagine and predict apocalypse, mostly within a Christian framework, attending specifically to how such fictions persist despite being repeatedly falsified. "Men in the middest make considerable imaginative investments in coherent patterns," he observes, "which, by the provision of an end, make possible a satisfying consonance with the origins and the middle."[14] Such investment helps explain, for Kermode, why our narratives are so often immune to reality: we are prone to a kind of ontological sunk-cost fallacy.

> Apocalypse can be disconfirmed without being discredited. This is part of its extraordinary resilience.... It is patient of change and of historiographical sophistications. It allows itself to be diffused, blended with other varieties of fiction—tragedy, for example, myths of Empire and Decadence—and yet it can survive in very naïve forms. Probably the most sophisticated of us is capable at times of naïve reactions to the End.[15]

The enduring appeal of apocalypse was not, for Kermode, a problem for medieval mystics alone. "Of course we have it now, the sense of an ending," he writes. "It has not diminished, and is as endemic to what we call modernism as apocalyptic utopianism is to political revolution."[16] Indeed, apocalyptic thought is neither specifically medieval nor specifically modern, but foundational to the Western, Christian conception of linear time. As Kermode puts it, "The apocalyptic types—empire, decadence and renovation, progress and catastrophe—are fed by history and underlie our ways of making sense of the world."[17] Political scientist Alison McQueen's study of apocalypse and the modern tradition of political realism supports Kermode's analysis. She defines apocalypse as a robust genre of political theodicy that "took shape gradually in response to the political crises that plagued ancient Palestine" and can be understood as "an attempt to situate . . . contemporary political circumstances within a sacred worldview, thereby endowing them with divine significance."[18]

As Kermode demonstrates, exploring the persistent appeal that the apocalyptic consonance between political and sacred time holds can help us make sense of the persistent gap between reality and the narratives we craft to make sense of it while also helping us understand our "considerable imaginative investments" in such narratives. Such study further helps us see how the stories we develop to organize human social life cannot be easily divided into true and false based on their correspondence to physical reality, because the schemata we use to organize physical reality are themselves socially constructed, and much of what we take to be "true" has only an accidental relation to that reality. Why is an apple an apple and not a *pomme* or *tafaha* or *píngguǒ* or *blork*? What is money? Blue? Race? Time? The number 2? How do you *prove* it? How are they *true*? These ponderous questions are of special concern to a literary critic such as Kermode, whose life's work is the study of manifestly false narratives. The ontological ungrounding effected by the fundamentally arbitrary quality of language in Kermode's fictionalism opens disturbing moral and ethical voids: "If literary fictions *are* related to all others, then it must be said that they have some dangerous relations."[19]

Kermode builds his fictionalism from Nietzschean perspectivism, Kantian epistemology, and German thinker Hans Vaihinger's philosophy of "as if," which argues that we live "as if" our fictions about reality were true.[20] Vaihinger holds that human thought is an organic function that adapts sensory input to its own purposive ends, and "however we may conceive the relation of thought and reality, it may be asserted from the empirical point of view, that the ways of thought are different from those of reality, [and] the subjective processes of thought concerned with any given external event or process have very rarely a demonstrable similarity to it."[21] For Vaihinger, reality-as-such and how we understand it are ontologically distinct: we make our way in the world by shaping our narratives to serve our goals, then chopping at reality to make them match up.[22]

What Kermode finds disturbing about this behavior is that while unconsciously committing to our fictions seems dangerous and self-deluding, doing so consciously isn't much better. Ontological commitment faces the same problem whether it's conscious or not, in part because our commitments are made in a moral abyss, since there are no fundamental grounds on which to adjudicate competing narratives. Kermode brings this issue into focus with reference to what was, for his audience, the most salient example of the horrific consequences that can follow from a zealous commitment to fiction, the Shoah. "The validity of one's opinion of the Jews," he writes, "can be proved by killing six million Jews." The philosophical assumptions behind Nazi antisemitism, Kermode points out, "were not generically different from those of the scientist." A fiction is a fiction is a fiction, and since the only way to know whether any given fiction is really true is to make it true by deed, we face an ethically paralyzing existential conundrum.

> How, in such a situation can our paradigm of concord, our beginnings and ends, our humanly ordered picture of the world satisfy us . . . ? How can apocalypse or tragedy make sense, or more sense than any arbitrary nonsense can be made to make sense? If *King Lear* is an image of the promised end, so is Buchenwald; and both stand under the accusation of being horrible, rootless fantasies, the one no more true or more false than the other, so that the best you can say is that *King Lear* does less harm.[23]

Kermode attempts to solve this conundrum by positing a distinction between "myth" and "fiction," defining myth as absolute, ritualistic, traditionalist, unselfconscious, and escapist, and fiction as conditional, self-conscious, and self-aware. According to his argument, the self-aware complexity of fiction offers an ethically preferable alternative to the escapist absolutism of myth, since for the conscious fictionalist there is always a certain irony and contingency in their efforts to find "what will suffice," which may at least secure the conscious fictionalist against catastrophic moral blunders.[24]

Yet Kermode fails to offer any persuasive criteria for determining how to decide between competing fictions, nor does he make a strong case for the value of self-consciously fictional ones. His attempt to distinguish between antisemitism and *King Lear*, for instance, founders on his specious distinction between myth and fiction, which seems plausible until you start asking how it might actually work. Is reading a seventeenth-century succession drama based on ancient British myth an "encounter with oneself" or "escapist"? How about watching the Shakespearean HBO series *Succession*? What about reading Hitler's *Mein Kampf*, Karl Ove Knausgaard's six-volume *My Struggle*, or Margaret Atwood's *The Handmaid's Tale*? How do we decide which is "fiction" and which "myth"? Does context matter? Does it matter who is reading or why? And what about not just explicitly narrative fictions but complex social ones, like Kermode's example of Nazi antisemitism? Were the Proud Boys who stormed the US Capitol on January 6, 2021, participating in self-conscious fiction, unselfconscious myth, or both at the same time? How about cosplayers at Comic-Con, "Swifties" flocking to their idol's latest concert, undergraduate activists chanting "From the river to sea," or anyone anywhere on social media? How do we tell the difference? A self-reflexive experience with fiction for one person may be unselfconscious escapism for another, and for yet a third may be escapism masquerading as self-reflection, vice versa, or both at once. For yet a fourth, the same fiction may involve the devout performance of religious ritual experienced as at once mannered repetition, self-reflexive fiction, and a deep communion with the divine. The truth is that humans inhabit multiple layers of symbolic and narrative reality in various ways at the same time, living in complex, ambiguous, and shifting currents of faith, doubt, irony, performance, enthusiasm, repression, deceit, imagination, reflection, mimesis, and indulgence, sometimes believing, sometimes merely echoing the belief of others, sometimes performing belief in spite of our doubt.[25] How self-conscious one may be about the fictional status of one's myths seems to be a question more about the sophistication of the believer than about some quality inherent in the fictions themselves.

Quite contrary to Kermode's efforts to establish a firm ethical binary between fiction and myth, robust fictionalism lacks any grounds for such a distinction. While we may grant practical, rhetorical, and even epistemological differences between "mere fictions," myths, social fictions, and empirical data, we can grant no metaphysical distinction, since there is no human access to reality unmediated by consciousness, language, and sociality. As Nietzsche famously put it, there are no facts, only interpretations.[26] We may commit ourselves to one interpretation over another, one perspective over another, but the question of how to choose among different fictions can be answered only by appealing to judgment, which means making decisions based on prior knowledge, experience, discussion, and debate.[27]

Yet on what basis do we judge judgment? The question offers no easy answer, as Kermode knew well. In the end, he gives up his simplistic distinction between "myths" and "fictions" and argues instead for skepticism, complexity, and a hopeful if tenuous faith in human solidarity. He writes:

> Fictions in the end fail under the pressure of what [Henry] James is said, in his last words, to have called "at last, the real distinguished thing"; but meanwhile we have our predictive games, our family jokes like *Lear*, our anthropomorphic paradigms of apocalypse; we have a common project, truth in poverty, and a common need, solidarity of plight in diversity of state. The free imagination makes endless plots on reality, attempts to make our proportionals convenient for our equations in everything; our common sense makes us see that without paradox and contradiction our parables will be too simple for a complex poverty, too consolatory to console. Our study . . . must have a certain complexity and a sense of failure. "I cannot do it; yet I'll hammer it out."[28]

This is an appealing formulation. It is restrained yet hopeful, earnest in its desire for concord, and generous in its humanity. Yet there is something hollow in it and indeed in Kermode's whole approach.

Handling the fictions through which we comprehend reality with sophisticated skepticism is prudent up to a point. But it fails to answer the existential human need for ethical grounding: we live in a world of struggle and want to believe our struggle is meaningful inherently, not only insofar as we attribute meaning to it. What's more, ironic sophistication can easily turn into another kind of escapism, a fiction of disinterested contemplation distracting itself with the endless play of endlessly fascinating language games at the expense of attending to the "blooming, buzzing confusion" those games evolved to address. Worrying the fictionality of our apocalyptic narratives conveniently lets us ignore the all too real apocalypse unfolding before our eyes.

Thinking back to the tumultuous violence of the late 1960s, it's not hard to see how Kermode's turn away from the ethical challenges posed by war, nuclear annihilation, decolonization, racial strife, and the terrifying future of mass technological society and toward elaborate meditations on the fictionality of apocalyptic stories so dense and labyrinthine they verge on sophistry might have appealed to his audience. *The Sense of an Ending* must have been a great comfort to English professors, literary critics, and university intellectuals facing apocalyptic change. And one cannot in good faith deny an aspect of practical quietism in all criticism and philosophy, no matter how revolutionary any given thinker's claims.[29] Kermode is aware of this and opens *The Sense of an Ending* by disavowing any responsibility for solving earthly problems: "It is not expected of critics as it is of poets that they should help us make sense of our lives; they are bound only to attempt the lesser feat of making sense of the ways we try to make sense of our lives."[30] Yet this disavowal is disingenuous, even for Kermode. He later contradicts himself: "The critics should know their duty," he writes, and defines that duty as "the justification of ideas of order," by which he means the critical analysis and evaluation of how we "make sense of our lives."[31]

Performing a delicate intellectual balance, Kermode holds simultaneously contrary positions. On the one hand, the critic is not responsible for solving the world's problems. On the other, it is the critic's express duty to analyze, explicate, evaluate, and judge the various fictions proposed by others for making reality sensible and bearable. An enviable position, high above the fray, never risking the embarrassment of a "naïve reaction," never being taken in by an erroneous myth because never making a commitment, yet burdened with the obligation to adjudicate between competing fictions. And here, of course, is the crux: the complex dynamic at work between the stories we live, our relations with other beings, and the ways we make sense of events—the tension between narrative, ethics, and politics at the heart of social life—cannot be resolved by bracketing it as a merely interesting aesthetic or philosophical puzzle. If we must live by fiction—and we must—then we are obliged to reflect on which fictions we impose on reality.

INTELLECTUALS AND IDEOLOGY

The problem is that choosing our fictions isn't a choice we get to make. The stories we live within are imprinted on us in the cradle, drunk in with our mother's milk, imposed by collective diktat, and enforced by the demands of social conformity, and not even the most sophisticated critic can ever fully free themselves from the stories they inhabit. It's not even clear that freedom is what critics are up to. After all, the traditional social role of the modern intellectual is not that of the revolutionary but the apparatchik. "Whatever structure of domination is reflected in, and served by, a given concept," writes sociologist Zygmunt Bauman in his history of the intellectual from the Enlightenment to the present, *Legislators and Interpreters*, "all such concepts are coined, or refined, or logically polished, not by the dominating side of the structure as a whole, but by the intellectual part of it"—by that class of self-appointed legislators who base their claims for social authority on the ostensibly transcendental critique and esoteric interpretation of the dominating structure they cannot help but serve to legitimate and strengthen.[32]

Knowledge is inextricable from power, an insight that goes back beyond Foucault and Adorno to at least Francis Bacon, and the problem turns out to be deeper than the one illuminated by Marxian insights into the material determination of culture. There can be no doubt that "the mode of production of material life conditions the social, political, and intellectual life-process," as Marx famously put it, that "it is not the consciousness of men that specifies their being, but on the contrary their social being that specifies their consciousness," nor that we live today in a global civilization fundamentally shaped by fossil fuel capitalism.[33] But we must also recognize that beyond and beneath the material conditions structuring social life, language and culture as such serve an essentially homeostatic function, predetermining the formations that condense meaning out of material reality and constraining the emergent rationalizations deployed by diverse factions competing for influence within any given order. Petrocapitalism is a form of social life emerging out of a specific cultural history, that is, not only its material conditions. That the history of ideas is the history of ideology must be granted not merely in the sense lampooned by Walter Benjamin in his image of the chess-playing Mechanical Turk at the beginning of his "Theses on the Philosophy of History"—in the sense of history being the manifest operation of a hidden providential agent—but rather, as Benjamin himself understood, in a theological, mythical, and ontological register, as a recognition that language is the field of signification through which what we call reality acquires meaning.[34] "The interpenetration of language and the social field and political problems lies at the deepest level of the abstract machine," Deleuze and Guattari write, "not at the surface."[35] The epistemological challenge of grounding human knowledge in some irreducible *ding an sich*—the problem of "philosophy and the mirror of nature," in Rortyean language, or grounding the Symbolic in the Real, in Lacanian terms, or whatever terms you like—is compounded, even foreclosed, not merely by the material determination of history and our partisan cultural blinders but also by grammatical constraints, cognitive biases, biological limits, diction, and status quo ontological commitments so foundational to our sense of existence they can scarcely be articulated, much less disentangled from any effort to achieve so-called critical intellectual distance.

We may press the point further: Not only does the intellectual make herself complicit in the dominating structures of society by articulating the conceptual and narrative superstructures that rationalize said domination, even under the sign of "critique," but there is no way to escape such complicity insofar as one performs the social role of the intellectual. There is no transcendental theoretical grasp on reality beyond the socially agreed-upon conceptual and narrative schemata that give our experience shape and meaning. We are all trapped within our "subject positions" and "standpoints," constrained by our ideological, cultural, and grammatical frameworks, locked in the model of reality that is language, this present author not excepted.[36] As W. V. Quine writes: "The totality of our so-called knowledge or beliefs, from the most casual matters of geography and history to the profoundest laws of atomic physics or even of pure mathematics and logic, is a man-made fabric which impinges on experience only along the edges."[37] And for better or for worse, there's no way to know whether the social fabric of which our world is woven is an accurate model or a veil of illusion *except* through experience. "Even great spirits have only their five fingers breadth of experience—" writes Nietzsche, "just beyond it their thinking ceases and their endless empty space and stupidity begins."[38] Or as Alfred North Whitehead put it more austerely, "Apart from the experiences of subjects there is nothing, nothing, nothing, bare nothingness."[39]

Confronted with such nothingness, one can only turn back to experience, mine and yours, lived and reported. Thus the intentional pursuit of "objective knowledge" leads in two directions at once: within and without. The first path follows the hoary Socratic injunction to know thyself, or as Whitehead put it in his definition of philosophy, "the self-correction by consciousness of its own initial excess of subjectivity."[40] The second follows a Nietzschean multiplication of perspectives, or in more contemporary language, epistemic diversity. Analysis of how various beings and competing factions represent the experience of reality through conflicting nodes and flows of power and are represented in turn by contradictory narratives and conceptual schemata may grant a tenuous though not insignificant purchase on reality-qua-reality. Even if we cannot ultimately *know* much about reality as it "actually" is, we can make more or less empirical claims and more or less reliable judgments, and we can also work to improve our understanding even as we recognize our epistemic limitations. As discussed in the introduction, this wager serves as the core methodological premise of this book.

Yet as we knowers invest ourselves in performing the role of "one who knows," which role is practically and by definition a position of authority invested in preserving the status quo social relations that are the very condition of its existence, we diminish the scope of our inquiry. Approached structurally, the social role of the intellectual is best understood as the rationalization of power, a "secularization of the pastoral and proselytizing techniques" once wielded by priests, which role is at best in tension with empirical reality but more often invested in repressing, occluding, and controlling it in order to better control the population at large.[41] Intellectuals were historically and are still *civilizers*, engaged in a "conscious proselytizing crusade . . . aimed at extirpating the vestiges of wild cultures—local, tradition-bound ways of life," a crusade "aimed above all at the administration of individual minds and bodies."[42]

The intellectual's first loyalty is to themselves and their caste. As Bauman writes,

> for the better part of their history, Western intellectuals drew the blueprints of a better, civilized or rational society by extrapolating their collective experience in general, and the counterfactual assumptions of their mode of life in particular. A "good society," all specific differences between numerous blueprints notwithstanding, invariably possessed one feature: it was a society well geared to the performance of the intellectual role and the flourishing of the intellectual mode of life.[43]

Setting aside genuine political divisions internal to their caste, intellectuals' fundamental social commitment is to a world amenable to intellectual activity—that is, to a world in which the use of reason improves the conditions of human existence—because only in a world wherein improving the human condition through the use of reason is a feasible project does the critical function of the intellectual serve a purpose. Otherwise what good are we? In pre-modern cultures, priests pass on inherited knowledge, yet the role of the intellectual as it emerged in the Enlightenment is to critique inherited knowledge and lift our eyes to the future. The intellectual-qua-intellectual—that is, in their role as cultural critic—is structurally committed to the idea of progress and all it entails.

Building on Bauman's critique, Jamaican novelist, playwright, and theorist Sylvia Wynter helps illustrate this point in her dense, suggestive essay "Unsettling the Coloniality of Being/Power/Truth/Freedom: Towards the Human, After Man, Its Overrepresentation—An Argument," where she writes that "we, as Western and westernized intellectuals, continue to articulate, in however radically oppositional a manner, the rules of the social order and its sanctioned theories."[44] Because social truth is always truth *for* the society that produces it, because "our varying . . . modes of being human, as inscribed in the terms of each culture's descriptive statement, will necessarily give rise to their varying respective modalities of adaptive truths-for . . . up to and including our contemporary own," we must reckon with the chastening implication that cultural production as such inescapably serves the status quo.[45]

Wynter makes her argument about the inherently conservative role of cultural production most pointedly through an analysis of the debate between two pre-Enlightenment proto-intellectuals, Dominican friar Bartolomeo de Las Casas and humanist scholar Juan Ginés de Sepúlveda, on the rights and treatment of indigenous peoples in the New World. This debate took place before a council established on behalf of the Holy Roman emperor and king of Spain Charles V to determine the legitimacy of the Spanish conquest and forced conversion of the Indians of the New World, in two sessions, beginning respectively August 1550 and April 1551, in the city of Valladolid, Spain, from which site the debate takes its name.[46] The council, the Consejo Real de las Indias, comprised fourteen learned jurists and theologians, and its composition was representative of the deeper issue at stake in the debate and embodied by the two key disputants—namely, the conflict between secular state power and clerical authority. Sepúlveda, a key early proponent of just war theory, argued that the Crown had the right to conquer and convert the natives of the New World because they were manifestly incapable of governing themselves. Las Casas held that Indians had the God-given right to self-determination, which right could not be justly abrogated by a foreign power.

"The Las Casas–Sepúlveda disputation of 1550 was the last important event in the controversy on Indian capacity that bitterly divided Spaniards in the sixteenth century," writes historian Lewis Hanke.[47] And though the debate ended equivocally, it nevertheless marked a distinct shift in colonial authority from church to state and signifies in retrospect an important turning point in modern history.[48] As Wynter shows, Valladolid illustrates how the secularization of Christian metaphysics was coeval with the emergence of biophysically grounded social hierarchies structured through concepts of racial difference; for Wynter, the debate helps us see the transition from an ontological binary predicated on the division between Christian and Heathen to one predicated on the division between the Human and the Savage.

Yet the debate itself was not taken up with such heady topics. Though the disputants were, at some level, arguing about what it means to be human, the manifest content of their dispute concerned rights claims framed within contemporaneous legal, cultural, and ontological norms. "Contrary to conventional history or popular accounts . . . ," writes historian David Lantigua, "the debate was . . . not about the humanity of the Indians. Rather, the ideological struggle concerned infidel rights."[49] What Bauman, Wynter, and Lantigua together help us see is what we might call, after Nietzsche, the "ignorance of the knowers": how those most passionately concerned with defining, articulating, and adjudicating the terms of justice and meaning in society—the priests, jurists, scholars, and intellectuals who elaborate the ideological superstructure of social life in any given period—are often those least capable of comprehending the deeper currents shaping their thought. The pessimism in Hegel's famous pronouncement that the owl of Minerva takes flight at dusk gives way to an even more pessimistic corollary: we spend the better part of our lives in passionate obscurity, stumbling in the blinding light of noonday. This in itself is humiliating for the intellectual's sense of authority and purpose, but what attending to the debate at Valladolid further illuminates is the emergence of the secular intellectual as the servant of state power. The figure Juan Ginés de

Sepúlveda presents us with, justifying conquest and enslavement in terms of social and moral progress—advocating a "conscious proselytizing crusade . . . aimed at extirpating the vestiges of wild cultures," in Bauman's phrase—is the progenitor of the modern self-appointed legislator described in Bauman's book, the ambivalent critic we see in Kermode, and the contemporary academic.[50]

Sepúlveda was a deeply learned and renowned humanist scholar, translator, and philosopher, who built upon the work of Scottish theologian John Mair, Dominican friar Francisco de Vitoria, and Jesuit priest Francisco Suárez to help develop just war theory. At Valladolid and in earlier writings (particularly *Democrates secundus sive de justis belli causis apud Indos*, ca. 1544), Sepúlveda argued that because the Indians had committed crimes against humanity (allegedly idolatry, sodomy, cannibalism, and human sacrifice), they were incapable of ruling themselves and thus "must naturally submit to wiser and more humane civilizations."[51] The question at stake was not whether the natives were human or whether they, as infidels, had a right to self-determination, both of which Sepúlveda granted, but whether more advanced nations had a right to intervene against criminal ones. "In a pivotal shift," writes Lantigua, "Sepúlveda transmuted a medieval papalist doctrine of war as a punishment for sins against the natural law into a thoroughly secular political idiom no longer dependent on the Church or the Pope for its execution."[52]

Las Casas, on the other hand, then bishop of Chiapas and designated Protector de Indios, argued that it was precisely because the native peoples of the New World lived outside the authority of the church that they held the right to self-determination, even if they engaged in what the Crown regarded as criminal acts; Las Casas's controversial theological argument was that neither the church nor the Crown had authority beyond the bounds of Christendom. Las Casas was also learned and quite as capable of quoting Aristotle as Sepúlveda, but he possessed in addition years of firsthand experience in the Americas as well as admirable moral courage. His *Short Account of the Destruction of the Indies* (written in 1542, published 1552) and his longer *History of the Indies* (published 1561), from which his *Apologética historia summaria de las gentes destas Indias* was excerpted, are striking and original works, and he was a founding thinker for modern ideas of human rights and anthropology, as well as an inspiration to later anti-colonial activists.[53] Yet despite the ethnographic and empirical knowledge informing his position, Las Casas's argument against Spanish rule in the New World and Indian enslavement was fundamentally theological: all humanity was granted reason by God, and thus every people had not only the right but the obligation to rule and defend itself.[54] Conquering people and converting them to Christianity by force was sinful, argued Las Casas, and undermined the evangelical Christian mission in the New World.[55]

To modern sensibilities, the conflict between Sepúlveda and Las Casas presents a conundrum: on the one side we see a progressive secular humanist advocating forced conversion; on the other, a devout religious arguing for cultural relativism. Yet this very paradox is what makes the debate so illuminating as a case study in social change and so valuable for Wynter's analysis of what she calls the "over-representation of Man." Sepúlveda and Las Casas were explicitly arguing about the rights of Indians and claims about state and church power but implicitly arguing about unarticulated ontological presuppositions, grounding metaphysics, and the very conceptual frameworks through which such rights and power relations were understood. They were arguing—even if they didn't and couldn't know it at the time—about the terms in which European culture understood what it meant to be human.[56]

One of Wynter's main points about the dispute between Las Casas and Sepúlveda is that neither participant had full awareness of the stakes and broader implications of the positions they were arguing, *nor could they have*, since ideological formations are *adaptive* truth forms that serve a homeostatic function for particular social modes of being. She writes:

> As human beings who live in society, and who must also produce society in order to live, we have hitherto always done so by producing, at the same time, the mechanism by means of which we have been able to invert cause and effect, allowing us to repress the recognition of our collective production of our modes of social reality. . . . Central to these mechanisms was the one by which we projected our own authorship of our societies onto the ostensible extrahuman agency of supernatural Imaginary Beings. This imperative has been total in the case of all human orders (even where in the case of our now purely secular order, the extrahuman agency on which our authorship is now projected is no longer supernatural, but rather that of Evolution/Natural Selection together with its imagined entity of "Race").[57]

It's important to point out that Wynter's argument is neither wholly relativistic nor a social constructionist critique of science, as some of her readers mistakenly seem to think. Despite the critical character of her approach, Wynter remains committed to empirical knowledge and even to the idea of progress. Her essay is a sophisticated if sometimes obscure analysis of the complex relationship between empirical science and ideological change, as well as a hopeful argument that the humanities may be able to use neuroscience and systems theory to develop a new, more "objective" epistemological framework. As Wynter puts it elsewhere, "The *Studia* must be reinvented as a higher order of human knowledge, able to provide an 'outer view' which takes the human rather than any one of its variations as Subject; must be reformulated as a science of human systems, which makes use of multiple frames of reference . . . to attain to the position of an external observer, at once inside/outside the figural domain of our order."[58]

Wynter argues, that is, for applying empirical science (or "nonadaptive knowledge," in her terms) to the humanities in order to develop "a science in which the 'study of the Word'—of our narratively inscribed, governing sociogenic principles, descriptive statement, or code of symbolic life/death, together with the overall symbolic, representational processes to which they give rise—will condition the 'study of nature.'"[59] Wynter's hope seems to be that such a scientific "study of the Word" may free us from the constraints of ideology as well as the limits of nature and thus enable what she calls autopoiesis, or self-making.[60] Yet the call for a "new science" with which Wynter ends her argument, reminiscent at once of Fanon's "new Humanity," Nietzsche's transvaluation of all values, the Hegelian Absolute, and E. O. Wilson's consilience, never addresses the question of how creative human action might escape the determinism of material existence on which empirical knowledge depends.[61] Nor is it clear how empirical understanding of neurophysiological mechanisms would lead to increased human autonomy, nor why whatever such mechanisms might be found would not be immediately exploited by the very forces their revelation ostensibly undermines. Given the profound recent restructuring of human attention under what Shoshana Zuboff calls surveillance capitalism, the hope that a "new science of the word" might lead to liberation rather than novel forms of repression seems dangerously optimistic.[62]

In any case, what Wynter helps us see, particularly through her attention to the transition from late Christian to early modern discursive formations, is how, in Fanon's words, "each generation must discover its mission, fulfill it or betray it, in relative opacity."[63] We do not know ourselves, we knowers, even in our most articulate, sophisticated, and passionate arguments about who we are. As Marx wrote in *The Eighteenth Brumaire of Louis Bonaparte*, making a point that has become so much a part of the contemporary intellectual's armature that we no longer remember it applies to us:

> Men make their own history, but they do not make it as they please; they do not make it under circumstances chosen by themselves, but under circumstances directly encountered, given and transmitted from the past. The tradition of all the dead generations weighs like a nightmare on the brains of the living. And just when they seem engaged in revolutionizing themselves and things, in creating something that has never yet existed, precisely in such epochs of revolutionary crisis they anxiously conjure up the spirits of the past to their service and borrow from them names, battle cries, and costumes in order to present the new scene of world history in this time-honored disguise and borrowed language.[64]

So we, too, translate the unprecedented impasse we face back into the borrowed language of apocalypse, despite its obvious inadequacy. Facing an intractable existential crisis destabilizing global civilization and undermining concepts central to modern social life, threatening to erase "Progress" and "Freedom" and "Man" as easily as rising oceans might erase "a face drawn in sand at the edge of the sea," we argue about nomenclature, messaging, and the need for optimism.[65] We strain to dress our catastrophe in the "names, battle cries, and costumes" of a dead world, straining all the more the less they fit.

The impulse to reduce planetary ecological crisis to one or another preexisting political or conceptual framework is unavoidable. When faced with a situation offering no solutions but only more or less risky coping strategies, the most robust of which lie well beyond probable or even realistically imaginable geopolitical affordances, it makes sense to grab whatever theoretical apparatus lies ready to hand and use it to force reality to yield, at least in our imagination.[66] Against an intractable problem, storytelling provides consolation and relief, particularly if the story has heroes to cheer and villains to jeer. Identifying a scapegoat turns a thorny analytic problem about how to make sense of an epochal transformation in the human lifeworld into a straightforward political problem that can be solved with simple aggression, but as philosopher and literary critic Claire Colebrook points out, to argue about who "caused" the Anthropocene is to grasp the problem by precisely the wrong end.[67] Such narrativization is nevertheless our first response in the face of existential catastrophe, inextricable from our "sense of an ending," and, at the end of the day, inescapable: we are *Homo narrans*, the storytelling ape, wired to simplify and reduce the reality we experience in order to make it make sense.[68] Contrary to what we knowers would like to believe, the challenge we face isn't that people have the wrong ideas, lack information, or even subscribe to the wrong narrative. The problem, deeper than false consciousness, is consciousness itself.

BIASES AND EMPIRICISM

"We are prone to exaggerate the consistency and coherence of what we see," writes psychologist Daniel Kahneman.[69] We regularly construct plausible, coherent stories of causality and intention from the most fragmentary intuitions, often at odds with objective reality.[70] As Kahneman and Amos Tvsersky write in their groundbreaking 1974 study "Judgment Under Uncertainty: Heuristics and Biases," "people rely on a limited number of heuristic principles which reduce the complex tasks of assessing probabilities and predicting values to simpler judgmental operations," and which, while necessary and useful, often "lead to severe and systematic errors."[71] Some of the best known cognitive biases and heuristics include *anchoring*, the tendency to make judgments based on prior information regardless of its relevance; *confirmation bias*, the tendency to filter for information confirming what one already knows; *hindsight bias*, the tendency to view past events as having been predictable; and the *optimism bias*, which we'll talk about more later. In general, the theory of cognitive bias is well supported by extensive evidence from numerous studies and has robust explanatory power for politics, history, and day-to-day human behavior.

The basic idea is that as our brains process sensory impressions, those impressions get interpreted, filtered, and sorted in various ways, and while these interpreting, filtering, and sorting adaptations have proven advantageous over the long run of Homo sapiens' time on Earth, they are also often empirically erroneous and sometimes maladaptive, especially under the recent (and from an evolutionary perspective anomalous) conditions of industrial capitalism.[72] Yet such interpreting, filtering, and sorting is not only advantageous but essential, since constantly processing the full range of sensory data to extract maximal content would lead to cognitive paralysis indistinguishable from psychosis. Living *without* the intuitions and snap judgments of our biases and heuristics would mean experiencing everything all the time.

Jorge Luis Borges describes just this in his story "Funes, His Memory," about "a street tough from Fray Bentos" named Ireneo Funes.[73] Funes first appears as a shy, mocking young man, who is later bucked from a horse and crippled, then spends the rest of his life in bed, immobilized not only by physical impairment but also by a miraculous and terrifying new gift. When Funes loses the use of his legs, he gains the ability to perceive and remember *everything*. "When he fell," Borges writes, "he'd been knocked unconscious; when he came to again, the present was so rich, so clear, that it was almost unbearable. . . . Now his perception and his memory were perfect." Borges inverts the problem of cognitive bias by imagining the horrific sublimity of consciousness with no filter:

> With one quick look, you and I perceive three wineglasses on a table; Funes perceived every grape that had been pressed into the wine and all the stalks and tendrils of its vineyard. He knew the forms of the clouds in the southern sky on the morning of April 30, 1882, and he could compare them in his memory with the veins in the marbled binding of a book he had seen only once, or with the feathers of spray lifted by an oar on the Río Negro on the eve of the Battle of Quebracho. Nor were those memories simple—every visual image was linked to muscular sensations, thermal sensations, and so on. . . . A circle drawn on a blackboard, a right triangle, a rhombus—all these are forms we can fully intuit; Ireneo could do the same with the stormy mane of a young colt, a small herd of cattle on a mountainside, a flickering fire and its uncountable ashes, and the many faces of a dead man at a wake. I have no idea how many stars he saw in the sky.[74]

Funes is cursed with absolute facticity: reality is fully and everlastingly present for him in its concrete totality.[75] You and I, on the other hand, are cursed with factitiousness. The worlds we move through every day are partial, invented, reconstructed, woven and rewoven out of stories and customs, imagination and error, hearsay, manipulation, and lies. Our day-to-day lives are conducted within and through various adaptive "truth-for" heuristics, schemata, practices, descriptive statements, paradigmatic examples, discursive and epistemic frameworks, narratives, and ontological commitments that have evolved because they have proven advantageous in preserving and reproducing the social order.[76] As novelist Samuel Delany puts it, "In the field of human endeavor language is a stabilizing mechanism, not a producing mechanism—regardless of what both artists and critics would prefer."[77]

Yet Funes is more than just a theoretical foil to show us how little reality we actually see. He's also an allegorical figure embodying the triumph of *scientia* over *sapientia*.[78] Funes's absolute facticity corresponds to the desideratum of positivist epistemology, the map containing every detail of the territory, wherein all reality is captured, correlated, and organized in a comprehensive whole, all of it searchable, usable, understood, and under control. Although this ideal has yet to be achieved, we are even now overwhelmed by data that no longer serve our needs but impose their own strange needs upon us. This Funesian condition goes back beyond the internet to the rupture that, in Sylvia Wynter's words, made possible "the rise of a nonadaptive . . . mode of cognition with respect to the 'objective set of facts' of the physical level of reality"—in short, the Scientific Revolution.[79]

The forging of systematic nonadaptive cognition, or modern science, was a profoundly ambiguous process. On the one hand, empirical reason was a sword with which to conquer and subdue, a new weapon in the arsenal of the emerging state. Foundational to modern science is the belief that understanding nature is a way to control it: "Human knowledge and human power come to the same thing," wrote Francis Bacon in 1620, setting out the underlying philosophical assumption of the scientific method. "For Nature is conquered only by obedience, and that which in thought is a cause, is a rule in practice."[80] At the same time, empirical reason produced knowledge that undermined traditional authority. Scientific cognition was "nonadaptive," to use Wynter's term, because it disconfirmed adaptive biases and heuristics, destabilized the social order, and undermined the metaphysical grounds of religious, cultural, and political truth.

Yet despite the revolutionary nonadaptive potential of the scientific method, science is still practiced by scientists who, however well trained in methodological skepticism, are as subject to cognitive bias as any other social primate and, what's more, powerfully susceptible to overvaluing their own judgment because of their expertise in what they tend to regard as the acme of human endeavor.[81] The more one knows about a single subject, the more one tends to confuse specific expertise with general reasoning ability: hence the hubris of climate scientists pronouncing on government policy (or literary scholars writing about climate change). Which is not to disparage scientists as a class, only to make the point already well argued by sociologists, historians, and philosophers of science, from Max Weber to Naomi Oreskes, that the purest nonadaptive cognition is still shaped, compromised, and corrupted by adaptive truth-for presuppositions, biases, and frameworks as well as grounded in metaphysical commitments that "cannot possibly be secular—open to choice—or provisional."[82] In Oreskes's words, "The idea of science as a value-neutral activity is a myth."[83] That the scientific method has so far offered human beings the most reliable way to mitigate the adaptive constraints of cultural tradition in no way guarantees that it can free us from them, nor that we would know what to do with ourselves if it did.

One might go so far as to argue, as some have, that the scientific domination of nature wasn't progress at all, but a mistake: the arrogant, self-destructive, alienating rejection of our unity with a greater ecological whole. Carolyn Merchant, for instance, argues that "the most far-reaching effect of the Scientific Revolution" was the "removal of animistic, organic assumptions about the cosmos" that she calls "the death of nature."[84] Merchant may be right, but we needn't grant her thesis to recognize that the consequences of the epistemological rupture we call the Scientific Revolution were, at best, mixed. One may as easily point to antibiotics, air conditioning, agricultural intensification, rapid intercontinental travel, refrigeration, near-universal literacy, the decline in infant mortality, new understandings of Earth and the cosmos, and the transformative connective potential of the internet as one may point to new antibiotic-resistant viruses, oceanic dead zones, hypersonic nuclear missiles, pervasive toxic and hormone-disrupting chemical and plastic waste, industrial-scale genocide, the precipitous decline in global biodiversity, unmanageably high levels of human population and consumption, the yet-unknown cognitive effects of the internet, destabilizing nihilism, and, of course, catastrophic global warming. The decades since the end of World War II, in particular, have been a period of radical transformation in the human lifeworld as a result of relentless scientific research and technological innovation, powered by an ever-increasing consumption of cheap energy, primarily from fossil fuels. Yet arguing whether or not "science" is to blame for our impasse, as Merchant and others might wish to do, is beside the point.

What we can be sure of is that science is a problem. Notwithstanding the undeniable successes and catastrophes of scientific rationality and its wanton helpmeet technology, the philosophical problem scientific epistemology poses goes to the core problem of philosophy itself—the limits of reason, which cannot ultimately distinguish between true and false. When it comes to positive claims about reality, the nonadaptive empirical sciences are limited to probabilistic generalizations about *what seems to have been* and *what might be*. Karl Popper famously argued that scientific theories, "if they are not falsified, forever remain hypotheses or conjectures," but the problem is even deeper, since we can't rely on falsification either.[85] As both Pierre Duhem and W. V. Quine argue in different ways, in what is now called the "Duhem-Quine thesis," no hypothesis about reality can be isolated from the implicit and ultimately overdetermined background knowledge and assumptions that give said hypothesis meaning, and thus "proving" a given hypothesis "false" means no more than showing that some aspect of it doesn't fit contemporary scientific paradigms.[86] As Duhem writes, "for the physiologist and chemist as well as for the physicist, the statement of the result of an experiment implies, in general, an act of faith in a whole group of theories," whether the result is positive or negative.[87]

The humanities are even less certain and of a fundamentally different character. Knowledge production in the humanities typically consists in adaptive truth-for descriptive statements stabilizing power relations, even when those statements are couched in terms of critique. Yet where the humanities do distinguish themselves and produce nonadaptive empirical insight is in their comparatist method. Historical, philological, anthropological, theological, and literary comparison demonstrate not only *that* but *how* humans are both the same and not the same over time and across different cultures, not only *that* but *how* human culture is both universal and polymorphous, not only *that* but *how* "the shifting self-interpretation of the changing historical subject" operates within a range of patterns and habits we can discern recurring again and again in different times and places.[88]

Science is relentlessly absolutist and presentist, while the humanities are essentially relativistic and historical. One has fed the world and landed humans on the moon; the other seems at times to offer nothing but pedantry and rancor. Conversely, scientific knowledge has made possible the development of world-destroying bombs and the disruption of long-term planetary cycles, while the humanities preserve and cultivate "the best that has been thought and said in the world."[89] Both are compromised by their implication in structures of oppression. Both are tools of domination, swords and shackles to conquer and enslave. Both offer at best partial authority, constrained and filtered through unavoidable biases and heuristics. Neither offers a royal road to wisdom nor a philosopher's stone transmuting expertise in one field into general intelligence or good judgment. Yet each on their own and both together offer leverage against the stories we find ourselves trapped in by connecting us to a reality beyond those stories, a reality existing independently of what we want the world to be: in the one case, physical energy and matter; in the other, historical text. Both science and the humanities are social, socially constructed, historical human practices, yet their methods and objects of study give them inestimable nonadaptive (though still not unmediated) purchase on reality. Burgess Shale doesn't care what you believe, and neither does the *Oresteia*.

"Why trust science?" Naomi Oreskes asks in her book of the same name. As she carefully explains, not only is the authority of science in public discourse tenuous, but long-held claims for scientific neutrality, objectivity, and methodological rigor have all been refuted or dismantled:

> There is now broad agreement among historians, philosophers, sociologists, and anthropologists of science that there is no (singular) scientific method, and that scientific practice consists of communities of people, making decisions for reasons that are both empirical and social, using diverse methods. But this leaves us with the question: If scientists are just people doing work, like plumbers or nurses or electricians, and if our scientific theories are fallible and subject to change, then what is the basis for trust in science?[90]

Oreskes accepts that science is contingent, partial, social, and diverse in its methods, and grants that scientists are subjective and limited. Yet we can still trust scientific knowledge, she argues, because of two core practices: "One is experience and observation of the natural world; another is the collective critical scrutiny of claims based on those experiences and observations."[91]

Scientific expertise is grounded in close, comparative attention to physical interactions and a process of communal reflection and adjudication of claims made about the insights of such study. Similarly, the humanities are grounded in close, comparative attention to historical texts and communal reflection and adjudication of claims made about them. At no point are we confronted with raw data. "As philosophers going back to Plato (and perhaps before) have long recognized," Oreskes writes, "we do not have independent, unmediated access to reality and therefore have no independent, unmediated means to judge the truth content of scientific claims. We can never be entirely *positive*."[92] Yet we can know *something* about reality beyond the stories we tell, beyond our heuristics and biases, and we can use that knowledge to make the world a better place.

At least that's the story we like to tell ourselves: a story of civilizational and moral improvement based in ever-increasing human knowledge leading to ever-increasing human power over nonhuman reality—a story, that is, of progress.

THE MYTH OF PROGRESS

The Myth of Progress is probably the most powerful story in the world today.[93] It tells us that human action can improve the general conditions of human life, that humanity has advanced from primitive societies through feudalism and industrialization and will continue to advance into the future, and that the "judgment of history" will condemn today's failures just as we condemn those of the past.[94] Although evidence is often produced for this myth, as logical proofs were once adduced for the existence of God, the Myth of Progress is a structure of belief, for which evidentiary claims are irrelevant. In the words of Finnish philosopher Georg Henrik von Wright:

> The modern Myth of Progress is a conjecture.... No facts about diminishing illiteracy, improved sanitary conditions or increased per capita income can, by themselves, prove this conjecture true. If it is inherent in the idea of modernity that there are no objective standards of goodness (value), then, in an enlightened view, belief in progress is just another article of faith.[95]

And it is an article of faith axiomatic to modern culture. As Wittgenstein put it, "Our civilization is characterized by the word 'progress.' Progress is its form. . . . It is occupied with building an ever more complicated structure."[96] In the words of Joseph Tainter, "The concepts of civilization and progress have a status in the cosmology of industrial societies that amounts to what anthropologists call 'ancestor myths.'"[97]

The Myth of Progress has its roots in that synthesis of Christian millenarianism and Baconian rationality we call the Enlightenment and has some of its most enthusiastic proponents in Enlightenment thinkers such as Kant, Condorcet, Adam Smith, William Godwin, and the founders of the American Republic. While often anti-religious from its first days ("*Écrasez l'infâme!*"), the Myth of Progress is best understood as a secularized Christian theology pursuing the "immanentization of the eschaton"—that is, as a faith that seeks to bring about the end of the world in the present.[98] Progress is, as Kermode points out, a type of apocalypse. As Karl Löwith writes, "We of today, concerned with the unity of universal history and with its progress toward an ultimate goal or at least toward a 'better world,' are still in the line of prophetic and messianic monotheism."[99] And while not necessarily or always utopian in the strict sense, the Myth of Progress is relentlessly, boundlessly optimistic. As Condorcet writes in his *Outlines of an Historical View of the Progress of the Human Mind*, an enlightened view of humankind holds "that no bounds have been fixed to the improvement of the human faculties; that the perfectibility of man is absolutely indefinite; [and] that the progress of this perfectibility . . . has no other limit than the duration of the globe upon which nature has placed us"—and if we believe Elon Musk and Ray Kurzweil, not even that.[100]

While the Myth of Progress is best understood as a secularized apocalypse that grew from the merging of Christian faith and scientific rationality, it is at the same time a repudiation of both. In its radical insistence on the human as the measure of all things, the Myth of Progress rejects both religious humility and empirical skepticism, and in its vision of temporality as an infinitely ascending curve, spurns both Christian judgment and Einsteinian relativity. As Christopher Lasch points out in his history of the idea of progress, *The True and Only Heaven*:

> Once we recognize the profound differences between the Christian view of history, prophetic or millenarian, and the modern conception of progress, we can understand what was so original about the latter: not the promise of

a secular utopia that would bring history to a happy ending but the promise of steady improvement with no foreseeable end at all.[101]

If and when God or science do appear in versions of the myth, they are beneficent agents of human material comfort, not terrifying forces revealing the mystery of existence. The God of Progress is not the God of Abraham, but the victorious Jesus of the prosperity gospel, just as the versions of science that progress favors lie in the technological realm of inventors and engineers, rather than the cosmic eons of geologists and astrophysicists.

Faith in progress pervades American political discourse from left to right, at least among middle- and upper-class college graduates, as progress is a core creed of modern liberal arts education. Your particular version of the myth may depend on the arc of history bending toward justice, or it may depend on human selfishness being "led by an invisible hand to promote an end which was no part of [its] intention"; it may take form in your rainbow bumper sticker proclaiming "Science is real," or it may take form in your commitment to innovation and entrepreneurship; it may depend on a moral commitment to universal equality or on a belief that rising tides lift all boats.[102] Whatever version of the myth one subscribes to as private creed, one cannot help participating in the public rituals and liturgies of the faith: the daily haruspicy of the S&P 500, flash sermons over the commercial break preaching the gospel of consumer joy, the latest product rollout bodying forth our newest communion, and so on.

What may be most striking about our faith in progress, however, is not its pervasiveness but rather its robustness in the face of contradictory evidence. Progress, to adapt Kermode's point about its master genre, apocalypse, "can be disconfirmed without being discredited. . . . It is patient of change and of historiographical sophistications. It allows itself to be diffused, blended with other varieties of fiction . . . and yet it can survive in very naïve forms."[103] Even many who see clearly the scope of our present impasse still espouse militant optimism regarding its resolution. From climate change and war to ecological collapse and existential risk, recognition of the intractable problems we confront often goes go hand in hand with unquestioned faith in our ability to solve them.

Witness the recurring waves of upbeat, optimistic, inspiring, hopeful, and transformative op-eds and commentaries and books we might categorize as "climate bright-siding," written with the ostensible goal of motivating public action on climate change but serving ultimately to justify a broader civilizational narrative of technocratic progress.[104] David Spratt, research director at the Breakthrough National Centre for Climate Restoration and co-author of *Climate Code Red: The Case for Emergency Action*, first wrote about climate bright-siding in 2012, on his blog climatecodered.org, where he criticized several greenhouse-gas mitigation efforts as being hampered by dogmatic optimism, including the Obama administration's failed effort to pass climate legislation under the banner of "energy independence."[105] Ten years later, Jem Bendell identified the same tendencies in a blog post on the website Brave New Europe.[106] The intervening decade saw repeated global and national political failures to construct or adopt substantive climate policy, an ongoing increase in greenhouse-gas emissions, accelerating ramifications from global warming, increasing global conflict and migration, the rise of increasingly powerful reactionary political movements, and an appalling failure of governance in response to the COVID-19 pandemic.[107] Rather than come to terms with the disastrous consequences of political and cultural failure, however, bright-siders keep on bright-siding, endlessly insisting on progress and optimism.

After all, as progressivist cheerleaders Stephen Pinker and Bjorn Lomborg argue, look at the data. "The world has been radically transformed for the better in the last century," writes Lomborg, "and it will continue to improve in the century to come. Analysis by experts shows that we are likely to become much, much better off in the future."[108] By almost any measure we might choose, argues Pinker, from infant mortality to absolute poverty, from violence to hunger, human life is materially and physically better today than at any time in recorded history.[109]

> Past decades had much worse poverty at home and in the developing world, more and deadlier wars, nuclear escalations, a homicide rate twice as high as what we have now, homosexuality illegal in many states, and much else. It's simply a mathematical fallacy to point to things going wrong now and to say life is getting worse: you have to compare it to life in the past.[110]

Most of this improvement has come since the Enlightenment, with a further surge in aggregate material prosperity occurring in the decades following World War II. One cannot help but admit that to Bartolomeo de Las Casas or

Immanuel Kant, even a modestly wealthy modern citizen today would seem a wizard or a god. Any given afternoon, millions if not billions of people cook frozen food in seconds with microwave radiation, fly around the world, talk to people living thousands of miles away, make the dead speak, conjure phantasmatic and hallucinatory stories of cosmic scope, light and heat vast structures with the flick of a switch, cure and prevent fatal diseases with pills, access seemingly infinite archives of human knowledge and culture on devices roughly the size of a piece of fruit, drive machines powered by tiny explosions faster than any land animal on Earth can run, remotely pilot battery-powered flying devices for sheer amusement, dispose of their feces with the push of a button, instantaneously find out about major events happening on the other side of the world, call up relatively accurate short-term weather reports for anywhere they might like to go, buy vegetables and fish delivered daily by air from other continents, and make visual and audio recordings of the whole spectacle to stream live on a global information network running twenty-four hours a day. That's not even talking about what we might call civilizational achievements such as nuclear weapons, aircraft carriers, space flight, hydroelectric dams, high-fructose corn syrup, in vitro fertilization, corrective eye surgery, and Wegovy. Steven Pinker and Bjorn Lomborg are right about this: as long as you're not poor, modern life is awesome.

The main problem with the Myth of Progress isn't that technological development doesn't exist. Nor is it that technological development is unequal, that it doesn't make us happy, that it's merely incremental, or that it generates whole new problems, though all these criticisms are valid. The main problem with the Myth of Progress isn't even that it conflates material well-being with moral development and biological evolution, weaving a Whiggish, Eurocentric, racist apologia for European and American imperialism, though this, too, is a legitimate criticism. It may be true that even sophisticated arguments for progress suffer from both presentism and cultural chauvinism, presuming that contemporary Euro-American liberal capitalist values are morally better than other values in other times and cultures; that they succumb to the post hoc ergo propter hoc fallacy, assuming causal relationships between moral values and other factors such as aggregate human wealth or infant mortality that cannot be established by evidence; and that they take for granted close causal connections between distinct civilizations widely separated in space and time.[111] It may also be true that, as Swedish philosopher Dag Prawitz points out, the "philosophy of science has no satisfactory notion of truth which allows us to say that science makes progress with respect to its main aim to give a true account of the world."[112] None of these quibbles really gets to the crux of the issue.

The fundamental problem with progress, it turns out, is physics. Contrary to the self-congratulatory tone adopted by boosters like Pinker and Lomborg, it's not reason, humanism, or even science that has brought about miraculous changes in the human condition over the past two hundred and fifty years, but cheap energy from fossil fuels. The fourfold increase in global energy consumption from approximately 5,654 terawatt-hours in 1800 to approximately 22,869 terawatt-hours in 1940, thanks mainly to coal, is dwarfed by the almost eightfold increase over the next eighty years (reaching approximately 176,431 terawatt-hours consumed in 2021) thanks mainly to oil.[113] It's no coincidence that this more than 3,000% spike in energy consumption happens to correlate with all the markers of modern development contemporary progressivists like to point to, such as declining infant mortality, an expanding franchise, decreasing everyday violence, and the widespread delegitimization of forced human labor.[114] Yet it wasn't any advance in human sentience or moral progress that led to this boom: it was the same ingenious bricolage that has led to all the most transformative human technological innovations, from the discovery of fire to the invention of agriculture, applied this time to easily accessible carbon stocks.

No appeal to the Enlightenment or "science" or "progress" that ignores the role cheap fossil fuel has played in increasing social complexity can be taken seriously, not least because all this "progress" has a cost—and a limit. As historians J. R. McNeill and Peter Engelke observe in their book *The Great Acceleration*:

> Within the last three human generations, three-quarters of the human-caused loading of the atmosphere with carbon dioxide took place. The number of motor vehicles on Earth increased from 40 million to 850 million. The number of people nearly tripled, and the number of city dwellers rose from about 700 million to 3.7 billion. In 1950 the world produced about 1 million tons of plastics but by 2015 that rose to nearly 300 million tons. . . . This period since 1945 corresponds roughly to the average life expectancy

of a human being.... **The entire life experience of almost everyone living now has taken place within the eccentric historical moment of the Great Acceleration, during what is certainly the most anomalous and unrepresentative period in the 200,000-year-long history of relations between our species and the biosphere.** That should make us all skeptical of expectations that any particular current trends will last for long.... One cannot say when the Great Acceleration will end, and one cannot say just how, but it is almost certainly a brief blip in human history, environmental history, and Earth history.[115]

As much as it's "simply a mathematical fallacy to point to things going wrong now and to say life is getting worse," it's equally fallacious to point to the last two hundred years of material progress as evidence for the infinite continuation of that trend. What the evidence actually shows is that humans have used a one-time energy surplus from fossilized carbon to create an incredibly complex global society (perhaps the most complex ever, if such a metric is even meaningful), to increase aggregate human wealth and general welfare beyond anything Kant or Condorcet might have imagined, to heedlessly dominate the planet to a degree that should horrify anyone with a love of wild nature or a sense of proportion, and to transform our environment beyond comprehension. This is unquestionable. What is questionable is the assumption that such a trajectory can go on. There are limits to the amount of oil than can be profitably extracted from the Earth, how much carbon waste can be dumped in the atmosphere before forcing abrupt and unstoppable climatic transformation, how much damage and exploitation the biosphere can bear, how quickly governments can transition to alternative energy sources, how much power those sources can provide, and how much instability global political and economic institutions can withstand before failing. Human hope may be infinite; the Earth is not.

Increasing technological and social complexity, growing human population, lengthened average human life expectancy, and a general increase in human material prosperity over the past two hundred years offer no evidence for civilizational or cultural progress but are rather the effects of a limited, one-time energy surplus from fossilized carbon. Things are better now for humans than they were in the past only because of this temporary affluence, which will soon come to an end. That human political and social arrangements have grown less violent and repressive in a time of immense prosperity is no surprise, but neither is it proof of moral improvement. Humans are so energy-rich today that we've grown fat, lazy, and magnanimous.[116] Shut off the tap and see what happens.

Progress is what Kermode would call an unselfconscious myth and it is a bad fiction for making sense of our impasse. Trusting in progress is not only complacent and self-serving, it's inadequate for anyone who takes seriously questions of ethics, politics, or the human future. Insisting on optimism, placing our faith in technological innovation, and asserting despite all evidence to the contrary that we can somehow *choose* to consume less energy are profoundly maladaptive strategies for dealing with the changing ecological conditions industrial society has created.[117]

THE GREAT SIMPLIFICATION

A better fiction for our moment, perhaps not as useful for maintaining the status quo but almost certainly more accurate, is the story of overshoot and collapse. We should see the destabilizing events of recent years not as momentary stumbles in the otherwise confident ascent of progress, but as a foreshadowing of what's to come as global capitalist civilization accelerates past overshoot "on an ascending curve of radical change" and plunges into uncontrollably rapid simplification.[118]

If Tainter's analysis of the increasing marginal costs of complexity is right, we don't even need climate change to push us over the edge: all we need are increasingly complex attempts to solve the problems we already face.[119] "Like the Romans, and perhaps also the Maya, we must increase the complexity and costliness of our way of life just to maintain the status quo," Tainter writes. "This will produce diminishing returns and serious consequences, including a reduction in our problem-solving abilities."[120] Carbon capitalism faces the Red Queen effect, as we are forced to run faster and faster just to stay in place. Sooner or later, we're going to fall.

Complexity scientist Peter Turchin and his team have used a massive database of historical information to demonstrate "that key aspects of human social organization tend to coevolve in predictable ways," supporting the hypothesis that "there are substantial commonalities in the ways that human societies evolve."[121] If this is accurate, then it is also likely that human societies experience simplification in predictable ways, as Joseph

Tainter and others have argued. It would be wise to use such insights to help us deal with our impasse, but our faith in progress disdains the very possibility of collapse. In the words of Miguel Centeno et al.:

> Our hubris lies not only in our overconfidence in our increasingly fragile systems, but more so in our belief that our 21st century civilization is immune to the tragic fate of fallen societies in history. While our modern societies and the systems on which they rely are at a scale, scope, and degree of complexity far greater than their historical counterparts, the mechanisms of failure and collapse remain the same.[122]

In other words, if modern fossil fuel civilization is not exempt from the tendencies, constraints, and natural limits that have governed every other complex society that has ever existed on Earth, then simplification is inevitable.

Many suspect a great simplification may have already begun.[123] As it accelerates, it will likely grow increasingly violent. It will also be global: if there is one way in which contemporary civilization differs substantially from previous civilizations, it is precisely in how it represents "the emergence for the first time in human history of a single, tightly coupled human social-ecological system of planetary scope," facing what political scientist Thomas Homer-Dixon calls "synchronous failure."[124] "No longer can any individual nation collapse," writes Tainter. "World civilization will disintegrate as a whole."[125]

In the end, no matter what unselfconscious myths we subscribe to, reality will have the final say. It is an inescapable condition of human life that we face limits, from the ultimate limit of mortality to more proximate limits of ecological dependence, epistemic constraint, social cohesion, narrative framing, linguistic indeterminacy, cognitive bias, and the radical unknowability of the future. Reckoning with such limits is a challenge under the best of conditions. Under conditions of existential crisis, ecological catastrophe, and political dissolution, the compulsion to act overwhelms judgment, and even the wise succumb to wishful thinking.

The Myth of Progress not only misrepresents our situation, but fosters a dangerous optimism. The challenge we face is not that of finding the right messaging nor making a heroic existential commitment in a moment of crisis, but rather negotiating the failing fictions we have inherited. How do we narrativize the failure of our narrative? How do we navigate the eroding metaphysical grounds of our world? How do we even talk about our impasse when it warps our sense of reality, and how might we articulate an ethical framework for dealing with the paradoxical unknowability of our forbidding future? How do we live ethically in a world of catastrophe?

Before turning to these questions in Part Two, I want to look more closely at some of the discursive, political, and conceptual problems we face. We turn in the next chapter to focus on a specific case study in climate narrative, David Wallace Wells's 2017 article "The Uninhabitable Earth," in order to explore negative messaging, problems in climate change communication, and the value of fear. Against the unfounded assertions of dogmatic optimists, evidence suggests that fear in the face of our impasse is not only justified but effective. It may well be the beginning of wisdom.

Two

THE FAILURE OF CLIMATE POLITICS

With its lurid title and splash art of a fossilized skull wearing sunglasses, David Wallace-Wells's 2017 *New York* article, "The Uninhabitable Earth," was clearly sensationalistic. Its opening sentence, "It is, I promise, worse than you think," struck a tone closer to horror fiction than journalism, and its subject matter was intentionally alarmist.[1] In spite of its obvious sensationalism, however, "The Uninhabitable Earth" was also pretty good science reporting—not flawless, but based in solid research. The article was "the result of dozens of interviews and exchanges with climatologists and researchers in related fields and reflects hundreds of scientific papers on the subject of climate change" and was explicitly framed by its author not as a "series of predictions of what will happen" but "a portrait of our best understanding of where the planet is heading absent aggressive action." That last bit—"absent aggressive action"—was key: Wallace-Wells's tone was monitory but neither fatalistic nor hopeless. The express intent was to warn people. In any case, the sensationalism worked: the piece went viral, becoming the most-read article in *New York* magazine history to that point, reaching hundreds of thousands if not millions of readers.[2]

Given the article's reach, one might suppose that scientists, activists, and journalists working on climate change would see the story as an opportunity to talk about the risks we face. Yet before the article had been online even twenty-four hours, Wallace-Wells was being attacked on the left and the right, most aggressively by the very journalists, activists, and scientists you'd think would be his allies. Anyone who observes climate change discourse will know that such hostility is no aberration.[3] Activists, climate scientists, and science journalists concerned with the threat climate change poses regularly attack popular articles that highlight just that, and the response to "The Uninhabitable Earth" was exemplary of a pervasive resistance not only to negative messaging and fear appeals, but also to public discussion of worst-case outcomes.[4] Why might this be the case? What might explain such fierce resistance to discussing worst-case scenarios? What does social science research actually tell us about negative messaging? And what do problems in climate change communication tell us about the long-term, ongoing failure of climate politics? Looking more closely at how "The Uninhabitable Earth" was received might help us begin to answer these questions.

The first attacks on Wallace-Wells's article came from Silicon Valley futurists, graduate students, and even a few actual scientists, the most significant of which was Distinguished Professor of Atmospheric Science at Pennsylvania State University Michael Mann, who offered a relatively restrained comment on Twitter, "Not a big fan of the doomist framing of new @NYMag article," then wrote a more thorough response on Facebook, which became a touchstone for the next round of discourse.[5] Mann began his Facebook post by explaining his motivation for commenting on the article, emphasizing how important it is to talk about the risks climate change poses, and noting how he himself criticized those who underplay such risks. He then turned to his main criticism of Wallace-Wells: "But there is also a danger in overstating the science in a way that presents the problem as unsolvable, and feeds a sense of doom, inevitability and hopelessness."[6]

Mann here conflates at least three distinct issues: first, "overstating the science"; second, presenting climate change as unsolvable; and third, nurturing an affective register of "doom, inevitability, and hopelessness." There is no necessary connection between these three issues, though their interdependence is often taken as axiomatic, as it is here. According to Mann's reasoning, presenting climate change as "unsolvable" necessarily overstates the science and feeds a sense of hopelessness. Yet this reasoning is neither logically valid nor empirically supported. One may overstate the science or not, and one may do it in either a pessimistic or optimistic way. Indeed, optimistic overstatement of scientific evidence in support of climate change solutions is so common that even neutral assessment, without what novelist Jenny Offill called the "obligatory note of hope," is often seen as overly pessimistic.[7] And it's possible to argue that climate change is unsolvable without overstating the science, since how societies respond to climate change is not a scientific question at all but a political one. Whether any given social problem involving disparate actors with competing interests can be solved in an abstract scientific sense has little bearing on whether it can be solved in a concrete political sense; scientifically speaking, we have the capacity to "solve" poverty, war, racism, crime, world hunger, and water scarcity, but that has nothing to do with whether or not these problems can actually be eliminated in the world in which we live. Finally, one can make an argument that presents evidence in an affective register of "doom, inevitability, and hopelessness" without either overstating the science *or* arguing that climate change is unsolvable, but Wallace-Wells didn't even do that. His article was spectacular, alarming, and frightening, but hardly lachrymose: his stated intention was to scare people into action.

The assumption that pessimistic assessments lead to paralysis is key to Mann's brief against Wallace-Wells, but the question of whether or not it's actually true is one Mann never bothers to ask. In a co-authored *Washington Post* op-ed attacking Wallace-Wells, for instance, Mann writes that "Such rhetoric is in many ways as pernicious as outright climate change denial, for it leads us down the same path of inaction," but the only support he offers for this claim is a single review article from 2014, "The Role of Emotion in Global Warming Policy Support and Opposition," which doesn't actually prove what Mann is claiming.[8] Regrettably, Mann's unfounded confidence in the conclusions of social science research he hasn't bothered to read is not unusual. In an article in *Mashable* criticizing David Wallace-Wells, for instance, science journalist Andrew Freedman wrote: "Doom and gloom only leads to fear and paralysis, studies have shown."[9] Which studies? Freedman doesn't say. Similarly, Robinson Meyer at *The Atlantic* wrote: "Over the past decade, most researchers have trended away from climate doomsdayism. They cite research suggesting that people respond better to hopeful messages, not fatalistic ones."[10] Yet Meyer doesn't bother to identify the research he refers to. Similar claims appear in an article from *Climate Feedback* titled "Scientists Explain What *New York Magazine* Article on 'The Uninhabitable Earth' Gets Wrong."[11]

Likewise, both Emily Atkin at *The New Republic* and Eric Holthaus at *Grist* argued that negative messaging about climate change is ineffective, but at least they attempted to back their claims up. The trouble was with their evidence. Atkin offers the most nuanced case, first noting that "the complaints about the science in Wallace-Wells's article are mostly quibbles," then observing that "the more common critique is that Wallace-Wells engaged in *some* hyperbole to describe what might happen, and then didn't present enough solutions or optimism to counter it. . . . Critics say such doom-and-gloom is unpersuasive and discouraging."[12] Atkin grants that "the notion that fear is a motivating force may be true in some political arenas" but asserts that "research suggests that's not quite true for climate change."

In support of her statement Atkin cites two studies and also quotes John Cook, a cognitive scientist at George Mason University's Center for Climate Change Communication, but her evidence doesn't add up. Cook, for instance, actually says (in Atkin's words) "that fear is effective." He also says that "If we communicate the solutions only, people lack the urgency that the situation requires." Not only does Atkin's own expert source say that negative messaging works, so do her two studies. The first study Atkin relies on is Saffron O'Neill and Sophie Nicholson-Cole's often-cited article, "'Fear Won't Do It': Promoting Positive Engagement with Climate Change Through Visual and Iconic Representations," which argues "that fear is generally an ineffective tool for motivating genuine personal engagement," based on survey results showing that negative representations tend to correlate with feelings of impotence, while images of small, personal actions, such as changing a lightbulb, correlate with a sense of political efficacy.[13] Yet this study is highly questionable in several respects. First, it relies on nonrepresentative sample groups from Norwich, England, one group of thirty young people ("young mothers from a deprived area, young professionals between the ages of 26 and 35, and high school students"), one group of twenty-seven parents and environmental activists, and one online survey of sixty-three members of the forum ClimatePrediction.net.[14] These sample sets are not only exceedingly small, they are also ideologically skewed toward people predisposed to think of climate change in terms of positive engagement—which is to say, activists. This matters because some research suggests that insiders fluent with communications strategies are affected less by negative emotional appeals and further, that liberals find negative messaging less persuasive than do conservatives, indicating that fear appeals may be significantly less effective among climate change activists than among the general public.[15] So while O'Neill and Nicholson-Cole found that "participants in the focus groups disagreed strongly with using fear as a communications tool," it's likely the participants were predisposed against the very strategy the survey purports to test. Nevertheless and in contradiction to its own hypothesis, O'Neill and Nicholson-Cole's study shows that negative messaging gets people's attention. As they write: "This research has shown that dramatic, sensational, fearful, shocking, and other climate change representations of a similar ilk can successfully capture people's attention to the issue of climate change and drive a general sense of importance on the issue."[16]

The second study Atkin cites is even less helpful for her argument. In their 2015 article in *Sustainability*, "Motivating Action through Fostering Climate Change Hope and Concern and Avoiding Despair among Adolescents," Kathryn Stevenson and Nils Peterson use a quantitative survey of a random sample of around 1,500 middle school students in North Carolina to see how "climate change hope, despair, and concern predict pro-environmental behavior." What Stevenson and Peterson found was equivocal and doesn't tell us much about negative messaging. They state in their abstract: "We did not find an interaction between climate change hope and concern or despair, but instead found climate change hope and concern independently and positively related to behavior and despair negatively related to behavior. These results suggest that climate change concern among K-12 students may be an important antecedent to behavior which does not dampen the positive impacts of hope."[17] That is to say, though Stevenson and Peterson found that despair independently correlated with decreased action, concern about climate change did not lead to despair, dampen hope, or inspire inaction, but rather correlated with pro-environmental behavior. As they write, "Avoiding despair is important, however climate literacy efforts should not shy away from communicating the seriousness of climate change as adolescents are likely capable of productively responding to concern." This conclusion directly contradicts Atkin's interpretation.

Finally, in an article for *Grist* titled "Stop Scaring People About Climate Change. It Doesn't Work," science journalist Eric Holthaus argues that "if you're trying to motivate people, scaring the shit out of them is a really bad strategy" and that "time and time and time again, psychology researchers have found that trying to scare people into action usually backfires."[18] Holthaus, like Atkin, includes links to references purportedly supporting his argument, but as with Atkin, when you look at his evidence, the argument falls apart. His first piece of evidence is a study from *Nature Climate Change* by Paul Bain et al., that argues, "Communicating co-benefits could motivate action on climate change where traditional approaches have stalled."[19] While Bain et al. do demonstrate the possible effectiveness of some positive messaging, they have nothing to say about the effectiveness of fear appeals or negative messaging. Holthaus's second piece of evidence is an article in *Time* magazine by Dean Ornish, a medical doctor best known as a popular advocate for treating and preventing heart disease with diet and lifestyle changes. Ornish writes, "Fear is not a sustainable motivator—in health or in politics. In the short run, fear is powerful, it gets our attention. . . . In the long run, though, it's too scary to think that something really bad may happen to us, so we usually don't, at least not for long."[20] Ornish offers various anecdotes to support his arguments but provides no references to empirical research.

The third piece of evidence Holthaus cites is a news summary by Paul Stern of more research by Paul Bain et al., again from *Nature Climate Change*.[21] Stern writes: "Scientists often expect fear of climate change and its impacts to motivate public support of climate policies. A study suggests that climate change deniers don't respond to this, but that positive appeals can change their views."[22] In fact, the article Stern discusses doesn't specifically address fear appeals. Rather, Bain et al. address the question of value-based messaging in addressing conservatives, or as their title puts it, "Promoting Pro-Environmental Action in Climate Change Deniers." Their study finds that "to motivate deniers' pro-environmental actions, communication should focus on how mitigation efforts can promote a better society, rather than focusing on the reality of climate change and averting its risks."[23] That is, Bain et al. found that arguments about values are more persuasive than arguments about facts, specifically among climate change deniers. This conclusion is important for climate messaging and is well supported by other research showing that fact-based messaging fails to convince conservatives regardless of whether the messaging is positive or negative, but it says nothing about the general effectiveness of negative messaging or fear appeals. In sum, Holthaus cites three supposedly reputable sources that show how research has discredited the effectiveness of negative messaging, but his only source that addresses fear appeals is an article by a pop science advocate who's made his career selling diet books.

What's going on here? One wants to assume that these journalists and scientists are serious people of good intentions, with professional commitments to their respective fields' standards of evidence. Why, then, would people who in other contexts are so concerned with evidence—who, indeed, criticize David Wallace-Wells for exaggerating his—be so negligent and tendentious when it comes to assembling their own evidence against negative messaging?[24] Are narrow, dubious studies and articles from diet fad promoters really the best evidence there is? What do studies actually show?

PROBLEMS IN CLIMATE CHANGE COMMUNICATION

To begin with, research suggests that there are significant challenges to effective public discussion of climate change in practically every respect.[25] Each audience brings its own concerns, biases, and beliefs to the discussion, and each messenger brings their own baggage, blind spots, and presumptions, all of which have consequences.[26] What's more, the standards and norms of scientific discourse differ significantly from those of public discourse, which is itself not monolithic but composed of many different discursive communities.[27] Because of this, scientists and science communicators face what Stephen Schneider famously called a "double ethical bind," a tension "between being an effective agent for change and being honest about the limitations of the state of knowledge."[28] As Chris Russill explains, "Climate scientists find it very difficult to adhere to the conventions of media discourse and scientific discourse at the same time; their capacity to reframe the interaction by commenting on the context via meta-discourse or meta-communication is limited; and they cannot avoid the interaction without losing all opportunity for public education."[29] Hence two reasons why Michael Mann might attack Wallace-Wells: first, Wallace-Wells doesn't talk like a scientist, and second, the standards Mann adheres to professionally (if not on social media or in *Washington Post* op-eds) preclude the kinds of simplification, generalization, and narrative framing necessary for broad public engagement.

Another major challenge from the communicator's point of view is that no one knows what they're doing. They might know the science, and maybe they've had executive-level TED-Talk training in "storytelling," but nobody actually knows how social change happens, much less how to motivate it. In spite of valuable research on nudging, framing, heuristics, and messaging, one of the most persistent failures of the social sciences in general is the lack of any robust understanding of the relationships between language, culture, and long-term social change. As communications researchers Brigitte Nerlich, Nelya Koteyko, and Brian Brown put it, "On the whole, there is no direct correlation between communication and behavior change."[30] While we may make general observations and draw inferences from the historical record, human society is too complex to effectively model or manage.[31] Which is not to dismiss the vast and ever-growing body of research on the subject from political science, psychology, marketing, social media companies, sociology, history, communications, and even philosophy and rhetoric, but only to note that what conclusions may be drawn from such research are limited, provisional, and often gloomy.

There is, for instance, good evidence that positive messaging on climate change doesn't motivate people.[32] One such study, by Yanni Ma et al., suggests that messaging focused on the basic facts about climate change actually turns people away.[33] Recent research by psychologists Matthew Hornsey and Kelly Fielding suggests that optimistic messaging may decrease mobilization because it fosters complacency. They write in their study "A Cautionary Note About Messages of Hope," "Relative to a pessimistic message, the message of hope *reduced* mitigation motivations."[34] Another study showed that activist-style persuasive messages "were either ineffective or demobilizing" for people who knew a little about climate change, and failed to convince people who knew a lot about climate change to engage in any behavior that demanded time, effort, or economic sacrifice.[35] Another study found that showing people either "hope" or "doom" videos about climate change had no noticeable effect on their beliefs or willingness to take action.[36] According to a 2021 meta-analysis from Jacob Rode et al., the general effect of climate messaging is equivocal at best: people are more willing to entertain doubts than they are to commit to a strong stance on the issue; negative messaging is generally more effective than positive; and it's harder to change people's support for actual policies than it is to change their beliefs.[37] More recently, a large international study with over 59,000 participants tested eleven different kinds of climate messaging interventions against various outcomes, including "completing . . . a tree-planting task . . . resulting in the real planting of a tree," with disappointing results.[38] While the study showed that some kinds of messaging correlated positively to increased belief and willingness to share information, none of the tested messages showed any positive effect on tree planting and half of them seemed to decrease it, suggesting that messaging on its own may be ineffective for motivating effortful action on climate change. Worse yet, an ongoing replication study on the effectiveness of various climate change messages has so far failed to replicate the findings of several highly cited studies, raising questions about the efficacy of climate change communication in general and the reliability of the social science supporting it.[39]

So much for messaging. What about the audience? The truth is, most people don't have the time or energy to think through complex issues that seem abstract and distant, and when they do, they tend to think more like lawyers than like scientists, amassing evidence for what they already believe rather than disinterestedly seeking objective truth.[40] The combination of motivated reasoning and confirmation bias means that people tend to actively resist data that challenge their beliefs and behaviors. Policymakers are as susceptible to such biases as anyone else, sometimes making decisions out of ignorance and confusion while at other times intentionally misrepresenting evidence and exaggerating existing uncertainties, in either case delegitimizing the political use of science and fostering warranted public skepticism.[41] People also tend to reason more from personal experience and group norms than from rigorous analysis, which can be especially debilitating with regard to climate change, since the phenomenon is relatively slow on individual human time scales and people quickly get used to changing temperatures: research on "shifting baseline syndrome" shows that abnormal temperature and weather patterns can become normalized in just five years.[42] Then there's the fact that climate change is scary and portends real losses, which people don't like to think about.[43]

Compounding these problems is the shaky state of scientific literacy in the US: according to 2019 Pew research, fewer than 40% of Americans surveyed showed even high-school-level scientific knowledge.[44] The situation is even worse with climate change specifically, since most people think they know more than they actually do. Research by Yale law professor Daniel Kahan and others in 2014 showed that while 67% of Republicans and 84% of Democrats surveyed believed they knew at least a moderate amount about global warming, significant majorities across the spectrum were mistaken about basic facts and mechanisms, and held erroneous beliefs about how global warming actually works (and more Democrats held false beliefs than Republicans).[45] Perhaps related to this general lack of scientific literacy in the US is a general lack of trust in scientific authority. Only 23% of Americans today have "a great deal of confidence" in the scientific community, a number that has been falling for the past several years, and while conservatives have significantly less confidence in science than do liberals, the overall decline has been bipartisan.[46] The number of Americans who believe science has a mostly positive effect on society has also declined. Many Americans, especially conservatives and people of color but also nearly half of all surveyed moderates and approximately a quarter of surveyed liberals believe that scientists adjust their findings to get the results they want.[47] And populist antipathy toward sources perceived as elitist—such as scientists or the pope—may be an even stronger motivator for resistance to climate change messaging than ideology.[48] Populist anti-elitism tends to correlate to climate skepticism and distrust toward environmental policies, as climate and the environment are perceived as elite-driven issues.[49] Numerous studies show that people tend to believe things that conform to what they already think, to be skeptical of information that challenges their beliefs and behaviors, and to resist authoritative messaging advocating personal change.[50]

While such resistance seems to hold true across the political spectrum, there's no denying that another major challenge in climate change communications is the stark partisan divide on the issue.[51] Many on the left consider this divide the result of ignorance and anti-science bias among conservatives, and research does show a strong correlation between conservative political identification and distrust of scientific experts, especially experts whose research "identifies environmental and public health impacts of economic production" or who are involved in public debate.[52]

Overall, however, the partisan divide on climate change is less about bias or ignorance and more about group identity. Research from Kent Liere, Riley Dunlap, and others shows that it's not science per se that conservatives reject when it comes to climate change, but the cultural identification they see belief in climate change as marking.[53] As Kahan puts it, "'Belief in human-caused global warming' items measure 'who one is, what side one is on' in an ugly and highly illiberal form of cultural status competition, one being fueled by the idioms of contempt that the most conspicuous spokespeople on both sides use."[54] This view is substantiated widely across several studies in different fields.[55] Such sectarianism has little to do with being closed-minded or uneducated. Kahan and Corbin found that critical, active, open-minded thinking in conservatives actually correlated with *increased* skepticism toward climate change, and another study led by Kahan found that "people with stronger analytic abilities are more likely to twist data at will than people with low reasoning ability," across the political spectrum.[56] In fact, partisan divisions on climate change tend to *widen* with education, rather than decrease.[57]

"What you 'believe' about global warming doesn't reflect what you 'know,'" argues Kahan; "it expresses 'who you are.'"[58] Or as Mathew Hornsey puts it, "People view climate science through the lens of their worldviews."[59] This even affects how people perceive their own firsthand experience of climate change: research by sociologist Lawrence Hamilton shows that politically conservative observers in areas of New Hampshire with increased flooding and in wildfire-prone areas of Oregon with significant warming simply did not see those trends.[60] Sectarian framing impedes climate change communication with both liberal and conservative audiences, as it nurtures confirmation bias and provokes resistance to evidence that doesn't align with cultural values, whether those values favor individualism, Christian identity, and the free market or social justice, progressive multicultural democracy, and meritocratic elitism.[61] Such polarization may be one of the key reasons why most Americans—around 60%—rarely or never talk to family and friends about climate change.[62] As Erik Nisbet et al. have shown, the very fact that climate change is politically controversial increases negative emotions, cognitive resistance, and distrust of the scientific community among *both* conservatives and liberals.[63]

And although polarization may be an organic feature of partisan political conflict, there is also good evidence that it has been actively fostered by a "network of political and financial actors," notably large corporations.[64] It's not too much to say that corporate influence pervasively corrupts climate change discourse. Research from Robert Brulle, Erik Conway, Geoff Dembicki, Naomi Oreskes, Geoffrey Supran, and others has conclusively established that motivated actors among the wealthy and powerful, particularly in the fossil fuel industry, have been working to spread misinformation, decrease public engagement, obscure key issues, dominate the overall framing, and focus attention on individual action rather than systemic change.[65] "The very notion of a personal 'carbon footprint,' for example," write Oreskes and Supran, "was first popularized in 2004–2006 by oil firm BP as part of its $100+ million per year 'beyond petroleum' US media campaign."[66]

Yet another problem is that the topic itself is forbiddingly complex. The basic elements of global warming can be explained simply, but the mechanisms of planetary climate transformation are stupendously intricate and, in many cases, still being worked out. Even highly educated laypeople can find it hard to keep up with the changing science and are easily confused by perplexities such as the role of paleoclimate proxy data, how various climate models differ, and jargon like RCPs, SSPs, AMOC, and CP–ENSO.[67] The issue isn't made any easier by climate change being what's called a "wicked problem," meaning there's no straightforward, one-and-done solution, like getting rid of chlorofluorocarbons (CFCs) to stop making a hole in the ozone layer. Rather, it's the kind of problem that can at best only be managed, and even that would take an array of strategic interventions and an unprecedented level of global cooperation. Indeed, talking about climate change "solutions" may be more fraught than talking about the problem, since in addition to all the other issues we've looked at, the discourse about solutions is rife with wishful thinking, political manipulation, greenwashing, predatory scams, and outright hucksterism.[68] Finally, there's the fact that climate change is an existential threat that entails real losses even if we get it under control, coupled with the fact that genuinely addressing it would mean a comprehensive transformation of the way modern society works. Climate change means *real change*, which humans tend to be skeptical and even frightened of, especially when it involves unpredictable costs.

FEAR WORKS (MOSTLY)

Meanwhile, leading voices in climate change communication not only disdain addressing the complicated, unpleasant emotions that arise from seriously considering the significant risks climate change poses but enthusiastically attack anyone they deem too pessimistic. Climate scientist Katharine Hayhoe put the charge to her followers on Twitter, writing, "We need to push back just as strongly on doomerism as we do on denial because they both accomplish exactly the same thing: inaction."[69] Michael Mann has similarly argued that pessimistic assessments are "as pernicious as outright climate change denial," as cited above, and devoted an entire chapter of his 2021 book *The New Climate War* to attacking what he calls "doomism."[70] Yet none of these claims against negative messaging and the discussion of worst-case scenarios are supported by evidence. As Joseph Romm wrote in 2013, "The two greatest myths about global warming communications are 1) constant repetition of doomsday messages has been a major, ongoing strategy and 2) that strategy doesn't work and indeed is actually counterproductive."[71]

Neither of these myths is true. In fact, climate communications and big science studies like the IPCC reports tend to *underrepresent* extreme possibilities, and numerous studies across disparate fields show that negative messaging emphasizing threats is highly effective for getting people's attention and influencing public behavior, at least over the short term.[72] Substantial research on negative messaging began with Ronald Rogers's "protection motivation theory," first posited in 1975, and the consensus scientific position on the subject was established in the early 2000s with two key meta-analyses in the fields of psychology and health education.[73] Roy Baumeister et al. found that "Negative information receives more processing and contributes more strongly to the final impression than does positive information," or, put another way, "bad is stronger than good, as a general principle across a broad range of psychological phenomena."[74] Another meta-analysis by Kim Witte and Mike Allen found that "strong fear appeals produce high levels of perceived severity and susceptibility, and are more persuasive than low or weak fear appeals."[75] More recent research has confirmed the effectiveness of fear appeals, with a large 2015 meta-analysis concluding that "(a) fear appeals are effective at positively influencing attitude, intentions, and behaviors; (b) there are very few circumstances under which they are not effective; and (c) there are no identified circumstances under which they backfire and lead to undesirable outcomes."[76] As public health and climate change communications expert Edward Maibach and co-authors put it, "The climate change literature contains frequent warnings to avoid fearful messages, yet the more general persuasive communication literature indicates that fear appeals are effective in motivating behavior change, especially if they are accompanied by efficacy-enhancing information."[77]

As Joseph P. Reser and Graham L. Bradley write in the *Oxford Encyclopedia of Climate Change Communication*, "there is majority agreement that fear appeals can be effective."[78] According to Witte and Allen, the scarier the better. "The stronger the fear appeal," they write, "the greater the attitude, intention, and behavior changes."[79] The reason for this is theorized to lie in what researchers call "negativity bias," or in the words of Roy Baumeister and John Tierney, "a universal tendency for negative events and emotions to affect us more strongly than positive ones."[80] Not only do studies in political science, psychology, marketing, and health education support the effectiveness of negative messaging, but so do studies specific to climate change. As Reser and Bradley write, while "relatively little experimental research has examined the impact of fear appeals on attitudes, intentions, and behaviors in the environmental domain, and even less has focused on messages that contain climate change related content," and this "modest body of experimental research ... has yielded mixed findings," a majority of studies nevertheless offer "at least partial or qualified support for the effectiveness of fear appeals in promoting environmentally responsible attitudes and behaviors."[81] Research by Thomas Lowe et al., showed that a number of people who saw the climate disaster film *The Day After Tomorrow* "expressed strong motivation to act on climate change; more so than prior to seeing the film" and that less than 5% of their sample "believed that there was no point in taking action."[82] This is striking because *The Day After Tomorrow* is not only ridiculously inaccurate but super cheesy.[83] Why psychologists would choose such a film to test the effectiveness of climate messaging is a real question, but the results of the study suggest that even bad negative messaging can be effective.

Contradicting the baseless assertions of the anti-doom tone police, substantial research suggests that fear can motivate action on climate change or at least increase people's intentions to act.[84] Research from the Centre for Climate Change and Social Transformations at the University of Bath shows that climate anxiety correlates to a range of pro-environmental actions, suggesting that it may be an "adaptive, motivating response to climate change that encourages effective action."[85] An international study from psychologist Charles A. Ogunbode et al., surveying more than 12,000 respondents across thirty-two countries, found that while climate anxiety was negatively linked to mental well-being, it was positively linked to the amount of attention people pay to climate change information, pro-environmental behavior, and activism, though the correlation between anxiety and activism was "largely confined to Western and relatively affluent countries."[86] Research from Wen Xue et al. on climate change messaging in China showed that high-threat messaging "elicited higher levels of perceived efficacy in viewers."[87] Another study from the Yale Program in Climate Change Communications found that Americans who had experienced climate distress were much more likely to have taken action, more willing to take future action, more willing to join a climate campaign, and more willing to talk about climate change than people who hadn't experienced climate distress, while multiple studies from Naoko and Kosuke Kaida suggest that "a pessimistic perspective of future subjective well-being facilitates pro-environmental behavior," and work by communications and media researchers Sol Feldman and Lauren Hart shows a positive correlation between fear and climate activism.[88] Moreover, Feldman and Hart found that fear messaging is not only effective at getting attention across the political spectrum but also "encourages opinion moderation among political ideologues, especially conservatives."[89]

Several thoughtful recent studies from Matthew Hornsey and Kelly Fielding et al. at the University of Queensland show not only that threat messaging works, but that it might actually *increase* a sense of collective efficacy.[90] That is, the more frightening the threat, the more people believe that society can effectively respond. Similar results were found by Brandi Morris et al. "In contrast to work positing that 'fear appeals' may hinder efficacy," Morris writes, "our results suggest that pessimistic endings actually *increase* people's belief that their own individual behavior matters for climate change, even in the face of fatalistic messaging."[91] That is, their research counterintuitively suggests that doomist messaging may actually *decrease* despair and *increase* agency.

Of course, motivating action isn't as simple as scaring people. Most research tends to suggest that negative messaging works better when it's combined with a sense of efficacy: you don't just show people the monster, you show them how to beat it. Kim Witte established this idea with her extended parallel process model [EPPM], which argues that messages combining high threat and high efficacy provoke people to address the threat, while messages combining high threat and low efficacy provoke people to avoid the message. As Witte puts it: "When both perceived threat and perceived efficacy are high . . . individuals respond to the danger, not to their fear. Conversely, when perceived threat is high, but perceived efficacy is low . . . individuals respond to their fear, not to the danger."[92] Witte's later meta-analysis supports her theory, as do a 2011 reassessment and numerous subsequent studies.[93]

Finally, although the majority of evidence supports the argument that fear works, some research complicates the picture, especially in relation to climate change.[94] Different people respond to fear in different ways. Some respond more assertively, others more aversively.[95] Some fight, some flee. Conservatives appear to find negative messaging more persuasive than do liberals.[96] And some research shows that insiders fluent with communications strategies are less affected by the emotional appeals in negative messaging, which suggests that repetition and familiarity may dull impact.[97] What's more, nobody can pay equal attention to everything, a simple truism that researchers refer to as the "finite pool of worry." Thus different people prioritize different threats differently at different times, and such assessments typically have more to do with salience than with the actual threat posed.[98] Some threats are going to be judged more pressing simply because they're more easily remembered or imagined, a phenomenon psychologists call the availability heuristic.[99] Getting people to see climate change as salient has been a persistent challenge, for all the reasons discussed previously, and the efforts of climate bright-siders like Hayhoe and Mann to dampen and marginalize negative messaging on the topic only makes that challenge harder.

Researchers have also brought up ethical concerns about negative messaging and methodological questions about the science, and a few studies have attempted to provide empirical evidence that negative messaging doesn't work.[100] Evidence against negative messaging is scarce, however, as representative studies and meta-analyses show, and it is often compromised by sampling error and other problems. A 2022 study by Canadian researchers Marjolaine Martel-Morin and Erick Lachapelle, which purports to show "that negative messaging can be less mobilizing than positive messaging," suffers from sampling error in that their sample set is limited to 308 environmental group supporters—that is, activists who are likely going to be biased against negative messaging where they are not inured to it by overfamiliarity.[101] Yet another study, by Sonia Merkel et al., claims that "preliminary findings suggest that hopeful, solution-based messaging may be more effective in facilitating pro-environmental behavior than either fear- or guilt-based appeals," but this survey had only twenty-four respondents, mostly young, liberal University of Oklahoma college students.[102] The authors themselves admit, "Our study was a locally recruited, non-representative sample, in which generalizability is ceded."

Another often-cited study, titled "Apocalypse Soon? Dire Messages Reduce Belief in Global Warming by Contradicting Just-World Beliefs," is just as flimsy. In this study, Matthew Feinberg and Robb Willer seek to explain stable or increasing disbelief in climate change among Americans by arguing that "information about the potentially dire consequences of global warming threatens deeply held beliefs that the world is just, orderly, and stable" and that "individuals overcome this threat by denying or discounting the existence of global warming."[103] The authors rely on two surveys to show that "less dire messaging could be more effective for promoting public understanding of climate-change research." One immediate problem with the study is, again, the narrowness of the sample: One survey uses a sample of less than one hundred college students, the other survey a sample of forty-five people solicited on Craigslist. How these tiny samples might be supposed to be representative of anything is hard to see. In any case, the study hasn't withstood the test of science. A recent attempt to replicate the Feinberg and Willer study using a much more robust sample set failed.[104] The authors of the replication study write, "There was no evidence that dire messages affected belief in climate change among participants low or high in belief in a just world and in the less or more solvable solution condition. In short, the Feinberg and Willer model was not supported."

Overall evidence suggests that fear works, mostly, but in ambiguous ways. Conversely, there is little evidence that positive messaging works, at least on its own, at least when it comes to climate change, and there is good evidence suggesting that it often promotes complacency and resistance. More broadly, current research into climate change communications and the general scientific literature on persuasion do not encourage an optimistic view of the human animal. They paint a picture of fearful, self-interested primates who care more about social belonging than they do about understanding reality, and they provide little evidence to inspire faith that finding the right message or crafting the right story is going to shift the conversation. Cultural conflict over climate change looks less like a war of maneuver than a war of attrition, with both sides digging in and trying to bombard each other into silence.

Which brings us back to the question with which we began: What might explain the zealous commitment among so many climate scientists, science journalists, and activists to the idea that negative messaging doesn't work, in spite of overwhelming evidence to the contrary? Why, if the problem of climate change is so serious and urgent, do so many of those who advocate addressing it seem hostile to open discussion of how serious and urgent the problem really is?

One reason may simply be that they're scared. After all, climate scientists, science journalists, and activists are among those most sensitive to the real danger unrestrained global warming poses, and numerous articles have drawn attention to the psychological costs of coping with this knowledge.[105] Dogmatic optimism and the rejection of negative messaging may simply be how some people manage their anxiety. Another reason why science journalists in particular might go along with a perceived consensus dismissing negative messaging could be changing conditions in science journalism that are making it harder for reporters to do their job. As Sharon Dunwoody and others have shown, structural changes in the media landscape over the past three decades, including digitization, consolidation, and the casualization of labor, pose serious challenges to the integrity and quality of science journalism.[106] Science journalists are often inadequately trained and poorly supported freelancers, who find themselves thrown into a chaotic and demanding digital media environment where they must compete for attention against and sometimes come into conflict with non-journalist science experts, amateur commentators, purveyors of misinformation, and occasionally hostile members of a distrustful public. They may also face pressure from the government to skew or censor their reporting, particularly on important public issues.[107] Such conditions are hardly likely to foster thorough, independent reporting and more likely to encourage fast and shallow "explainers" that avoid controversial issues.

Another possible reason for irrational resistance to negative messaging could be anxiety about scientific authority, including the fear of being labeled alarmist or extremist. Scientists tend to be highly conservative in their assessments, even underestimating the effects of climate change compared to conservative projections from the IPCC, and can be very quick to attack journalists they perceive as encroaching on their territory, as we've seen.[108] In his critique of Wallace-Wells, Robinson Meyer argues that one reason scientists resist negative messaging is that "they want to be exceptionally careful with facts for such a vital issue," but they might also be jealously guarding their declining public status.[109]

Consider one of the most significant and illustrative events in the brief history of climate change communication, the 2009 Climate Research Unit email controversy known as Climategate. Thousands of emails and other documents were stolen by hackers from the Climate Research Unit at the University of East Anglia and disseminated across the internet, where they were picked up by right-wing media and climate denialists, who alleged that the documents showed scientists involved at the CRU manipulating and misrepresenting data in a conspiracy to overstate the evidence for climate change. By and large, Climategate was a manufactured scandal, an unfortunately successful smear attack: what the documents actually illustrated was how difficult it is to do good science, how data don't exist independently of interpretation, and how scientists are fallible human beings who can be as territorial, biased, and rude as anybody else. In sociologist Reiner Grundmann's assessment, the scientists involved in Climategate "did not act in a disinterested way" but rather acted "strategically, showing self-interest and zeal" and actively "tried to stifle skeptical voices."[110] Science journalist Andrew Revkin noted at the time, "In one e-mail exchange, a scientist writes of using a statistical 'trick' in a chart illustrating a recent sharp warming trend. In another, a scientist refers to climate skeptics as 'idiots.'"[111] The scandal was a major topic in the mainstream media for several weeks, led to a House of Commons Science and Technology Select Committee inquiry in Britain, and was likely traumatizing for the scientists involved, if not for the profession. It certainly changed the conversation about climate change, leading to "significant declines in Americans' climate change beliefs, risk perceptions, and trust in scientists" between 2008 and 2010.[112] It's possible that the knee-jerk antipathy toward negative messaging shown by Michael Mann and others is a consequence of persistent conservative attacks on their integrity and research, in which case the petroleum flacks will have succeeded beyond their wildest dreams by inspiring climate scientists to minimize the salience of their own work.[113]

Another possible motivation for reasoning against the evidence on negative messaging could be the pervasive human bias toward optimism and, specifically, American cultural optimism. Pessimism on *any* issue is chastised as cynical and nihilistic, especially but not exclusively by progressives, a topic to which we'll return. This is related to yet another reason why people might make arguments against negative messaging, which is that focusing on negative outcomes may support adaptative strategies, while many climate journalists and scientists appear to favor mitigation. That is to say, dismissing negative messaging might be a way of foreclosing a more complicated discussion about what kind of action to prioritize. As John Holdren, former president of the American Association for the Advancement of Science, told the *New York Times* in 2007, in what has become a touchstone summary: "We basically have three choices: mitigation, adaptation and suffering. We're going to do some of each. The question is what the mix is."[114] Many proponents of mitigation dismiss adaptation as defeatist, not only immoral but un-American, "neo-Malthusian," pro-corporate, and even ecofascist.[115] Al Gore himself characterized adaptation as a "kind of laziness, an arrogant faith in our ability to react in time."[116] Critics often conflate arguments for adaptation with arguments for inaction.[117] And as Debra Javeline and Sophia Chau argue, adaptation is poorly understood and "riddled with uncertainty and controversy."[118]

Although it seems clear that from an abstract viewpoint society should prioritize mitigation, in the face of a Category 5 hurricane or wildfire, it's not "counseling despair" to advocate flight rather than "choosing hope." Similarly, in the face of 2°C or greater warming, now regarded by many scientists as inevitable, mitigation on its own is clearly inadequate: any responsible climate policy must prioritize adaptation.[119] Nor is it "giving up" to express skepticism about the good will or capability of governing elites to show responsible leadership on the issue after more than three decades of outrageous negligence. In any case, sometimes giving up makes sense: according to psychologists Charles S. Carver and Michael F. Scheier, "giving up is a functional and adaptive response when it leads to the taking up of other goals."[120] The question of messaging in no way answers the question of what our goals should be, which is not a question that can be resolved by science. At the end of the day, arguing about messaging seems like an instance of the substitution heuristic, a reasoning fallacy that substitutes an easier problem for a harder one.[121] It's far simpler to talk about how we should talk about climate change than it is to talk about how we're going to live with it. Perhaps this suggests yet another reason why climate scientists, journalists, and activists ignore the evidence on negative messaging: they're afraid scaring people might motivate them, but in the wrong way.

RIGHT-WING CLIMATE POLITICS AND ECOFASCISM

After all, as we saw with COVID, recognizing a global threat to "our way of life" doesn't necessarily mean there's any clear consensus on what the threat is, how "we" should respond, or even who "we" are. The progressive left, liberals, conservatives, and the far-right disagree profoundly with each other and amongst themselves about what risks and costs are acceptable for what goals and ideals. These disagreements are just what distinguish such moieties. Appeals to "the science" are inadequate: despite its ideals of neutrality and objectivity, science is shaped by politics and identity, and how people understand science is itself political. Even more fundamentally, our ideas about who and what science is talking about, about "nature" and the "human," aren't scientific at all but cultural and even ontological. They are also under profound stress from the emergent crises constituting our impasse. Expecting crisis to resolve cultural and political conflict is to expect the dissolution of political difference as such; it is a fantasy of unification through trauma, a "state of exception" where the sovereign will of the collective transcends dissensus.[122] As history shows, however, long-term crises tend not to resolve political conflicts but rather exacerbate them, even and perhaps especially when such conflicts are masked behind coercive ideological unification.[123]

In response to recent crises such as COVID, we have seen just this: not unification, but polarization. One effect of such polarization is increased insularity within ideological groups and decreased willingness to engage across disagreement. Concerned climate scientists, academics, activists, and journalists, for instance, often seem less interested in persuading their opponents than in reassuring their supporters.[124] Climate protests are a telling example: sociologist Dylan Bugden found that civil disobedience and peaceful marches increase public support for climate activism, but only among Democrats and independents who already believe in climate change.[125] Rebecca Solnit explains the appeal of such activities in an essay celebrating the self-congratulatory comfort of "preaching to the choir": "Why else do we go to church but to sing," she writes, "to pray a little, to ease our souls, to see our friends, and to hear the sermon?"[126]

Not all researchers, scholars, and activists concerned with climate change have been content to preach to the choir. Some have explored framing climate change in terms that speak to conservative values, most publicly the evangelical Christian atmospheric scientist Katherine Hayhoe. Such efforts may seem, at first glance, promising. A study by Graham Dixon et al. found that "highlighting free market solutions to climate change . . . appears effective in shifting beliefs about climate change among conservatives."[127] Another study by Matthew Feinberg and Robb Willer found that "framing environmental messages in terms of the moral value of purity fosters an increase in pro-environmental attitudes in conservatives."[128] Similarly, a study by Christopher Wolsko et al. found that depicting environmental protection "as a matter of obeying authority, defending the purity of nature, and demonstrating one's patriotism to the United States" persuaded conservatives to favor environmental values.[129]

On reflection, however, there are real problems with such strategies. The trouble with highlighting free-market solutions to climate change, for instance, is that it's not clear that free-market solutions are viable or that free-market capitalism can be reconciled with ecologically sound and sustainable human social organization. Similarly, while framing climate change and environmental issues in terms of purity, authority, and patriotism may be effective, such strategies are doubtful and possibly dangerous.

First, such framing faces an ideological contradiction in that effective climate change mitigation must necessarily be internationalist rather than nationalist in approach—that is, successful carbon emissions reductions depend on international regulatory cooperation, rather than on nationalist economic and political competition. One might imagine something like a Carbon Olympics, where nations compete for status based on their emissions reductions, but such fanciful ideas elide the central role fossil fuels play in production, trade, and military power. Second, the values of purity, patriotism, and authority fit ambiguously with core environmental concerns about environmental racism and climate justice. Defending one's own nation and identity aligns uneasily with demands for distributive justice, the needs of people displaced by climate disaster, and calls to recognize the long history of "slow violence" directed against ethnic and racial minorities, as well as the poor.[130]

What's more, while research shows that values matter when it comes to general attitudes about climate change, the question of whether value reframing works to shift support for specific policies is much more equivocal.[131] In a 2016 research letter for *Nature Climate Change*, Thomas Bernauer and Liam McGrath argue that there is no "robust empirical evidence for alternative framing (justification) of climate policy being able to increase public support for GHG mitigation."[132] According to a 2021 study by Hilary Byerly et al., while narrative framing about climate change is likely to persuade liberals, it is much less likely to persuade conservatives and may even provoke resistance.[133] A study by Lukas Fesenfield et al. on the role and limits of strategic framing similarly showed no significant impact on the consumption of fossil fuels, fish, or meat.[134] Likewise, a study from Laura Arpan et al. found that moral framing had no effect on people's support for green pricing programs or renewable energy, and Samantha Stanley et al. found that "comparing the current state of the environment with the past (and not the future)," or what they call "temporal framing," was ineffective and that, in general, "current attempts to promote action by appealing to conservative values have returned mixed results."[135]

In an article summarizing Bernauer and McGrath's research, science journalist David Roberts posits three main reasons why reframing fails. First, "people's lives and identities have great inertia." Second, people are guided by confirmation bias and group norms. Finally, "source and repetition matter more than cleverness."¹³⁶ These problems are compounded across the political spectrum by the severe threat climate change poses to modern cultural values of freedom, individualism, consumerism, and, of course, progress. As Robert Brulle and Kari Norgaard put it, "climate change constitutes a profound challenge to established ways of life in Western nations and constitutes the emergence of an ongoing and expanding cultural trauma."¹³⁷ The cultural and conceptual destabilization our impasse poses not only precludes but actively undermines sociopolitical unity. Thus, even if clever reframing, national security, and message saturation were able to turn climate change into a viable conservative issue, the likely outcome wouldn't be national or international consensus but rather new partisan divisions.

Given reactionary and anti-democratic tendencies on the right, we might want to think through the possible consequences of framing climate change as a conservative issue at all. We've seen what reactionary responses to climate catastrophe look like, for instance in the work of biologist Garrett Hardin, who gave us the idea of the tragedy of the commons and the concept of lifeboat ethics. Many are familiar with some version of the tragedy of the commons: the typical argument goes that since no one is personally responsible for public resources, such resources tend to be exploited to the detriment of all, and thus private property offers the best strategy for managing resources. There are numerous problems with this argument, many of them articulated in Elinor Ostrom's 1999 refutation, "Revisiting the Commons," but the metaphor remains lodged in American culture, probably because, despite good arguments against it, "the tragedy of the commons" captures something about our individualistic consumer society.¹³⁸

What fewer people know is that the article introducing the idea isn't actually making an argument for private property, but for global population control. "No technical solution can rescue us from the misery of overpopulation," Hardin writes. "Freedom to breed will bring ruin to all. . . . The only way we can preserve and nurture other and more precious freedoms is by relinquishing the freedom to breed."¹³⁹ As Hardin implies, it's primarily the poor who need to have their "freedom to breed" curtailed, by whatever means necessary. Hardin expanded his brief against the world's poor in his subsequent essay, "Living on a Lifeboat," published in *Bioscience* in 1974.

Here Hardin argues against Kenneth Boulding's and Buckminster Fuller's metaphor "Spaceship Earth," which envisions planet Earth as a spaceship with all humanity as its crew, and asserts that thinking of the Earth in such a way will lead to "suicidal" immigration policies.[140] Instead, he contends, we should think of the world's nations as lifeboats with limited carrying capacity:

> Metaphorically, each rich nation amounts to a lifeboat full of comparatively rich people. The poor of the world are in other, much more crowded lifeboats. Continuously, so to speak, the poor fall out of their lifeboats and swim for a while in the water outside, hoping to be admitted to a rich lifeboat, or in some other way to benefit from the "goodies" on board. What should the passengers on a rich lifeboat do? This is the central problem of "the ethics of the lifeboat."[141]

The answer to this quandary, Hardin argues, is to keep the poor out of your boat. According to Hardin, this entails not only strict immigration but curtailing foreign aid and perhaps even limiting democratic governance.

Hardin's concept of lifeboat ethics reappears in the astringent work of Finnish deep ecologist Pentti Linkola, who wrote, "When the lifeboat is full, those who hate life will try to pull more people onto it, thus drowning everyone. Those who love life and respect life will instead grab an ax and sever the hands clinging to the gunwales."[142] Similarly reactionary responses to overpopulation and ecological crisis can be seen in the work of French thinker Guillaume Faye, who writes, "The planet Earth is simply not capable of answering the needs of an excessively large humanity that always wants more. Notions of 'justice' and 'injustice' are no longer relevant. Morality is disappearing in the face of physical obstructions, while ideologies are stuttering in the void."[143] Such vatic brutality can be seen even in *Time* magazine, in Parag Khanna's ecomodernist argument that although hundreds of millions may die from climate catastrophe, "billions of people will survive" in nomadic, adaptive, 3D-printing, renewable-energy-powered cities.[144]

There is a line from reactionary arguments like Hardin's and Faye's to ecofascist violence like that of March 15, 2019, when Brenton Tarrant killed fifty-one people in two mosques in Christchurch, New Zealand; August 3, 2019, when Patrick Crusius killed twenty-two people in a Walmart in El

Paso, Texas; and May 14, 2022, when Peyton Gendron killed ten people in a Tops supermarket in Buffalo, New York. All three shooters disseminated explicitly ecofascist manifestos before their attacks, which weave together environmentalism, racism, population anxiety, and calls for revolutionary violence. In his manifesto "The Great Replacement," Tarrant identifies himself as an "Ethno-nationalist Eco-fascist" and articulates concerns with racial demography, comparative population levels, and the decline of what he identifies as "white" or "Western" culture. The "green" side of Tarrant's argument is an eco-nationalist concern for environmental degradation, grounded in anxieties about overpopulation.[145] He writes: "Each nation and each ethnicity was melded by their own environment and if they are to be protected so must their own environments." Crusius's screed, on the other hand, is primarily anti-corporate and anti-immigration, fixing on immigration and automation as the two greatest threats to what he sees as the American way of life, while also arguing for decreasing US population because "our lifestyle is destroying the environment of our country." Echoing the rhetoric of Bill McKibben and Greta Thunberg (perhaps satirically), Crusius writes, "The decimation of the environment is creating a massive burden for future generations. Corporations are heading the destruction of our environment by shamelessly overharvesting resources. This has been a problem for decades. For example, this phenomenon is brilliantly portrayed in the decades old classic 'The Lorax.'"[146] Peyton Gendron copied large sections of Tarrant's manifesto verbatim, including his argument that "green nationalism is the only true nationalism." Gendron also writes, again echoing Tarrant: "I would prefer to call myself a populist. But you can call me an ethno-nationalist eco-fascist national socialist if you want, I wouldn't disagree with you."[147]

Broadly construed, ecofascism combines a commitment to environmental sustainability with a commitment to political separatism based on race, ethnicity, or nationality, along with a willingness to use violence to make good on those commitments.[148] The term is often deployed by journalists and scholars to refer to a range of reactionary European political movements that combine environmentalism and anti-immigration politics, such as France's Rassemblement National, Hungary's Fidesz, Germany's AfD, and Italy's 5 Star Movement, as well as more disparate American phenomena such as the Pine Tree Party, neo-pagans, men's rights groups, white supremacists, the Green Tea Coalition, accelerationists, and anti-immigration groups like the Federation for American Immigration Reform (FAIR).[149] According to criminologist Imogen Richards et al., ecofascism is "a belief system, trope, or extremist sensibility rather than a coherent ideology."[150] Similarly, in their authoritative overview of the online "far-right/ecology nexus," Brian Hughes et al. write that ecofascism is "less a coherent ideology, still less a political movement per se," and "first and foremost an imaginary and cultural expression of mystical, anti-humanist Romanticism."[151]

Despite its conceptual bagginess, ecofascism is still "capable of inspiring political action, from propaganda to lone actor violence to parliamentary politics."[152] Indeed, the violent acts of self-identified ecofascists such as Tarrant, Crusius, and Gendron meet the criteria for forensic psychiatrist Marc Sageman's social identity perspective theory of terrorism.[153] Sageman writes: "In the presence of an escalating conflict with a contrasting outgroup (often the state), disillusionment with peaceful protest, and moral outrage at out-group aggression, some militants start thinking of themselves as soldiers protecting their political community."[154] Tarrant, Crusius, and Gendron saw themselves in just this way, joining Theodore Kaczynski and Anders Behring Breivik in an emerging ecofascist pantheon of anti-state, white supremacist martyrs.[155]

While contemporary ecofascism is distinct from historic European fascism, the two are related. Hitler's National Socialist movement was both political and ecological, and its followers conflated politics and science just as readily as do both liberals and conservatives today.[156] And journalists and scholars have traced the roots of ecofascism not only to ecologically oriented aspects of Nazism and its successors but also within American environmentalism itself, particularly to persistent anti-indigenous and anti-immigration tendencies in wilderness conservation, anti-humanist tendencies in deep ecology, and racist strains in American environmental thought, typified by early conservationist and white supremacist Madison Grant.[157] Sometimes these connections are described with nuance and care; other times they are painted in broad strokes.[158] The main point to take away is the one made by Janet Biehl and Peter Staudenmeier in *Ecofascism Revisited*: "'Ecology' alone does not prescribe a politics; it must be interpreted, mediated through some theory of society in order to acquire political meaning. Failure to heed this mediated interrelationship between the social and the ecological is the hallmark of reactionary ecology."[159] The same holds true, *mutatis mutandis*, for global warming and climate change. Climate anxiety, population anxiety, and anxiety about social disruption, just like concern about environmental degradation, are not "progressive" or "reactionary" issues but complex responses to emergent phenomena. How these phenomena are narrativized and politicized are questions that must be foregrounded, not conflated with scientific modeling, buried under partisan slogans, or squelched by dogmatic optimism.

The relationship between "nature" and politics is complex, contentious, and full of contradictions not only across the political spectrum but within the groups at each end.[160] As Sam Moore and Alex Roberts write in their study *The Rise of Ecofascism*, "Articulations of 'nature' on the far right now are multifaceted, chaotic even," and core tendencies of right-wing politics such as nationalism, anti-elitist populism, and extractivism may conflict with ecofascist strains, particularly because of strategic conservative commitments to climate change denial.[161] Similar conflicts characterize how nature and climate change are articulated on the political left, particularly between liberal advocates of green growth and more radical proponents of degrowth, eco-Marxism, and cultural revolution. Further, while ecofascism is a genuine threat, focusing too much on individual ecofascists may miss the broader alignment between ecological concerns and conservative politics that researcher Balsa Lubarda identifies as Far Right Ecologism (FRE), as well as overlook the peculiarities of American politics, namely the bipartisan confluence of techno-utopian progress, Protestant Christianity, petrocapitalism, racialism, and militarism at the core of American imperial ideology.[162]

Were the US to drift into reactionary ecofascism, that is, it may take on a distinctly liberal cast. Since World War II at least, American cultural unity has centered on a "way of life" characterized by consumer gratification, personal autonomy, economic abundance, faith in progress, and technological mastery. Modern American environmentalism has always been most successful when it has been able to tap into core postwar values (as with Earth Day and the *Whole Earth Catalog*), and so has modern American nativism. It's not difficult to imagine a version of reactionary American climate politics combining nativism, environmentalism, consumerism, and interventionism. Such an ideological constellation would look less like ecofascism as people currently understand it and more like the kind of American chauvinism familiar from the Spanish-American War to the proxy war in Ukraine—which is to say, center-right, pro-business, flag-waving interventionist liberalism—the very thing activists like McKibben, Ocasio-Cortez, and others seem to be calling for in their repeated demand that we "declare war on climate change."[163]

American ecofascism might actually look a lot like the developing merger between green growth and the military-industrial complex that Mona Ali calls "militarized adaptation"; Pierre Charbonnier, "war ecology"; and Thea Riofrancos, "the security-sustainability nexus."[164] Even if "climate realism" demands recognizing the role fossil fuels play in maintaining US geopolitical power and managing the security disruptions that green energy poses, given Chinese dominance in both critical mineral supply chains and green energy technology, there is at the same time no question of American power ceding the field of green growth to Europe and China: American strategic interest lies in a both-and or all-of-the-above energy strategy.[165] Thus, as Ali writes, "given what American legal scholar Cass Sunstein calls 'the dark cloud that now looms over the administrative state,' and the nonpartisan nature of the US defense spending, it is likely that climate finance will in future be folded into the US Department of Defense budget."[166]

It's hard to decide which imagined future seems more awful: one in which partisan division continues to paralyze national action on climate change until it's too late or one in which the admonition to "save the planet" merges with apocalyptic militarism. It would be all too human to excuse horrific policies overseas and at the border as the hard choices imposed by the moral obligation to fight climate change. After all, it wasn't so long ago that American politicians, lawyers, intellectuals, and filmmakers justified torture, assassination, indiscriminate bombing, and the aggressive invasion of a sovereign nation as the hard necessities of the Global War on Terror. The risk is so great as to put into question the wisdom of communicating the reality of our impasse at all. Historian Timothy Snyder suggests, "A sound policy for our world . . . would be one that keeps the fear of planetary catastrophe as far away as possible," but unfortunately for us, there may be no avoiding it.[167]

INDEPENDENTS, ELITES, AND LEFTIST ECOTERRORISM

Instead of preaching to the liberal choir or reaching across the aisle, some argue for persuading independents. One of techno-utopian thinker Ramez Naam's main criticisms of Wallace-Wells's article "The Uninhabitable Earth," for instance, was that its alleged "combo of exaggeration and hopelessness" would repulse "those in the middle we need to persuade."[168] Who is this imagined middle? About 7% to 9% of the voting population, it turns out.[169] Gallup Poll trends show a robust, decades-long tripartite division within the American body politic, with between a quarter and a third of Americans identifying as Republicans, around a third identifying as Democrats, and the remainder (between a third and half) identifying as Independents, though most Independents consistently lean one way or the other.[170] Although exact numbers have fluctuated over the past two decades, the deeper trend is robustly binary, with almost half of Independents leaning Democrat, just under half leaning Republican, and only a small minority tending toward non-affiliation. These unaffiliated independents also tend not to vote. Most Americans "pick a team" when it comes to politics, and those who identify with neither party tend not to play at all, making the goal of "persuading those in the middle" seem doubtful. While it's true that converting 7% of the population into climate activists would be a huge political shift, the chances seem slim that anyone is going to be able to motivate this disengaged segment of unaffiliated independent voters after decades of failed effort to do so. Yet contra dogmatic optimists like Naam, a fear-based approach like "The Uninhabitable Earth" may actually be the best way to get their attention.

Given the partisan character of climate change politics and antidemocratic structures built into American government such as the Senate and the Electoral College, however, the strategy of democratic "pressure from below" might need to be re-evaluated. There are numerous issues, after all, where significant citizen majorities consistently diverge from government policy priorities (often bipartisan), such as gun control and immigration, and climate change is one such issue. We would do well to take into account robust evidence that it's not actually democratic majorities that drive government policy but rather multinational corporations and wealthy elites.[171] Considering this, perhaps it would be more effective for climate activists to shift their target from the hoi polloi to the rich and powerful.

The trouble with this strategy, of course, is that economic and political elites are precisely those with the greatest stake in the status quo and the most to lose in any radical restructuring of society. They also have the least to gain from saving poor people from climate disaster. Saving the world costs money, after all. And even if we do convince elites that climate change is a priority, their material interests are going to incline them toward technocratic mitigation and "green growth" (à la Bill Gates), lifeboat ethics and climate apartheid (à la Peter Thiel), or an empty politics of hope. It may be that the gravest social threat climate change poses, as sociologist Daniel Aldana Cohen argues, is "a vicious right-wing minority imposing the privilege of the affluent few over everyone else," but the vicious, affluent minority currently imposing its privilege is actually quite bipartisan.[172] Indeed, wealthy elites are increasingly concentrated in the Democratic Party.[173]

Looking at Democratic mayor Michael Bloomberg's New York in the wake of Hurricane Sandy, Cohen found an "adaptation-focused defensive parochialism" he called a "fortress of solitude."[174] In response to climate disaster, Cohen writes, "Elites forged new, adaptation-focused climate policies to take the path of least political resistance, while maintaining their favored model of economic development."[175] This is not, Cohen argues, an exceptional case: "Urban and disaster sociologists find that, after crises, elites pursue new strategies to preserve prior power arrangements, prioritizing a 'growth machine' form of urban governance that puts short-term recovery and (unequal) economic development before long term ecological concerns."[176] What Naomi Klein and others have called "disaster capitalism," it turns out, is highly adaptable to climate disasters.[177]

The problem again is that convincing someone to respond to a threat doesn't mean they'll agree with you about what to do or even about what's at stake. Different social and political groups have different interests and goals. As Cohen puts it, "Extreme weather cannot do the work of politics."[178] While ruling elites may grant limited, targeted patronage to valued subalterns, in the big picture they are rather more likely to double down on defending their economic privilege than make sacrifices for an abstract global good. "Capitalism is clearly able to continue and even thrive in the face of vast human suffering," Alyssa Battistoni and Geoff Mann write, and "the idea that capitalism will ultimately collapse under the weight of its ecological contradictions is too simplistic, and may be too optimistic."[179] As Gilles Deleuze and Félix Guattari write, "No one has ever died from contradictions. And the more it breaks down, the more it schizophrenizes, the better it works, the American way."[180] In the end, persuading ruling elites to take climate change seriously runs the same risk as persuading conservatives, which is to say, reactionary eco-nationalism: Cohen's "fortress of solitude," or what Christian Parenti calls "the politics of the armed lifeboat."[181]

Given the doubtfulness of persuading elites, some advocate waging war against them. Revolutionary Leninism has found adherents on both the far left and the far right, though only one side has managed to get a president elected or storm the Capitol. From such a vanguardist position, the challenge is less to convince voters than it is to incite a dedicated cadre to commit propaganda by the deed and prepare the rank and file for the "tiger's leap" of violent revolt. Take, for instance, Andreas Malm's argument in his 2021 book *How to Blow Up a Pipeline* that the climate movement should "announce and enforce [a] prohibition" on all new carbon-emitting machines in a global campaign of industrial sabotage, or what the FBI would call ecoterrorism: "Damage and destroy new CO_2-emitting devices. Put them out of commission, pick them apart, demolish them, burn them, blow them up. Let the capitalists who keep on investing in the fire know that their properties will be trashed."[182] Adam Tooze generously reads Malm not as "making a proposal for action so much as undertaking a radical thought experiment," but there is every reason to believe Malm is deadly serious.[183] He presents a strong case against Erica Chenoweth and Maria J. Stephan's well-known but rather shaky argument for the effectiveness of nonviolent civil resistance.[184] He is correct in his assessment that "strategic pacifism is sanitized history, bereft of realistic appraisals of what has happened and what hasn't."[185] And he's also correct that the increasingly rapid emergence of ecological crisis "imposes tight constraints on those who want to fight."[186] Surveying the brief history of oil pipeline sabotage, Malm suggests extending that strategy to include luxury emissions, focusing explicitly on SUVs but also implying that air travel may make a good target.[187] But why stop at blowing up pipelines, deflating SUV tires, and sabotaging airplanes? Why not target refineries? Tankers? Gas stations?

Ultimately, Malm's argument for ecoterrorism is based on a series of misunderstandings. Most centrally, Malm doesn't understand political violence, probably because his idea of violence seems to derive mainly from Saturday afternoon vandalism.[188] "If destroying fences was an act of violence, it was violence of the sweetest kind," he writes, valorizing the "cleansing force" and "collective empowerment" of action.[189] Yet although Malm seeks to declare war on fossil-fuel capital, he doesn't comprehend that "war has its own logic": a logic of fragmentation, transformation, metamorphosis, and reciprocal mimetic intensification René Girard called "escalation to extremes."[190]

What "violence" means is itself a complicated question, which Malm does not adequately address. Some argue that destruction of property should not be considered violence and that only acts directly harming human persons or animals should count. Others argue that violence includes speech, discourse, and images that cause psychological or emotional distress. I cannot offer a comprehensive argument on the topic here, but I have experienced violence as both an agent and a target, reflected on the question at some length, and feel confident in adopting this working definition: Violence is physical force deliberately intended to harm a person, being, or social collective, whether directly or indirectly, e.g., through the destruction of property.[191] A full investigation of the concept would be the project of another book, or rather several, as the literature on the topic is extensive.[192]

The experience of committing violence, even against property, activates some of our most powerful emotions, and those who commit violence often find themselves transformed, sometimes destroyed by the trauma, sometimes empowered.[193] These transformations cannot be controlled except through rigorous discipline and training. The successful long-term deployment of Malm's "controlled political violence" would require highly motivated groups of people willing and able to carry out dangerous, coordinated acts of physical destruction, and developing such a force would require not only intensive training in violence but also a culture that valorizes group-belonging and aggression.[194] Small, elite units like military special forces, guerilla bands, and terrorist cadres all nurture strong unit cohesion and the ability to act with speed and initiative. As Marc Sageman writes about the formation of terrorist cadres: "The self-conceived soldiers start feeling and thinking of themselves as special, different from the rest of the community, believing they are creating history as they go along. They believe they are the vanguard of the revolution and, like soldiers everywhere, develop a strong esprit de corps."[195] No doubt this is exactly what Malm has in mind. He simply has no conception of what such a transformation entails.

"The soldier who has yielded himself to the fortunes of war . . . is no longer what he was," writes veteran and philosopher J. Glenn Gray. "He becomes in some sense a fighter, whether he wills it or not. . . . In a real sense he becomes a fighting man, a *Homo furens*."[196] The transformation into *Homo furens* is a physical, mental, moral, and narrative process.[197] The soldier learns in training and battle to objectify the world, himself, and others. He learns "to serve a different deity" than he served in peace, a deity concerned "with death and not life, destruction and not construction."[198] He learns to love war, to "delight in destruction," and to master pain.[199] Likewise, Malm's imagined militant would have to learn to love terrorism. To successfully prosecute an insurgent campaign like the one Malm advocates, the militant would also have to learn to love secrecy, deceit, and ruthlessness. Further, for all Malm's finicky insistence on "intelligent sabotage," violence against property is all but certain to lead to arguments for violence against people or at least for accepting collateral damage as inevitable, as happened with both the Environmental Liberation Front (ELF) and the Weather Underground.[200] As Sageman observes, "The history of Western political violence shows this pattern of evolution from narrowly targeted to indiscriminate violence against the population."[201] Self-isolating violent extremists hunted by the state tend not to favor moderation. Is this the hero Malm imagines saving us from climate change?

Malm also fails to comprehend that violence begets violence. As ready as he claims to be to lead his guerilla campaign against fossil capital, at least from behind his laptop, the fact is that you can never know in advance where violence will take you, except into more violence. Terrorism directed against core infrastructure such as pipelines, cars, and planes would surely inspire state repression, and the incommensurability of force between Malm's fantasy-league ecoterrorists and the US security apparatus is vast. Attacks on the energy system supporting American consumerism would not only bring down the hammer of the state, but they would also alienate and repulse the people who depend on that infrastructure to get to work, take their kids to school, buy Mexican blueberries in February, and visit their family on Thanksgiving. As Bue Rübner Hansen points out, "Malm does not ask how effective disruptions of fossil infrastructure will affect ordinary people through price rises on essential goods. And so, he does not pose the question of how to avoid resistances and ambivalences stemming therefrom."[202] Most people would surely reject such violence and see themselves as its victims, rather than identify with the terrorists against their ambiguous enemy, now identified as oil companies, now as the rich, now as "infrastructure."

In part, this is because terrorism tends to bring attention not to the cause behind it but to the tactics used. Sageman identifies this fact as perhaps "the fundamental paradox of domestic political violence as a strategy: it seldom has its intended effect—bringing attention to a set of grievances—but focuses attention, including that of scholars, on the violence and its perpetrators and on how they differ from the rest of society."[203] Consider, for instance, the lugubrious history of environmentalist sabotage in the US and Europe in the last decades of the twentieth century, in which thousands of acts of vandalism and sabotage not only failed to achieve the goals for which they were undertaken but actually turned people against environmentalism.[204] Malm gestures briefly toward this history in *How to Blow Up a Pipeline* yet draws from it the puzzling conclusion that American ecoterrorists failed because they had the wrong ideology.[205]

Malm seems to be arguing that we have reached the point of extremity where violence is the only option left, just as Theodore Kaczynski and Dave Foreman argued in the 1980s, but I doubt that very many people would agree. Although general confidence in public institutions in the United States is at an all-time low, public opinion remains favorable to state power and opposed to anti-state political violence.[206] Malm's imagined revolutionary future, in which climate catastrophes inspire an increasingly broad and radicalized social movement against petroleum infrastructure, seems founded on nothing more than wishful thinking.[207] As Bue Hansen writes:

> The problem of how a disciplined mass movement may come into being is nowhere discussed in the book, as if the size and disciplined non-violence of the 2019 climate movement made such reflections superfluous. But as radical environmentalists have learned in the past, the discipline and secrecy required to engage in sabotage in the face of anti-terror legislation, mass surveillance, and militarized policing—and the following media

backlash—can be very hard to combine with mass appeal and mass organization.[208]

Arguments for changing society by violent means certainly have their proponents, but they have little public support. And as someone who has witnessed and participated in political violence and dealt with its long-term costs, I cannot help but see such arguments as ethically and intellectually irresponsible. At best they represent the desperation of frustrated idealists, at worst the dangerous ravings of zealots.

Malm's fantasies of leftist political violence are ultimately just fantasies, based in simplistic ideas of how mass movements and political violence work, unconnected to social reality, as are the similarly fanciful representations of bloodless terrorism in Kim Stanley Robinson's *Ministry for the Future*.[209] "Time and again," writes Max Ajl in *Brooklyn Rail*, "Malm's recipes ignore the people, ignore our demands, and ignore how those demands enfold severe criticisms of the techno-fixes he embraces."[210] What's more, Malm's approach is both carbon-centric and Eurocentric. As Hansen points out, "Malm shows little interest in other drivers of extinction and eco-system breakdown such as colonialism, plantation agriculture, slavery, extractivism, landgrabs, pesticides and fossil-based fertilizers, etc."[211]

Malm's fervid dreams of righteous violence not only show a lack of concern for real-world complexity but imagine a violent eco-militant who does not exist, misread existing dynamics between state power and climate activism, and ignore the role group dynamics play in political violence. Substantive research suggests that left-wing extremists are much less likely to engage in political violence than right-wing ones.[212] In illustrative contrast to ecofascism, which has produced real-life versions of Malm's idealized fantasy ecoterrorists, the climate activist left does not appear to fulfill the conditions Marc Sageman argues are necessary for the emergence of political violence. Sageman's theory, based in extensive research and decades of experience working on terrorism, argues that the self-categorization of martial identity must meet three conditions: "escalation of the conflict between two groups, including a cumulative radicalization of discourse; protestors' disillusionment with nonviolent tactics; and moral outrage at state aggression against the community."[213] While climate activism does form a distinct "protest community" and has experienced isolated moments of state aggression, such as at the Dakota Access Pipeline and Stop Cop City protests, the climate activist left seems unlikely to produce the necessary "self-categorization into soldiers defending their community."[214]

Malm expresses disillusionment with nonviolent tactics and works to radicalize the discourse, but while this radicalization and disillusionment have found a welcome audience among leftists frustrated with government inaction on climate change, its target remains diffuse; the problem is that the target of their violence is not a group or even the government but *infrastructure*. Without a clear conflict between two opposing groups, there can be no real radicalization, and the climate activist left has so far been unable or unwilling to identify and demonize a clear target. "Oil companies" don't make a distinct enough imagined community, "people who use oil" is everybody, and "the rich" is relative and taboo anyway, because NGO climate activism depends on elite charity. The problem is that the problem is "us," meaning modern civilization and the people who thrive in it, such as Andreas Malm. Anyone who enjoys the comforts of the system is implicated and thus will find it hard to participate in oppositional violence, especially as immediate costs are high. The lack of clear in-group/out-group dynamics not only inhibits the necessary escalation but also mitigates "moral outrage at state aggression against the community," which is fairly minimal in any case. It's hard to be morally outraged about state repression when climate activists are giving speeches at the United Nations.

Finally, Malm's argument is grounded in an essentially voluntaristic understanding of political action. In Malm's view, willpower is preeminent, more important than intellect, ethics, or practical concerns. "If someone seeks to affect the ways of the world by acting in one way rather than another," he writes, "it must be because she holds an outcome to be desirable and wants to contribute to its realization."[215] Thus every action, for Malm, must align with the reality one wills, regardless of whether that reality is likely to come about or is even probable. As he writes, "to act politically is to reject probability assessment as a ground for action."[216] It's difficult to know what to say about a conception of politics that axiomatically rejects practical concerns as irrelevant and demands complete congruence between will and action, ends and means. At the very least, some skepticism seems warranted. One might also note with concern the totalitarian spirit of such an approach. It is a conception of politics that abjures compromise, nuance, irony, rhetorical positioning, empathy, moderation, prudence, and empirical evidence.

But why not indulge in violent fantasies? It's easy to sympathize with Malm's desperation. Raising awareness has failed, democracy has failed, protest has failed, the market has failed, communism has failed, hope and optimism have failed, even ecoterrorism has failed. Sheer will has failed, and so has compromise. On a deeper level, we must face yet again the humiliating fact that we don't actually know how to change society because we don't understand how long-term social change happens. There are too many variables, most of which operate below or beyond the level of analytic discernment, and the situation we face is unprecedented. Writes environmental philosopher Michael Zimmerman:

> No one knows whether local actions, global actions, or a combination of the two will generate a new social constellation capable of forestalling the grim social and ecological consequences predicted by many different studies; no one really understands how major social changes occur; no one really knows whether the inertia of industrial modernity is so great that it will inevitably lead to ecological catastrophe, or whether a combination of luck, unexpected developments, and concerted action can avert disaster.[217]

How much conscious influence we might exert over our own social evolution is a question to which no one knows the answer, especially since we don't even know how much conscious control we have over our own thoughts and deeds. Did Malm choose to tear down the Schwarze Pumpe fences with his Ende Gelände comrades, or did a constellation of unconscious desires and social forces put him there? I doubt he knows, and I suspect he would find the question impertinent. But the view from psychology and history is chastening: often what feels like intentional design or robust agency dissolves on closer inspection into a confluence of contingency, unconscious drives, inertia, biases, and environmental determinants. We do not know ourselves. We don't really understand what we're doing.

As I've been arguing, the impasse we face exceeds our ability to make sense of it, and elaborating violent, apocalyptic fictions is likely to make things worse. Before we can articulate an ethical framework for acting in a world of catastrophe, we must first come to terms with our situation, dispense with our illusions and unselfconscious myths, and accept that there are limits to civilization, to our knowledge, and to what we can do. The task we face is a paradoxical one: navigating the collapsing ethical, ontological, and cultural frameworks that give our lives meaning, under conditions of increasing extremity and duress. The next two chapters directly face this challenge, looking first at the problem of ethics in the age of acceleration, then turning back to the end of the world.

Three
THE AGE OF ACCELERATION

The conceptual outlines of our predicament have been clear for decades, articulated by a range of thinkers grappling with the dialectic of technology and nature in an age of accelerating global transformation, but the rapidly emerging consequences of climate change are putting unbearable pressure on even these deep-rooted and far-sighted theoretical formations. We have so far considered how the limits of narrative, inherent human biases and heuristics, the Myth of Progress, problems in climate change communication, and political polarization frustrate clear understanding of our predicament and hamper effective action in response to it. In this chapter, I will focus on how four key interrelated theoretical problems in the age of acceleration further impede such efforts: the ethics of globalization, thinking the human, the problem of time, and climate justice.

THE ETHICS OF GLOBALIZATION

The fundamental ethical problem globalization poses is how to articulate the relationship between the individual and the world. What "the world" means changed profoundly in the 1960s and 1970s, in ways we are still struggling to comprehend. In 1972, Peter Singer posed the problem in the following terms: Imagine you pass a child drowning in a shallow pond and could, at no more risk than getting your shoes and clothes muddy, save the child. Should you? Most people would say yes. The intuitive moral obligation to save a drowning child entailed, for Singer, an analogous obligation to the rest of the world. If you agree you should save a child drowning in a pond, he argued, then you have the same obligation to a child starving in Bangladesh. As Singer put it, channeling the spirit of British progressivist William Godwin, "If it is in our power to prevent something bad from happening, without thereby sacrificing anything of comparable moral importance, we ought, morally, to do it."[1] The mechanisms for care in Singer's argument—top-heavy NGOs, potentially corrupt governments, ignorance about where best and most efficiently to give—may complicate the practice but fail to touch the principle. If we are responsible for each other, he argues, then we are responsible for each other, and geographical or cultural distance should make no difference. "From the moral point of view," Singer writes, "the development of the world into a 'global village' has made an important, though still unrecognized, difference to our moral situation."[2]

Two years later, biologist Garret Hardin posed a contrary thought experiment, discussed previously.[3] Hardin asked readers to imagine they were among the survivors of a shipwreck lucky enough to have made it onto a lifeboat. The boat can only hold sixty people, and the more people in the boat, the higher the risk of capsizing. In Hardin's story problem, you've already got fifty survivors in your boat, and there are a hundred more around you treading water. Hardin sees three choices: take on as many as you can and capsize your boat, save ten and increase your risk of tipping, or let the hundred drown. This grim situation, Hardin argues, is analogous to the situation affluent nation-states face in dealing with growing global population, primarily in what was then called the Third World but today would be called the Global South. For Hardin, the basic claim in his thought experiment—that we have a right to self-preservation even if it means others will suffer—entails that we should support strict limits on immigration and oppose foreign aid to the poor.

Singer formulates an ethics of affluence, Hardin an ethics of scarcity, and both formulations resonate because they articulate key philosophical problems in a globalized world. What ethical obligations do we have to one another? At what scale do these obligations operate? How do we make sense of claims made on us by strangers we've never met and never will? How do we balance self-interest and compassion? How do we understand our competing obligations as individuals to the nation-states of which we're citizens, to the whole of humanity, and to any given stranger? Most people would probably agree with Singer that we ought to stop bad things from happening where we can do so without causing ourselves harm. And most people would probably also agree with Hardin that we have a right to self-preservation even if it means others will suffer. Where both Singer and Hardin pose a real challenge, however, is in projecting individual ethical agency onto a global framework.[4] The abstract concept of a universal rights-bearing subject, first articulated in the Enlightenment, takes on painfully concrete particularity when you can watch that subject drowning off the coast of Lampedusa on your TV—or your phone.

The problem with Singer's universalism is that, notwithstanding the undeniable fact that practically everyone on Earth today is entangled in the same planet-spanning civilization, not one single human being has yet become a "global citizen." We each inhabit a specific geographic, political, economic, ecological, and cultural situation with a specific history, and that specificity bears on our moral decisions. We may value tolerance, diversity, and cosmopolitanism, just as we may value hospitality, charity, and compassion, but such values are meaningful only if they emerge from and within a specific, concrete existence. Charity from nowhere has no way to discriminate: Why "Support the Yargle" rather than "Save the Blargle"? There are approximately 1.54 million registered nonprofits in the United States alone.[5] If you are in the lucky position of having $100,000 to give away without muddying your shoes, so to speak, perhaps the most equitable solution would be to give six cents to each nonprofit. But this would be absurd. You would want to choose which nonprofits to support, based on your values and commitments, and since presumably some of your commitments are to the community in which you live and the groups to which you belong, geography and identity matter. Singer has no compelling reason to feed a starving child in Bangladesh rather than a starving child in Camden, New Jersey, and since Camden is geographically closer to his own home in Princeton, I think there would be something morally suspect if he overlooked the suffering child at home in favor of feeding someone halfway around the planet.

With Hardin, the conflation of individual moral decisions with the political decisions of nation-states is an out-and-out category confusion. States are not people, and we err in arguing by analogy from individual moral agents acting in a crisis to the role of imagined communities in international politics. States have greater responsibilities and fewer justifications for self-preservation than do individual human beings or even small groups of humans, and Hardin's model lifeboats are unrealistically autonomous, existing in a cartoon world of "individually responsible sovereign nations," not the actual world of competing hegemonic and regional powers, client states, and transnational corporations interconnected in a single global civilization. War in Ukraine and drought in Kansas affect the price of bread in Soweto, public health policy in Italy impacts Indonesia, and labor policies in Germany reverberate in South Korea. Furthermore, Hardin's crude economic model elides important phenomena like differential labor markets and resource extraction: Hardin's "rich" lifeboats are rich not because they keep the poor out, but because they use the poor for cheap labor while extracting resources from their sinking boats.[6] We may grant some of Hardin's premises—that human overpopulation and ecological exploitation risk undermining the long-term viability of our habitat and that individual human beings have a right to self-preservation—without accepting the validity of his lifeboat metaphor.

Despite Singer's and Hardin's efforts to rethink ethics in a globalized world and despite their gruesome images of drowning children and capsizing lifeboats, neither thought experiment adequately represents the challenges we face in coping with global climate change.[7] In a time of widespread ecological transformation and social destabilization, the affluent savior imagined in Singer's story is nearly as vulnerable as the drowning Bangladeshi they are supposed to be obliged to save. Wealthy individuals may have better insurance or better access to disaster relief services, or they might not: either way, relative affluence is little comfort when your house burns down or the economy collapses. This is not to say that wealth makes no difference in such cases, only that risk changes the ethical calculus for the agent in question. If you yourself are caught in a flood, your obligation to save the drowning child doesn't supersede your right to save yourself. On the other hand, imagining states as autonomously sovereign "rich" and "poor" lifeboats offers a crudely simplistic model of actual global economic and political inequality.

Given how influential Singer's and Hardin's thought experiments have proven, however, it might be worthwhile asking what kind of metaphor might better represent our impasse. We might consider Kenneth Boulding's "Spaceship Earth," which provocatively posits a "closed economy of the future . . . in which the earth has become a single spaceship, without unlimited reservoirs of anything . . . and in which, therefore, man must find his place in a cyclical ecological system."[8] All well and good, except that Hardin's critique punches a hole in the spaceship's bulkhead:

> For the metaphor of a spaceship to be correct the aggregate of people on board would have to be under unitary sovereign control. A true ship always has a captain. . . . What about Spaceship Earth? It certainly has no captain, and no executive committee. The United Nations is a toothless tiger, because the signatories of its charter wanted it that way. The spaceship metaphor is used only to justify spaceship demands without acknowledging corresponding spaceship responsibilities.[9]

We can recognize the Earth as a closed system in which we all depend upon each other, but the political reality within that system resembles gang warfare more than it does a unified crew. Until Americans, Mexicans, Israelis, Palestinian Arabs, Ukrainians, Russians, Rohingya, Chinese, Inuit, Serbians, Bosnians, Syrians, Hutus, Tutsis, Somalis, Houthis, Hindus, Pakistanis, and everybody else is governed by one executive body with a monopoly on force, Spaceship Earth will continue its rudderless drift.

Even if we did all act like we were living together on one big spaceship, it happens be a spaceship that's full of garbage, overheating, and falling apart as it accelerates towards some ominous celestial event horizon. Yet even this thought-picture misses important aspects of our impasse, such as the fact that seemingly neutral daily choices contribute to widespread suffering, the use of race and ethnicity to divide the crew, and the issue that we're not talking about a man-made machine at all but a planet, with its own geological and energetic cycles operating across millions of years. A better allegory for our situation might be found in Jeff VanderMeer's novel *Annihilation*, in which a headstrong biologist finds herself trapped in a strange new world that looks like the Gulf Coast but may be an alien organism, mysterious storms surge and ebb in seasons that transcend comprehension, and the line between human and nonhuman blurs.[10] Like VanderMeer's lost biologist, we're caught in a bewilderingly uncanny reality, uncertain about important facts, coping with traumatic and terrifying changes, and forced to make poorly informed decisions not about spending surpluses or even minimizing loss but about what is even happening.

The big moral question we face is no longer how to increase the aggregate good in a globalized world, as progressivists from Godwin to Singer have argued, but how to live in a fragmenting, increasingly incomprehensible world of accelerating catastrophe. The obligation to "save the planet" is a messianic conceit progressivism inherited from medieval Christianity, a conceit implicated in power politics from Juan de Sepúlveda's justification for the conquest of New Spain down to whatever passes for moral reasoning today in Washington, D.C. In reality, there is little reason to believe that individual action makes a difference when it comes to climate change, we can't correlate the incommensurable scales involved in what's happening, and we have no universal basis on which to adjudicate competing claims.[11] Singer's argument about our obligations to strangers is grounded in a fundamentally utilitarian morality that's proven incapable of making sense of global warming.[12] Hardin's lifeboat nationalism rejects the empirically established interdependent character of our ecological relations. Boulding's Spaceship is set for self-destruction and its crew are at each other's throats. And while VanderMeer's novel captures something essential about the phenomenological strangeness of our ontological ungrounding, his biologist is the lone survivor of a failed expedition, with no ethical obligation to other human beings except her lost husband (who may have turned into a dolphin). As philosophers Dale Jamieson, Stephen Gardiner, John Broome, and others have been arguing since the 1990s, it is extremely difficult to make sense of our impasse through Kantian ethics, contractualism, commonsense pluralism, pragmatism, liberalism, and most other individualist ethical systems.[13] Indeed, under conditions of accelerating anthropogenic ecological crisis, we're hard pressed to even say what it means to be human.

THINKING THE HUMAN

As we apprehend climate change and ecological overshoot from the appropriate geophysical and temporal scales (i.e., across the planet from the depths of the oceans to the thinnest stretches of the atmosphere, from Antarctica to the Sahel, over thousands and even millions of years), the figure of the human dwindles to an abstraction, and the history of civilization shrinks to a very recent perturbation in a minor climatological plateau. Yet thanks to our disruption of the carbon cycle, "humanity" has emerged as a planetary actor on the geological stage, not as conscious agent but as Lovecraftian monstrosity, a world-destroying force without worldhood, a body without will, intention, or consciousness. As Dipesh Chakrabarty writes, "If we, collectively, have . . . become a geophysical force, then we also have a collective mode of existence that is justice-blind. Call that mode of being a 'species' or something else, but it has no ontology, it is beyond biology, and it acts as a limit to what we also are in the ontological mode."[14] At the same time, the greatest anthropogenic impacts on the Earth system, including most of the Great Acceleration and more than 75% of all human greenhouse gas emissions, have occurred within the span of one human life, largely under the hegemony of one empire, primarily as a result of choices made by a handful of oil company executives and US government officials. The staggering disjunction here between cause and effect, actors and act, inspires a kind of inverse sublime in which we shudder to behold the world-shattering consequences of run-of-the-mill human behavior: moral reason experiences the shock of a demonic Narcissus, withdrawing in repulsion from its incomprehensible power. Perhaps the most difficult theoretical challenge we face today is conceptualizing or representing the human as at once an ethical agent, ontological being, political actor, and biophysical phenomenon within the Earth system operating on geological time scales. The incommensurability is at once manifold and vast.

No modern conception of human ethics is adequate to make sense of the global, long-term consequences we face from simple hedonism, greed, and laziness, or from altruism, hard work, and sacrifice. After all, many of those who've worked to "develop" the world believed their efforts contributed to the progress of humankind, and many still do. Meanwhile, the millions of people driving cars and buying gas and taking flights and using cell phones and air conditioning and central heating and Amazon Prime delivery may feel twinges of guilt, but they're mostly thinking about going to work, taking care of their kids, and having a good time. Who can justly be held responsible for transforming the global climate and on what basis? Neither an Arendtian Eichmann nor a Conradian Kurtz seems adequate since what's at stake is at once broader and more banal than the ambitions of a Nazi bureaucrat or the reckless brutality of primitive accumulation. And while it might be comforting to blame the ninety or one hundred companies deemed responsible for two-thirds of all historic carbon emissions, such finger-pointing fails to name the goat we can sacrifice to cleanse our sins or the wrongdoer from whom we can extract compensation to make things right.[15] It's easy enough to shake a finger at a corporation, and a company can even be sued and held responsible in court, but this in no way answers the demand of justice. Corporate entities, though persons under the law, remain social abstractions; the question of agency is intentionally diffused by articles of incorporation, and it was precisely to disperse individual responsibility that the social fiction of the corporation was constructed. Would it be the CEOs who are to blame for climate change, their corporate boards, their shareholders, their customers, the law-making governments that cater to their whims, or the voters who sanction such governance? Should we blame every American who ever voted Republican, especially the 43 million who voted for Ronald Reagan in 1980 and thus reversed the limited, tentative steps Jimmy Carter had taken toward curtailing American energy consumption? Should we blame the Clinton-Gore administration, who could have acted, knew they should have acted, and didn't? Should we blame the Supreme Court justices who handed victory to George W. Bush in the contested 2000 election? Should we blame Barack Obama for his administration's unwavering support for fracking and big oil? And where does China fit in? Currently responsible for the largest global portion of ongoing greenhouse gas emissions and the second-largest portion of cumulative emissions, neither genuinely capitalist nor even ostensibly democratic, does the People's Republic of China bear the same moral burden as Western corporations for their industrialization, fossil fuel use, and support for global development, or are the Chinese somehow exempt from such an obligation because of past harms inflicted on their people by European and American imperial powers?

Any serious attempt to locate responsibility or even a fulcrum that could be used to leverage global political and economic transformation must soon recognize that such efforts founder against the unrepresentability and hence incomprehensibility of the total human world, even seen through the reductive lens of politics. As Claire Colebrook suggests, our very conception of "the political" may be at stake:

> Where the political tended, once, to be determined by way of socioeconomic relations, it is perhaps—today—shifting to a different register of affect and corporeality. . . . But what if those practices of political theory were themselves dependent upon an epoch of suspension, in which the earth, the globe, nature, affect or corporeality—or even humanity—could appear as an object of stable knowledge only with certain practices and formations that would precipitate the destruction of the milieu on which they depend? . . . What if what we know as politics—the practice of tracing what appears as contingent, universal, or natural back to human forces—were possible only in a brief era of the taming of human history?[16]

The limits of human cognition are stark, and the widely held belief that we can comprehend how global civilization works and how to change it is not so much hubris as naïveté, for it is to see the totality of the symbolic social imagination and individual desires of eight billion human beings in interaction with a swiftly changing planetary environment as nothing more than a complicated puzzle to be mastered. It is here that a responsible thinker must put down their foot and say, "E pur si muove!"—the world is more complex than any one of us can possibly imagine. To go back to the metaphor of Spaceship Earth, no one is steering the ship, no one really knows how it works, and the ship itself is too big and complicated for anyone to comprehend, much less explain.

This problem is different from the Arendtian one we face any time we act, which is that not only can we never act in full anticipation of the consequences of our actions, but our actions change the world so as to generate previously nonexistent choices and potential consequences.[17] Rather, we face the historically specific problem that in the nineteenth and twentieth centuries human civilization grew too large and complex to effectively model.[18] Putting aside the question of how accurate the models of less complex societies might have been in the past, we can observe with rueful irony that the very moment when diverse societies across the planet were being forcibly integrated into a single civilizational network through Euro-American imperialism in the early 1800s may well have been the same moment that the complexity of our social organization transcended the representational limits of our reason. It is in struggling with just this watershed that Marx's unfinished *Capital* achieves its tragic monumentality.

It's not the case that we haven't tried. On the contrary, we persist in the effort to make sense of reality and gain some measure of control over it, collecting data and generating narrative schemata even as the gap between the world's complexity and our cognition grows. The development of global digital networks, for instance, running on high-speed computers and managed by repressive government agencies and entrepreneurs interested primarily in harvesting human attention for profit, has opened incredible new vistas in data gathering and modeling at the same time as it has generated new levels of human social complexity further outstripping our abilities to understand who we are or what we are doing.[19]

As numerous commentators have pointed out, the very word *we* is effectively unutterable except in specifically marked situations, with profound consequences for any notion of human understanding or social progress. The dialectical movement of social ethics from *I* to *we* and back again is only possible today, it seems, within self-selected communities defining themselves against some "Big Other" figuration of the social whole. This is in itself another version of the problem of the human, as any universalizing representation emerges always already under erasure—Vitruvian Man canceled, *Homo sapiens sapiens* rejected as an instrumentalizing abstraction, *Homo economicus* rightly spurned as a tendentious fabulation—and thus any potentially liberated "New Human" is foreclosed even as it takes shape, not so much stillborn as aborted, an embryonic human future subject to permanent critical curettage.[20]

The problem of the human takes another form as well. The threat posed by climate change and ecological crisis is inescapably global, and addressing it demands a kind of collective solidarity and commitment to a "planetary good" never yet witnessed in international politics nor to be found in the demands for justice issued by those who claim to speak out of ongoing histories of oppression identified with various racial, ethnic, or national identities. Indeed, since at least the global economic collapse of 2008, if not since George W. Bush delegitimized the post-1945 world order and destabilized the Middle East with the illegal American invasion of Iraq, we have been witness to a surge of nationalist and identitarian politics, from the rise of ISIS to Charlottesville, from Viktor Orban to Black Lives Matter, all opposed to the universalist premises of the globalist order and the abstraction of a single Anthropos, sometimes to the point of opposing climate science itself as either a fabrication of the Marxist left or a racist manifestation of white supremacist epistemology, depending on your politics.[21] However we apprehend this impasse, no resolution toward one or the other extreme is able to reckon with its countervailing antithesis as anything but a negation, thus forcing a choice between, on the one hand, aporetic double-consciousness, like that articulated by Dipesh Chakrabarty in *The Climate of History in a Planetary Age* and elsewhere (and not unlike what I'm articulating here), in which incommensurate realities stand opposed and subject to all the attendant risks of self-contradiction, and on the other hand a polemical forgetting that reduces the complexity of our impasse to a more recognizable and factitiously manageable problem—such as racism, settler-colonialism, or CO_2 emissions—mistaking a symptom for the illness and usually taking the lesser problem's ongoing resistance to resolution not as indicating its embeddedness within a larger complexity, which is in fact the case, but rather as a merely temporary frustration more aptly blamed on whoever can be identified as the relevant scapegoat.[22] Some choose both, alternating between false consciousness and false double-consciousness, in a strategy tantamount to defending the status quo.

Finally, as postcolonial and Afropessimist scholars such as Sylvia Wynter, Anthony Pagden, and Frank Wilderson have persuasively argued, the progressivist understanding of the human is, at its core, a racialist narrative of moral and ethical evolution from the primitive to the modern. Although nineteenth-century biological explanations of racial evolution and degeneration, such as that articulated in conservationist Madison Grant's *The Passing of the Great Race* (1916), have been superseded by cultural explanations, the core narrative remains the same, the very logic we saw in Sepúlveda's argument for conquering and torturing the natives of the New World: there is such a thing as moral progress and we are its paragons, woke-liberal-civilized-enlightened-rational-developed-"WEIRD" or what have you, and are thus obliged to "educate" all those who have yet to catch up.[23]

What does being "Human" mean without the idea of progress? What would it mean to be "Human" without the various Others who have served as the Human's negative determination, be they Indians, Savages, Slaves, Women, Blacks, or the Deplorable Poor? And how can we make sense of ourselves as individual consciences, political agents, vectors of progress, biological organisms, and geological forces all at once? These questions are fundamentally unanswerable within our current civilizational framework. There is no "Human race" compatible with progress that doesn't depend on the simultaneous subjugation and redemption of an abject, subhuman Other from which it has evolved. That's what "progress" means. As Howard Winant writes, "The five-hundred-year domination of the globe by Europe and its inheritors is the historical context in which racial concepts of difference have attained their present status as fundamental components of human identity and inequality. To imagine the end of race is thus to contemplate the liquidation of Western civilization."[24] The one seemingly viable resolution, that of declaring climate change to be an "everything problem," which no honest analysis can but show that it is, nevertheless collapses under its own vagueness and bad conscience, as such abstraction is in practice nothing but the renomination of liberal universalism in the desperate hope of a resuscitated techno-utopian dispensation. An "everything problem," that is, begs for an "everything solution," but until we come to terms with the inescapable physical limits of human life on Earth, even the most pious vision of human justice is "pie in the sky, by and by."

THE PROBLEM OF TIME

Thus articulating the problem of the human helps bring out the third major conceptual challenge climate change poses, which we might call the problem of time. The problem of time confronts us with the intertwined challenges of making sense of vastly different temporal scales (from geological epochs to the twenty-four-hour news cycle), deciding how much to weigh the future (often called in economistic terms "the discount problem"), and deciding how future and past relate. The vatic riddle posed by T. S. Eliot in the opening lines to "Burnt Norton" takes on shades of horror vacui when read in the context of planetary ecological transformation:

> Time present and time past
> Are both perhaps present in time future,
> And time future contained in time past.
> If all time is eternally present
> All time is unredeemable.[25]

Redemption, of course, is key: to deliver from consequences, discharge a debt, or recover a loss. It's not enough to simply say that time exists. Time must be made meaningful, full of meaning—empty time must be filled, and not only filled but brought into concord, as Kermode argues, with our needs in the present. Unless, like Eliot, we are able to take comfort in an eternal present, the past must be distinguished from the future, and both must be stretched to fit the pegs of our narrative. Yet as the pressure of impasse fuses and explodes geology, biology, ontology, climate models, politics, and myth, we struggle to weave a coherent timeline and locate ourselves upon it.

Where did we come from and where are we going? What do we owe yesterday, today, or tomorrow? Is our preeminent ethical obligation to construct a plausible human future, to care for the living in the present, or to pay the debts of the past? Can we somehow, under the dubious aegis of "intergenerational justice," prioritize at once the needs of humans yet to be born, the needs of those who are alive today, and the demands of historical suffering? To argue that a plausible human future can only be constructed on justly accounting for the past is to refuse the question. To say there can be no peace without justice is to degrade peace to a pleasing side effect of the real goal, which is the payment of a blood debt. On the other hand, to argue for a tabula rasa commitment to human futurity amidst grotesque and historically conditioned inequality, as if human culture could ever be anything but the issue of its own past, is to argue from arrogance, ignorance, or naïveté. The revolutionary moment when we might have hoped to sweep history clean and begin anew was a delusion to begin with and a nightmare in execution: anyone who holds out hope for a new Reign of Terror or Great Leap Forward should be condemned out of hand as a fool.

One concrete way the problem of time emerges is with the question of carbon emissions. The best possible choice for human flourishing in the future would be to bring global carbon emissions immediately to zero. Yet as numerous observers have argued, this would preserve historically conditioned global inequality, as such a course of action would favor already affluent countries more easily able to transition to carbon-free energy systems, while at the same time retarding development in countries that have, so to speak, been "left behind." In response, many call for allowing less wealthy countries to pollute more, which would be globally disastrous. A significant portion of recent global carbon dioxide emissions have come from China, for instance, which has insisted on its right to achieve economic parity with the US and Europe, in part to redeem losses inflicted over the past two hundred years, even if doing so puts the future flourishing of the entire human world at risk.

The problem emerges on more quotidian scales as well. Should an impoverished local school district spend its limited budget on solar panels, transitioning to high-efficiency electric HVAC equipment, implementing "smart technology" and automation, adopting electric school buses, and replacing grass with drought-tolerant plants? Should it prioritize hiring diversity specialists, decolonizing its curriculum, and removing symbolic markers of the United States' imperialist past? Or should it prioritize teacher salaries, security, school breakfasts, making extracurricular activities accessible, and providing computers? Similarly, should a small city government focus on subsidizing "green" construction, preserving carbon sinks, purchasing carbon offsets, and investing in renewable energy? Should it focus on diversity initiatives, equitable development, eliminating racial disparities, and social justice? Or should it focus on renovating schools, housing the homeless, and promoting investment? To answer "all of the above" is to refuse the question. All human endeavor is limited. We must choose which problems to address, in what order, and with what resources.

My point here is not that addressing climate change must necessarily come at the cost of social justice or economic investment, but more simply that policymakers choose between conflicting priorities, and the policies and actions that would redress historical inequities or lead to a more just society in the present are not necessarily those that are likely to address climate change and secure human flourishing in the future. On an abstract level, the best way to address global poverty is to generate more wealth and distribute it more equitably. Yet decreasing emissions means decreasing global economic activity: the green dream of decoupling economic growth from CO_2 emissions has no basis in reality. As Giorgios Kallis et al. write: "There is no empirical evidence of absolute decoupling of throughput from economic growth."[26] And while economic contraction doesn't in itself imply increased inequality, it's hard to see what might persuade elites to give up their wealth and power for the sake of addressing historical injustice, saving strangers, or preserving the future of abstract millions yet to be born. Imagine Peter Singer's affluent Westerner having to choose not between saving a drowning child and keeping their shoes clean, but between saving someone else's drowning child and their own.

When we "think of the children" as actual children, that is, the problem of time resolves again into the dialectic of the universal and the particular, and the logic of the lifeboat whispers its ghoulish call. What kind of ethical system would prioritize distant, abstract, or potential suffering over local, concrete, and actual suffering? What kind of saint or zealot would be willing to sacrifice the particular future of a particular child, even within a collapsing system, to pay the debts of the past or to secure a universal good whose future emergence is scarcely plausible? Who would choose to kick their own child out of the lifeboat, so to speak, in order to give her place to another because of the hardships their ancestors had to endure—or less imaginable yet, sacrifice their own a child to save a future child who hadn't yet been conceived?[27]

Such a choice is repugnant on its face, but recognizing the moral horror in the fiction cannot assuage the anxiety to which the dramatic situation gives form. What the image of the lifeboat embodies is the problem of scarcity, anathema to the Myth of Progress from its inception, since progress fills time from the future back. According to the Myth of Progress, that is, all human problems can be solved in the infinity of time, so long as we keep our eyes resolutely on the future. Progress thus redeems the past by repudiating it and secures its faith in the future by postponing demands for justice in the present. The unwillingness and even incapacity of progressivism to answer the problem of scarcity has left it incapable of coping with the idea of ecological limits, which incapacity poses profound challenges to developing an ethical approach for addressing ecological crisis within a progressivist frame. Nationalist decolonization, Franz Fanon's New Human, Sylvia Wynter's "new science," Murray Bookchin's post-scarcity bioregionalism, the idea of climate reparations, Ecomodernism, popular versions of ecological Marxism, and the Green New Deal all share a basic Prometheanism: an Enlightenment faith in human reason and resource abundance, taking for granted the root logic of colonialism's extractionist matrix, in which nonhuman nature is an infinitely fecund resource to be plundered for the sake of infinitely expanding human potential. The future is thus the final frontier to be conquered and exploited, if only in our imagination, as in the related movements for "effective altruism" and "longtermism," which offer a secularized prosperity gospel.[28] Equitable distribution is thus a question of investment, development, and future allotments in a timeline that bends inevitably toward redemption.

CLIMATE JUSTICE

The problems of globalization, the human, and time are thus inextricably bound up in the problem of justice. Any ethically serious person must recognize that climate change is going to increase human suffering, especially for those whose lack of resources throws them on the mercy of increasingly dysfunctional states, and that this situation is inherently unjust by any credible meaning of the word. Yet the way "climate justice" has hitherto been articulated offers at best a deeply compromised framing of the problem. Multiple issues are conflated and elided, and moral arguments are made in an intellectual vacuum, disconnected from realities on the ground and the self-identified needs of particular communities.[29] Difficult questions around sovereignty, property, restitution, and access are obscured by a narrow, technocratic focus on carbon emissions, while important questions about managing adaptation, developing resilience, and sustaining social cohesion are muffled by demands for identity-based political recognition, charges of racial guilt, and condemnatory rhetoric. And the whole conversation takes place against a cynical political backdrop in which imperial hegemony, capitalist exploitation, and militarism are described as global cooperation, international development, and defending democracy.

As a topic of academic philosophical and political inquiry, climate justice encompasses a range of approaches that attempt to resolve, in one way or another, the numerous ethical issues climate change poses or exacerbates. In important ways, many of the problems in question are not new, having already been taken up in the fields of environmental ethics and environmental justice, though our impasse puts new pressures on such problems, at times to the point of conceptual deformation.[30] The question of how to adjudicate the competing needs of human society and nonhuman biota, for instance, is forced into strange new shape by the pressures of accelerating climate change and at the same time complicated by urgent concerns about human rights and human vulnerability. What role should conservation biology play in a time when the habitat ranges of countless plants and animals are being rapidly pushed toward higher altitudes and the poles or destroyed completely not only by direct human development, still the prime driver of habitat loss, but by global warming and climate change?[31] How do we adjudicate between the demands of conservation in such conditions and meeting the needs of impoverished communities who traditionally depend on such ecosystem services to survive? Should the Baka of southeastern Cameroon continue to hunt chimpanzees, the Inuit of Grise Fiord hunt polar bears, and the Makah of the Pacific Northwest hunt gray whales?[32] The answers are not clear, nor is it clear who should decide or how. The question of how to render unto each their due can only be answered within a coherent and agreed-upon framework of collective values, and insofar as any such collective framework of values exists in the world today—that is to say, under progressivist techno-utopian capitalism—it is maladaptive to long-term human thriving and incapable of coping with climate change.

Furthermore, our existing modern framework of individual rights, economic growth, and social progress presupposes a stark division between the human and the nonhuman, in which the human is a conscious agent, and nonhuman nature a concatenation of depersonalized, deterministic forces existing beyond transient human concerns. Somewhat ironically, mid-twentieth-century nature poet Robinson Jeffers gave vivid form to the modern materialist view of nature in his poem "Carmel Point," where he dramatizes nature's "extraordinary patience" with regard to human development:

> Now the spoiler has come: does it care?
> Not faintly. It has all time. It knows the people are a tide
> That swells and in time will ebb, and all
> Their works dissolve.[33]

Jeffers wrote his poem in 1954, in the dawn of the Great Acceleration, criticizing human development on the cliffs above Carmel Point, and he wasn't thinking about anthropogenic planetary warming, which can give the impression that nature does "care" what "the spoiler" does. On a cosmic scale, Jeffers's point holds: a materialist view of the universe diminishes human culture to an idiosyncratic and fundamentally irrelevant minor phenomenon. This was a comfort to Jeffers in his disdain for capitalist development, but his concept of nature as a vast, eternal resource is the same one that subtends capitalist extraction.

Jeffers advocated a philosophy of ecocentrism, or what he called "inhumanism": "a shifting of emphasis and significance from man to not-man; the rejection of human solipsism and recognition of the transhuman magnificence."³⁴ This is a sublime perspective—"There is grandeur in this view of life," as Darwin wrote of human evolution—yet it offers no grounds for ethical action.³⁵ As philosophers and thinkers such as Richard Watson, George Bradford, and William Cronon have pointed out, ecocentrism is ethically incoherent, since the claim that *Homo sapiens* is just another species undermines any claims that human beings should behave with special consideration toward the rest of the planet.³⁶ You cannot argue that humans are morally equivalent to spotted owls, whales, and sequoias while also arguing at the same time that humans have a moral obligation to behave with special care toward spotted owls, whales, and sequoias. "If we think there is nothing morally wrong with one species taking over the habitat of another and eventually causing the extinction of the dispossessed species—as has happened millions of times in the history of the Earth," writes Richard Watson, "then we should not think that there is anything morally or ecosophically wrong with the human species dispossessing and causing the extinction of other species."³⁷

On the other hand, taking a strong anthropocentric stance toward nonhuman nature offers, at best, a stewardship of utilitarian calculation, prone to the same shortsighted arrogance that put us on the trajectory of self-destruction we find ourselves pursuing today. From indigenous hunting practices to contemporary conservation, even the most ecologically sensitive approaches are inevitably constrained by human cognition, shaped by human needs, and liable to human error. When the hard limits of human understanding come up against the endlessly complex interdependence between organisms and their environment—an interdependence so deep that it brings into question that very distinction—one cannot avoid the implication that we will never fully understand nature and our relation to it, which means we will never fully understand ourselves and thus will inevitably fail in any attempt to do justice to the nonhuman or the human.

The notion of extending justice to the entire biotic community, as in Aldo Leopold's Land Ethic, renders the question of what is due to whom or what unanswerable. Truly "thinking like a mountain," as Leopold advocates, or "thinking like a planet," as J. Baird Callicott argues in the same vein, is to transcend the concept of justice entirely, at least given how we understand mountains and planets and justice.³⁸ On what scale could we possibly weigh competing demands from different individuals, political groups, ethnicities, states, geological formations, ecosystems, and species, comparing them against the long-term health of the biosphere? Is the individual organism even a meaningful ethical unit? If not, what is? Do we have a right to our self-preservation? How could any single being or genetic line compare against the needs of the planet? Imagine the classic trolley problem, played not with individual lives but with species. And what are the "needs of the planet" anyway, and how could we possibly know them? The answers to such questions are not easily clarified.

Many would consider such heady questions a distraction from more immediate concerns. In public discussion, climate justice tends to have a narrower meaning, though one no less unsound. Climate justice in contemporary discourse makes the explicit claim that "those who have contributed the least will suffer the most" and entails an implicit claim that this situation presents a moral obligation. As UN Under-Secretary-General Gyan Chandra Acharya put it in 2015: "It is a sad reality that while the world's most vulnerable countries have contributed the least to climate change, they are most at risk from its negative effects."[39] This argument is so widely repeated that it has achieved the status of truism, and although charts showing historical national emissions and evidence of disparate impacts are often brought out in support, the ostensible value of the argument is not in its empirical portrayal of reality—or in its imagination of the future, since it is not descriptive but predictive—but rather in how the argument justifies political claims: namely, that equity, diversity, inclusion, redress of racial disparities, redistribution, and reparations must be at the center of any conversation about climate change and indeed precede and frame (where they do not preempt) discussion of how we are to deal with and understand this unprecedented planetary process.[40] The idea of climate justice may seem and is often taken as self-evident insofar as it's obvious that the world's poor, who already bear the brunt of global capitalist exploitation, neocolonialism, and environmental violence, will be further impoverished and harmed by increasingly frequent and extreme climate disasters. It is further taken as obvious and thus in no need of explanation that the affluent have a moral obligation to protect poor people everywhere from such suffering, care for them when disasters do happen, and work to mitigate the numerous inequalities and disparities that lead to their vulnerability in the first place.

But while climate justice has its self-evident side, against which only an unrepentantly cynical interlocutor would advocate in favor of lifeboat ethics, it also has a subterranean structural logic that complicates its self-evident altruism. The structural logic of the climate justice argument, contrary to the avowed intentions of its well-meaning advocates, is striking precisely in how it obscures the scope of our impasse, strengthens status quo tendencies toward climate apartheid, sustains racial violence, and fosters elite complacency. In many respects, climate justice exemplifies the kind of "elite capture" philosopher Olúfẹ́mi Táíwò identifies as characteristic of "woke capitalism," combining social justice "wokewashing" with eco-capitalist "greenwashing," superficially attending to the manifest character of discriminatory social scapegoat mechanisms without transforming their underlying dynamics.[41]

The argument for climate justice, that is, works *within* the narrative structure of racism and economic inequality to map the threat climate change poses onto those groups already identified as the abject victims of global inequality, reassuringly reinforcing existing social relations.[42] As Amitav Ghosh writes:

> Activists have long sought to appeal to the conscience of the privileged by emphasizing the message that the costs of climate change will largely be borne by the world's poor, mainly Black and brown people. It now needs to be considered whether these appeals to the conscience may not have had exactly the opposite of the intended effect. Is it possible that this message has actually persuaded the privileged to think they need to do nothing about climate change because they will be insulated from the worst impacts of global warming by their affluence, and indeed by their bodily advantages?[43]

Ghosh's question articulates the troubling contradictions within the ostensibly self-evident claims of climate justice. Most centrally, Ghosh's question reveals that the connection between suffering and moral obligation assumed by the argument for climate justice is just an assumption and like most moral claims made for the impoverished upon the powerful amounts to little more than a plea for alms. To actually make the argument that the suffering caused by climate change poses a universal moral obligation would require something more substantial than affluent utilitarian magnanimity, or the techno-utopian "pie in the sky" of effective altruism.[44] Such an argument would need to give grounds for prioritizing the suffering of people and groups outside one's own under conditions of increasing social instability. It would need to be an argument, that is, for real sacrifice. Making such an argument within a progressivist liberal framework of perpetual economic growth is exceedingly difficult, as can be seen from the repeated failures of UN Conference of the Parties meetings to substantively address this very issue. But to make a persuasive moral argument for such self-sacrifice in a framework of ecological limits and economic scarcity is something else entirely.

Another aspect of the problem is the question of precisely what kind of moral obligation human suffering from climate change would impose, if one were to grant that such suffering does indeed present an obligation.[45] Does such suffering entail only minimal charity, forgoing for instance what Henry Shue calls "luxury emissions" for the affluent, while protecting "subsistence emissions" for the poor?[46] Or does it entail something more? Does it mean that populations in countries such as the United States and the United Kingdom, usually considered to have the greatest historic responsibility for carbon emissions, would be obliged to make whatever sacrifices necessary to bring their emissions to zero, regardless of the actions of increasingly high-emitting countries such as China and India? Or is it rather an argument that countries like the US and UK must take radical adaptive measures, including increasing immigration quotas or doing away with restraints on immigration entirely, relocating threatened populations, and committing to robust military and civil interventions in response to climate-driven conflicts and disasters? Or is it more simply an argument for radical economic redistribution, perhaps through climate reparations, as advocated by Táíwò and others?

Many would probably say all of the above, as indeed Táíwò does. He argues for climate reparations as "a systemic approach to redistributing resources and changing policies and institutions that have perpetuated harm," which "would address two distinct but interconnected issues: climate change mitigation . . . and just climate migration policy."[47] All it would take to achieve would be for global elites to "broadly redistribute funds across states to respond to inequalities," along with "an overhaul of the existing international refugee regime," including "the recognition of rights of movement and resettlement, and a steady liberalization of rich-country border policies." As Táíwò himself admits, this is no incrementalist adjustment to the status quo, but an "extreme . . . renegotiation of state sovereignty and citizenship," a radical demand for revolutionary political and cultural transformation. "Existing theories of international relations cannot provide policymakers with the intellectual resources to respond to the crisis," Táíwò writes. Corporations, governments, and private investors must "abandon financial and political self-interest in favor of the greater good." As he writes elsewhere, "Adapting to climate change is . . . a project that can involve no less than the reordering of the globe."[48]

What's important to recognize here is that Táíwò's argument is not a policy brief at all but rather a call for spiritual and moral revolution, in the tradition of Christian and later Communist metaphysics in which "the last shall be first."[49] As Táíwò recognizes, this demand is antithetical to liberal capitalism, its property owners, its corporate fictions, and its faith in progress: it asks the wealthy, powerful, and privileged not only to abjure selfishness and ambition but also to dismantle the institutions and practices that give their desire form and meaning, shape their identities, reward their ambition, and continue to structure their lives and the lives of their children.[50]

How might we go about working to achieve such a goal in purely practical terms? What are the first steps in dissolving the nation state, ending capitalism, and eradicating desire—not only in the elite 1% who own 38% of the world's wealth (which would still amount to some 81 million people, globally), but in the more than 317 million corporate shareholders around the world plus the small-business owners, clerks, and middle-class property owners implicated in global capitalism and hoping for a better life for their children—which is to say, in that half of the world's population that owns nearly all of its wealth?[51] Would this be a grassroots campaign demanding that the wealthy sell their stocks (*to whom*?) and give away their money and property for the promise of a "more sustainable future"? Or would this be a top-down campaign forcing governments around the world to tax their own patrons out of their privilege and legislate themselves out of existence? The paralyzing gap between the ends desired and the means available illuminates in its blinding emptiness the real stakes of claims for climate justice, which are nonexistent. One may insist that the last should be first, down should be up, and night should be day without being in any way accountable for making it so, since the demand is impossible on its face.

What's more, for those 2.5 billion people who currently enjoy the benefits of property ownership (however unscrupulously it may have been originally secured) and are being admonished to "abandon financial and political self-interest in favor of the greater good," the statement that "those who have contributed the least will suffer the most" also implies its logical obverse: "those who have contributed the *most* will suffer the *least*." Táíwò rightly argues that "compared to the horrors of climate apartheid and colonialism, having more neighbors is a small price to pay," but his argument comes up against one of the central challenges in climate ethics, that of discounting the future.[52] Táíwò is pitting distant, abstract, future dangers against a demand for immediate, concrete sacrifice in the present. The problem is only made more difficult when the dangers in question are explicitly framed as threatening *other people*, far away, who have already been written off as the pitiable victims of circumstance, where they are not feared as a potential threat. The privileged millions and billions concerned first of all with "getting and spending" are well practiced in ignoring the suffering of the other millions and billions who cannot get enough to eat, do not have reliable access to clean water or adequate sanitation, and have been driven from their homes by war, famine, and floods. The privileged have had plenty of opportunity to give up their security and wealth to help the millions who are destitute, displaced, and starving today, and they have not done so. It's hard to see how increasing scarcity, increasing instability, increasing stress and fear and chaos are going to inspire them to lift up the wretched of the Earth tomorrow.[53]

The unpleasant truth here revealed is that the actual moral code most self-conscious subjects of liberal capitalism live by is at odds with the values most of us want to think of ourselves as holding. We may profess progressive democratic solidarity or effective altruism, but mostly we act like we're on a lifeboat. We believe we're good people who care about others, are committed to the common good, and seek justice and equity, yet every day we ignore the cries of the drowning, not least because we don't feel all that safe ourselves. One big wave, one malicious colleague, one medical emergency, one unlucky accident, and we may suddenly find ourselves in the drink with all those *others*, treading water in a rough and rising sea.

Hence the real utility in avowing a commitment to climate justice and reiterating the empty mantra that "those who have contributed the least will suffer the most": it professes innocence at the same time as it assuages what Spanish philosopher Adela Cortina identifies as aporophobia, or fear of the poor.[54] That we pity the "marginalized" shows them that we care, while reassuring ourselves that we're safe enough to condescend. Yet as the increasingly unpredictable present increasingly undermines the promise of a just future, those committed to such an ideal are forced to appeal to increasingly dissociative and abstract fictions to sustain their faith, and as the gap between such fictions and reality grows, we are forced to confront not only our own deaths and the suffering of millions, but the very end of the world.

Four

THE END OF THE WORLD

In the fantasy version it starts with an event—a missile launch, a hacked virus, or a speeding comet hurtling toward the Earth—then moves swiftly to a new normal: you're scavenging canned beans from a plundered Costco, pushing a wobbly shopping cart down a highway littered with abandoned Teslas, on the lookout for mutant cannibals. You travel alone or in a small band, because "the Event" has shredded the social fabric, leaving isolated chiefdoms and freeholds but no bureaucracy, no administrative state, no police, no postal service, no Whole Foods. It's Jim and Huck on the raft again, in a Gothic-cum-Romantic quest through the Waste Land of the Old West, a confluence made obvious in Stephen King's metamythic genre mash-up, *The Dark Tower*, as the story of Robinson Crusoe and the legend of Parsifal merge with the millenarian revelation of St. John of Patmos in a promise of spiritual renewal through total devastation, regeneration through violence on a planetary scale.[1] As Achille Mbembe writes:

> The era's paranoid dispositions crystallize around the grand narratives of the (re)commencement and of the end—Apocalypse. Very few things seem to distinguish the time of (re)commencement and that of the end, since enabling both are destruction, catastrophe, and devastation. From this point of view, domination is exerted by modulating the thresholds of catastro-

phe.... So it goes that, to reach its apogee, Being must pass through a phase of purification by fire.[2]

We wait for such a baptism today, wondering what form it will take—yet perhaps the most intractable aspect of the global cataclysm we've unleashed is that it's not an event at all.

The peculiar character of our impasse is that it is unprecedented, obscure, and banal as the weather. Not a day of reckoning, but Apocalypse 24/7. Not a doomsday you can prep for, hack your way out of, or hide from, but your world dissolving around you. Not something with a beginning and end, but the prelude of a new form of life to come, which those of us alive today will never live to see: a promised land not of milk and honey but fire and flood, veiled in ashes and dust, more felt than seen. From an ecological perspective, we have created a new habitat less hospitable to the large bipedal primate we call *Homo sapiens*. On the ontological level, we have crossed a threshold into a new environing gestalt yet to coalesce, a new world not yet visible from where we stand. From an existential point of view, we face an impasse we cannot see through or beyond. It's not too much to say that the end of the world is at this point a given and that the question we face is what sort of world comes next, but as with the world we're born into, what world comes after we're gone is not something we get to choose. Our struggle is at once more direct and more obscure: how to live out our end.

In their book *The Ends of the World*, philosopher Déborah Danowski and anthropologist Eduardo Viveiros de Castro explore the ontological and metaphysical implications that emerge from contemporary discourses of the end. Among their numerous insights and observations, they make the Kermodian (and Heideggerian) point that the end of the world brings "the world" into sight in a new way. By pondering the failure and dissolution of our lifeworld, that is, we see the world anew, as a whole, and can construct a plausible fiction of its beginning, middle, and end. The denouement gives narrative shape to what came before. Danowski and de Castro write:

> *The end of the world,* then. Let us start from the *"end."* The formula places us in a paradoxical situation.... In a double movement, it drags us in two opposing directions, toward a past and a future that are equally double, each with an "empirical" and a "transcendental" face: the obscure and vi-

olent past of generation (cosmogenesis, anthropogenesis) and the painful future of decadence and corruption, or of the expectation of death; but also a past of pure existential plentitude (which was never present as present, as it is the present's regulative idea and therefore its mythical inversion) and a future of absolute inexistence (which has so to speak already happened, since absolute nothingness is transcendentally retroactive).[3]

Perhaps it is just this double movement that explains why disputes about the origins of our contemporary world so often take form around neologisms denominating the future, as with the tired controversy about whether the era in which we live should be called the Anthropocene, the Chthulucene, the Gynecene, the Plantationocene, the Capitalocene, the Trumpocene, or something else entirely. How are we to make sense of these labels except as desperate attempts to constrain the incomprehensibly complex disarrangement of modern temporality and disguise it as a recognizable problem with a recognizable solution, even when, as in the case of Haraway's Chthulucene, it gestures toward the incomprehensibility it occludes?[4] We may quibble about where and when this world began, which is one way of having the argument about what this world is, but whatever we call it, we must distinguish this world from all the others that have come and gone. Our world is not the world of Gilgamesh, Ahknaton, Abraham, or Moctezuma. It is not the world of Arjuna, Genghis Khan, Boudica, or Saxo Grammaticus. It's not even the world of Frank Kermode.

Common usage of the term *world* confuses, and it's important to distinguish, following Dipesh Chakrabarty, at least four different ways of talking about our relationship with the spinning rock on which our species' life depends.[5] When we say "the world" we typically mean the human world, our world, the world as we know it, the collectively imagined global chronotope of the present. This sense of "world" is not an objective description but a subjective one, denoting a matrix of space, time, and meaning inhabited by the beings with and through whom it exists. Our world is the world in which *we* live, a world that would be unimaginable without *us*. A "world without us"—a world without humans, that is—would be no world at all.

But our world is also a planet, a globe, and the Earth. We might say, somewhat reductively and with due respect to Chakrabarty's finely textured analysis, that the globe is political, the planet scientific, the Earth phenomenological, and the world ontological. Thus when we talk about "the end of the world," we speak almost exclusively in an ontological register. The globe is a spatial conceptualization, not a temporal one, and can only be said to "end" where political collectives no longer have reach—at one time the blank edges of the map marked "Here Be Dragons," now maybe somewhere between the magnetosphere and the moon. The planet on which the globe is mapped has limits, too, spatially at the Kármán line beyond the exosphere, temporally some five or six billion years hence, when it will be vaporized by the cooling and expanding sun. The Earth, in contrast, has no end. Our ecological entanglement with the thin layer of the biosphere, being a phenomenological affair—as Heidegger put it, "the serving bearer, blossoming and fruiting, spreading out in rock and water, rising up into plant and animal" dwelling-ground of Man—exists as long as does that being for whom the environing phenomena exists, which is to say as long as there are people dwelling in it.[6] We are Earthlings, as Bruno Latour has argued, in some sense perhaps inescapably Earthbound, despite our adventures in the vacuum of space.[7] Only the ontological world, which is to say the socially constructed world, existing through human culture in time, can truly be said to come to an end *for us* in any meaningful way.[8]

And what does it mean for something to end? To end is to finish, to come to completion, to cease, to no longer exist in the same form. Physics tells us that while mass and energy change into each other, nothing is ever lost, so from a cosmic point of view, to end means merely to change. But we do not live our lives from a cosmic point of view: We live in mortal bodies, among mortal bodies, forced to contend with the mystery of death, which is almost certainly our profoundest sense of the end of anything. We might even say that "the end," *any end*, is a metaphor for our own, as Kermode argues.[9]

So what could it possibly mean to say "the world is ending"? For Kermode, steeped in medieval apocalypticism and Wallace Stevens, such a proposition is not so much an attempt to describe reality as it is an aesthetic effort to establish a "fictive concord" between one's own mortality and some larger pattern of existence—an attempt, we might say, to feel a little less lonely about dying. We could push Kermode's subjectivism further and argue that the claim that "the world is ending" is no more than a narcissistic projection of one's own despair on a planet-sized screen. Yet such an interpretation feels hopelessly inadequate to a moment when the end of civilization as we know seems all too plausible. If we want to make sense of the proposition that "the world is ending" as a statement about reality, we must reckon with what it would mean for "the world" to "end." If we take "the end" as a euphemism for "death," then we could take "the end of the world" to mean human extinction or at least significant mass death. If we take a more nuanced view, "the end of the world" might indicate nothing more or less than the transmutation of our collectively constituted sense of the present: a "new world," a new now, a different collective sense of being.[10]

The end of the world is a fiction insofar as it's a judgment in the form of a narrative, constructed with symbols and synthesized from experience, emotion, and sense data. It expresses, as Kermode would say, a concordance between the speaker's transient being in the present and their sense of participating in a larger pattern, overlapping with any number of historical, geological, and religious stories about the past, and stamped today with the image of the planet, the "pale blue dot" we know so well from pictures, the globe whirling on its brass spindle. Which is to say that a being and its world are forever wrapped up in each other: "there is no person without a world," and there is conversely no world without a person.[11] A world exists insofar as someone gives it life, even if only in their imagination. Thus, as Danowski and de Castro write, "'End of the world' only has a determinate meaning . . . on the condition that one determines at the same time *for whom* this world that ends is a *world*, who is the worldly or 'worlded' being who *defines the end*."[12]

My "world" differs from yours, yours from Nancy Pelosi's, and Pelosi's from that of a twenty-seven-year-old Bangladeshi lab tech.[13] Nevertheless and despite what are likely significant differences between each person's sense of their "world," "the world" also points to something at least notionally shared by every living human being, a mutually constructed and concurrent coexistence happening within recognizable bounds. No one says "the world" to mean Neptune or the Incan Empire. One might say "the world" to mean specifically the planet Earth but only within a shared understanding of planetary history: what a dogmatic Creationist means by "the world" will differ from what a Gaia-worshipping neopagan means by the same phrase. Likewise, one could say "the world" to mean "the natural world," but that depends on what you mean by nature, and the very need for the adjectival modifier suggests that the two terms are not synonymous.

My point is that as much as the end of the world is a fiction, it is also the end of a fiction, or rather the end of fiction as such: in the individual case, the narrative of a life; in a collective sense, the weave of narrative, concepts, metaphysics, and myth that gives shape and meaning to reality. The "end of the world" signifies the conclusion of a story and in its determinateness indexes the empty possibility outside and beyond narrative.[14] As much as the end of our world allows us to conceive of that world as a discrete whole, it also forces us to see it as a monad in the void—as one transient world among many. Thus we can hardly make sense of the idea of the end of the world without attending to all the other worlds that ended before our own. The worlds of various paleolithic hunter-gatherer, fisher-folk, and cultivator tribes ended with the emergence of imperial cities and large-scale organized agriculture. So too ended the preliterate world of the eastern Mediterranean brought to life in Homer's *Iliad*, long before Plato was himself witness to the end of the world that followed Homer's, as literacy transformed Greek conceptions of meaning and being.[15] The world of Laozi and Confucius ended with the tenth-century collapse of the Tang dynasty. Countless worlds were destroyed forever through the so-called Columbian Exchange, including the feudal world of medieval European Apocalypticism. The world of the Umayyad Caliphate, the Zoroastrian world, the Aboriginal world, Chaco World, the world of the Tokugawa shogunate, the Mayan world, even the patriarchal, Eurocentric world of donnish sophistication in which Frank Kermode felt so at home—all these worlds are gone, even as they live on in the ruins of the past and the inherited lineaments of the present.

DECOLONIZATION, UNCIVILIZATION, BECOMING INDIGENOUS

Thus although our impasse is unprecedented in scope and consequences, it is not without historical analogue. This is not the first time the world has ended, nor the first time a people or culture has had to deal with the collapse of its lifeworld and the loss of its concepts.[16] Human culture has recorded and pondered such catastrophes again and again. The ancient Sumerian *Epic of Gilgamesh* tells the story of human survival beyond civilizational and ecological collapse: Gilgamesh brings back secrets from before the flood. Virgil's *Aeneid* tells the story of the end of Troy but the survival of the Trojans. The Hebrew *Nevi'im* and *Ketuvim* memorialize the Israelites' conquest and exile under foreign rule: the Babylonian captivity provided a powerful narrative of cultural survival for later generations struggling to maintain a sense of meaning after the destruction of the Second Temple under Rome, the Inquisition, and the Shoah.

One historical analogy stands out with particular force: the European conquest and genocide of the indigenous peoples of North and South America.[17] Here truly a world ended—many worlds, for each civilization, each culture, each tribe lived within their own sense of cosmos, reality, meaning, and worldhood. The Potawatomi were not Aztecs, the Nambikwara were not Inuit. Yet over five centuries, all of these peoples saw their lifeworlds destroyed as they struggled for cultural continuity beyond mere survival, in an utterly new and hostile environment.[18] As indigenous studies scholar Dina Gilio-Whitaker puts it, "To be a person of direct Indigenous descent in the US today is to have survived a genocide of cataclysmic proportions."[19] In the words of Peruvian sociologist Aníbal Quijano, speaking primarily of the devastation in what we now call Latin America:

> Between the Aztec-Maya-Caribbean and the Tawantisinsuyana (or Inca) areas, about 65 million inhabitants were exterminated in a period of less than 50 years. The scale of this extermination was so huge that it involved not only a demographic catastrophe, but also the destruction of societies and cultures. The cultural repression and the massive genocide together turned the previous high cultures of America into illiterate, peasant subcultures condemned to orality; that is, deprived of their own patterns of formalized, objectivized, intellectual, and plastic or visual expression. Henceforth, the survivors would have no other modes of intellectual and plastic or visual formalized and objectivized expressions, but through the cultural patterns of the rulers, even if subverting them in certain cases to transmit other needs of expression.[20]

We may thus turn to this cataclysmic encounter between Europeans and the people of the so-called New World to see what it means to lose your concepts, grapple for footing in the abyss, and struggle to articulate a meaningful life after the end of the world.[21]

Pursuing this insight, many look to Traditional Ecological Knowledge, indigenous epistemologies, or decolonizing the Anthropocene as sources of liberatory, salvific, and regenerative potential in the face of our impasse.[22] As Aboriginal scholar Tyson Yunkaporta writes in his book *Sand Talk*,

"Perhaps we need to revisit the brilliant thought-paths of our Paleolithic Ancestors and recover enough cognitive function to correct the impossible messes civilization has created."[23] Philosopher Kyle Whyte suggests that "renewing Indigenous knowledges can bring together Indigenous communities to strengthen their self-determined planning for climate change."[24] Building on the work of Walter Mignolo, British geographer Mark Jackson argues that decolonial critique, "epistemic disobedience," and "border gnosis" grounded in indigenous knowledge can open a liberatory cultural politics focused on "care, attention to flourishing, and . . . disobedience to hegemony."[25] Heather Davis and Zoe Todd assert that "decolonizing the Anthropocene" through indigenous knowledge can "begin to address the root of the problem, which is the severing of relations through the brutality of colonialism. . . . [and] the deeper questions of the need to acknowledge our embedded and embodied relations with our other-than-human kin and the land itself."[26] Perhaps the most popular version of this argument comes from biologist Robin Wall Kimmerer in her evocative book *Braiding Sweetgrass*, which assures us that by "becoming indigenous" and "learning the grammar of animacy," we may "choose the green path," defeat the "Windigo" of extractive capitalism, and come to live in sustainable reciprocity with nature.[27]

The terms deployed here are complex, used in sometimes contradictory or controversial ways, and rarely clearly defined. Firket Berkes offers one definition of Traditional Ecological Knowledge as "a cumulative body of knowledge, belief, and practice . . . handed down through generations through traditional songs, stories and beliefs" that is "concerned with the relationship of living beings (including human) with their traditional groups and with their environment."[28] Yet as Kyle Whyte points out, "a scan of environmental science and policy literatures reveals there to be sufficiently large differences in definitions of TEK" as to "obstruct the possibility of moving toward a consensus on the best definition."[29] One problem anthropologist Joseph Gone and others have identified is that the concept of indigenous epistemology gathers distinctly different cultures under a single broad category, *Indigenous*, defined primarily in its opposition to another broad category, variously invoked as White, European, American, Colonial, or Western, and begs important questions about how we might identify indigeneity on its own terms, how and whether indigenous epistemologies survive colonization and the imposition of literacy, and how to understand the relations of so-called indigenous peoples to non-European civilizing conquerors from southeast Asia to Islamized Africa to the Aztec Empire.[30]

Decolonization is an equally complex and controversial term. Critical race studies researcher Vanessa de Oliveira Andreotti and co-authors write that "decolonization has multiple meanings, and the desires and investments that animate it are diverse, contested, and, at times, at odds with one another."[31] The historically specific meaning of the term, having to do primarily with nationalist independence movements in formerly colonial empires in the two decades following World War II, is distinct from the broader usage characteristic of its contemporary ideological deployment, as sternly pointed out by education and critical race studies researchers Eve Tuck and K. Wayne Yang: "Decolonization brings about the repatriation of Indigenous land and life; it is not a metaphor for other things we want to do to improve our societies and schools."[32] Yet even in the specific historical context of its emergence, *decolonization* was more than a term of art restricted to political theory. Frantz Fanon, for instance, famously defined decolonization as "an agenda for total disorder" and "the substitution of one 'species' of mankind for another."[33] Distinct from Fanon as well as Tuck and Yang, the 2018 "Keele Manifesto for Decolonizing the Curriculum" defines decolonization as a process of "identifying colonial systems, structures, and relationships and working to challenge those systems," as well as "a culture shift to think more widely about why common knowledge is what it is, and in so doing, adjusting cultural perceptions and power relations in real and significant ways."[34]

In the face of gross injustice, ecological catastrophe, and civilizational collapse, there is a powerful appeal in returning to indigenous ways of knowing. Such arguments align with those of other thinkers, such as anthropologists James Scott, David Graeber, and David Wengrow, ecologist Paul Shepard, philosopher John Zerzan, and writers Paul Kingsnorth and Dougald Hine, all of whom assert in their different ways that modern humanity lives in a fallen state and that prelapsarian human life was qualitatively better, more sustainable, and more spiritually whole. Some, like Kingsnorth, tend to see humanity's fall from grace in industrialization, while others, like Scott and Zerzan, argue that humanity's original sin was the development of large-scale agriculture.³⁵ Graeber and Wengrow make a more sophisticated argument that cultural identity as such, emerging out of political differentiation between competing social groups, generated violent conflict that led to fixed hierarchies, slavery, and the erosion of what they identify as the three "basic forms of social liberty."³⁶ And while not all of these thinkers explicitly make the anarcho-primitivist argument that we should dismantle modern civilization, many of them do, and the ideal is implicit throughout.

"We can go back to nature . . . ," Shepard writes in *Coming Home to the Pleistocene*, "because we never left it."³⁷ Shepard argues that the hunting and foraging cultures that evolved over many thousands of years in the Pleistocene remain genetically predominant and accessible today. "White European/Americans cannot become Hopis or Kalahari Bushmen or Magdalenian bison hunters," Shepard grants, "but elements in those cultures can be recovered or re-created because they fit the heritage and predilection of the human genome everywhere."³⁸ In a more Heideggerian register, the anarchist philosopher John Zerzan writes, "What we have forgotten may be recovered. Unfolding origin, a journey to origins, is possible. Every authentic choice takes us nearer."³⁹ Confronting the challenge of climate change and the possibility of civilizational collapse through their well-known Dark Mountain Project, Paul Kingsnorth and Dougald Hine advocate "uncivilization," a kind of cultural renewal based in British Romanticism and the work of Robinson Jeffers, and specifically "uncivilized writing," which they define as "writing which attempts to stand outside the human bubble and see us as we are: highly evolved apes with an array of talents and abilities which we are unleashing without sufficient thought, control, compassion or intelligence."⁴⁰ More recently, Graeber and Wengrow argue somewhat paradoxically that increased scientific knowledge about the human past can help us recuperate our primordial "freedom to create new and different forms of social reality."⁴¹

Heideggerian anarcho-primitivists, paleo-ecologists, Romantic pessimists, and activist anthropologists may seem like strange allies for critical race theorists, self-appointed caretakers of indigenous knowledge, and advocates of decolonization, but as Robin Wall Kimmerer points out, "Traditional ecological knowledge is not unique to Native American culture but exists all over the world, independent of ethnicity"—even in Europe.⁴² Indeed, as numerous scholars have argued, including political scientist

Cedric Robinson, the first indigenous people conquered by European colonizers were European peasants, and some of the first Traditional Ecological Knowledges colonized by "Western epistemologies" were the "cumulative bod[ies] of knowledge, belief, and practice . . . handed down through generations through traditional songs, stories and beliefs" of pagan Britons, Gauls, Teutons, and Slavs.[43] Expropriation of peasant land, enclosure, and colonialism were inextricably bound up together in early capitalism in a reciprocal dynamic Foucault called the "imperial boomerang."[44] The conception of land as individual property rather than a locus of negotiated privileges and obligations emerged in England only in the sixteenth century, didn't take firm hold for another hundred and fifty years, and was bound up in contemporaneous phenomena like written recordkeeping, the spread of literacy, colonial surveying, New World land claims, financial risk management, the Protestant Reformation, disputes over ancestral land rights, witch trials, and dispossession.[45] In the words of Belgian medievalist Henri Pirenne, "Europe 'colonized' herself."[46] Thus alongside the Savage and the Slave stand the Pagan and the Peasant, a consonance given provocative narrative form in Paul Kingsnorth's postcolonial cli-fi Brexit novel, *The Wake*, which tells the story of an Anglo-Saxon yeoman farmer's failed resistance to the Norman conquest of 1066.[47]

"And this also . . . has been one of the dark places of the earth," observes Charles Marlow of Victorian London, then capital of the civilized world, in Joseph Conrad's *Heart of Darkness*, "thinking of very old times, when the Romans first came here, nineteen hundred years ago—the other day."[48] Marlow paints a vivid picture of a Roman commander going up the Thames, feeling "the savagery, the utter savagery" of the ancient Britons closing around him—

> all that mysterious life of the wilderness that stirs in the forest, in the jungles, in the hearts of wild men. There's no initiation either into such mysteries. He has to live in the midst of the incomprehensible, which is also detestable. And it has a fascination, too, that goes to work upon him. The fascination of the abomination—you know, imagine the growing regrets, the longing to escape, the powerless disgust, the surrender, the hate.[49]

These opening lines of *The Heart of Darkness*—or rather, the beginning of the story within the story—cast an ambivalent but illuminating light on the ideas under discussion, particularly the hope that if we could only escape back to indigenous, pre-modern, pre-agricultural ways of being, we might find redemption.[50] The novel's central intellectual drama depicts the civilizational double bind thoughtfully explored by Claude Lévi-Strauss in *Tristes Tropiques* and astutely diagnosed by Sylvia Wynter and others: the tragic awareness that the recognition of our common humanity cannot bridge the ontological divisions structuring social relations. Just as "White European/Americans cannot become Hopis," Kurtz cannot become a Congolese, Gilgamesh cannot become Enkidu, Roman cannot become Barbarian, Christian cannot become Pagan, Settler cannot become Savage, White cannot become Black, and Civilized cannot become Indigenous. As Marlow put it, "There's no initiation . . . into such mysteries."

THE MYTH OF RENEWAL

Those of us who live in modern, literate, global civilization are incapable of rewilding ourselves, uncivilizing, going native, or decolonizing the Anthropocene. We cannot change at will our "embedded and embodied relations with our other-than-human kin and the land itself," consciously transform "our understandings of ourselves as human," or rip up our civilization by the roots and degrowth it into a pastoral Eden.[51] The issue is not whether such a choice is preferable, merely whether it is possible. There can be no doubt that so-called indigenous ways of relating to the land and coexisting with the nonhuman were sustainable, so to speak, for hundreds of thousands of years before the development of large-scale agriculture, empire, literacy, and fossil-fueled industrial capitalism, even if, as archaeologist Steven LeBlanc argues, "the idea that Native Americans, as well as early humans the world over, lived in ecological harmony is pure fantasy."[52] Despite the clear limits of the racializing stereotype of the "ecological Indian," animist thought framing human and nonhuman relations through intersubjective kinship indubitably offers a more integrative ecological-cultural matrix than modern extractivism.[53] Further, it may be granted that *any* imagined alternatives to fossil-fuel capitalism may inspire novel social formations, no matter how impossible they might be in practice. In the final analysis, however, solutions to catastrophic global warming and ecological collapse that depend on Traditional Ecological Knowledges, becoming indigenous, decolonizing the Anthropocene, uncivilization, or anarcho-primitivism—all of which might be seen as versions of an apocalyptic Myth of Renewal—offer no concrete programs or effective tools and face insurmountable conceptual difficulties of which these four are salient: we have never been modern; indigeneity is always local; we cannot become illiterate; and the total violence required for such a vision to be achieved is ethically unsupportable.

The pithily phrased idea that "we have never been modern" we owe to Bruno Latour, from his book of the same name. His point was that modernity has not made humans "rational" but merely shifted the coordinates of our metaphysics. As I've argued throughout this book, we "moderns" not only believe in unseen forces that shape our lives but appeal to them through intricate rituals with as much care and passion as any Chacoan sky priest tending his kiva. We are not secular rationalists, that is, but the devout followers of deities like "the market," "the nation," "democracy," and "race," committed to the ancestral Myth of Progress, and perennially engaged in the ritualistic performance of complex ceremonies and sacrifices to sustain material relations around wholly metaphysical conceits. The first problem with the Myth of Renewal is that there is no question of "going back" to pre-modern embeddedness in an animate lifeworld, since we are all already embedded in an animate lifeworld today and thus "becoming indigenous" wouldn't mean simply returning to the old gods but killing the new ones. Such a spiritual revolt would no doubt find proponents, but it would also spark opposition, and the question of how to deal with any progressivist rump offers no easy solution. Even if you solve that problem, perhaps through conversion by the sword, you still face the question of which gods should rule in the new dispensation: Noqoìlpi the Gambler? Cōātlīcue with her skirt of snakes and necklace of hearts and skulls? One-eyed Odin? Samhara Kali? Crom Cruach? Babalú-Ayé? There is, after all, no universal indigenous pantheon, nor do indigenous beliefs coexist in some postmodern cosmopolitan multiverse, where Thor and K'uk'ulkan can team up to fight Dr. Doom. There were particular peoples with particular beliefs in particular deities, who did "not regard their world-founding myths and rituals as one of many countless narratives but instead as the 'Truth.'"[54] This point brings us back again to the stubborn fact that "Indigenous" is a racializing abstraction in no way adequate to the concrete lived realities of pre-modern peoples and only coherent within a racialized teleology weaving biology, history, ontology, and culture into a single axis leading from past to future, from primitive to modern.

Which brings us to the second point, which is that indigeneity is always local, and not merely local but embedded in specific ecological and geophysical affordances. Potawatomi are not Aztecs, Nambikwara are not Inuit, and Paniya are not Celts. To flip Latour's famous apothegm: *we have never been indigenous*. The abstract universality in the idea of indigeneity criticized by Joseph Gone and others not only conflates distinct peoples and times, but flattens and elides the specific ecological relations that make a Shoshone a Shoshone and not a Naskapi Cree, or a neolithic Majiayao farmer who they are and not a Thuringii of the Central European wald. This problem is made even more complex as climate change transforms ecosystems, habitat ranges, and weather patterns, and local ecological knowledge slips out of sync with lived reality. Tragically, the massive ecological perturbations caused by global warming are making indigenous ways of knowing *less* adaptive and *less* sustainable, because the lands to which particular ecological knowledges have adapted are changing so rapidly. This phenomenon can be most clearly seen today among Arctic peoples, whose traditional hunting grounds are changing or disappearing and whose traditional ways of living on the land are becoming impracticable.[55]

What's more, the geophysically specific character of particular indigenous identities is not only epistemological but ontological. As David Abram shows in *Spell of the Sensuous*, synthesizing the work of anthropologists who have looked closely at Aboriginal songlines and Apache narratives, the languages and stories of indigenous peoples are embedded directly in the landscape.[56] "For the Amahuaca, the Koyukon, the Western Apache, and the diverse Aboriginal peoples of Australia—as for numerous indigenous, oral cultures—the coherence of human language is inseparable from the coherence of the surrounding ecology, from the expressive vitality of the more-than-human terrain."[57] Comprehending this fact, writes Abram, helps us "understand the destitution of indigenous, oral persons who have been forcibly displaced from their traditional lands." He explains:

> The local earth is, for them, the very matrix of discursive meaning; to force them from their native ecology (for whatever political or economic purpose) is to render them speechless—or to render their speech mean-

ingless—to dislodge them from the very ground of coherence. It is, quite simply, to force them out of their mind. The massive "relocation" or "transmigration" projects underway in numerous parts of the world today in the name of "progress" (for example, the forced "relocation" of oral peoples in Indonesia and Malaysia in order to make way for the commercial clearcutting of forests) must be understood, in this light, as instances of cultural genocide.[58]

This insight helps us comprehend our destitution. The various lifeworlds of various indigenous peoples emerged through meaningful relations to specific ecosystem features, particular species of flora and fauna, and reliable regional climatological patterns. All these aspects of reality are now in flux, sometimes catastrophically. We cannot simply return to an animist world rich in meaning when that world has been stripped of the diversity and richness of nonhuman life in relation to which said meaning emerged, while our own daily practices remain embedded in an anthropogenic world of industrial technology, specialized labor, and complex social hierarchy.

Which is to say that modern humans are alienated from nonhuman nature today not by bad ideology or the wrong narratives but by the material structures of industrial life. We don't hunt our own food. We don't grow our own corn. We don't make our own clothing, tools, or homes. We don't produce our own heat. We are protected from most climatic variation. Our doctors treat illnesses with sophisticated pharmaceuticals and technological apparatuses. Our survival does not depend on close attention to the land, the plants around us, our animal kin, or the weather, but on attention to traffic, the market, and social media. And the meaning of our world is mediated not by the language of animals, plants, earth spirits, and sky gods but by text and images on pages, screens, doors, walls, billboards, and gas pumps.

My third point follows from the preceding. Just as our metaphysics inhere in our environment, which for modern humans is the built human world, so our ontological and epistemological commitments are structured by the symbolic world in which they take shape and the material media through which that symbolic world is constructed. Indigeneity is typically and historically associated with symbolic structures of existence sustained without phonetic literacy, which is to say that indigenous "ways of knowing" are typically those of primary oral cultures. The advent of literacy, forced or adopted, transforms not only indigenous epistemologies but indigenous ontologies as well. As Anne Carson writes, discussing the emergence of literacy in Greek culture, "Reading and writing change people and change societies.... Oral cultures and literate cultures do not think, perceive, or fall in love in the same way."[59] Indeed, "change" may be a euphemism. Literacy may obliterate indigenous ways of knowing and being.

The transition from orality to literacy is too complex to adequately address in this limited space, but in brief there is good evidence that the technological development of phonetic literacy was a major event in the process of human self-domestication, with profound consequences for human culture and patterns of cognitive practice, including specifically a transition from metonymic relationality to metaphoric categorization, alienation from the physical lifeworld, and the emergence of conceptions of discrete identity analogous to the bounded identity of a written word or name, thus giving rise to philosophical axioms such as the law of identity, the law of the excluded middle, and the law of non-contradiction, among other consequences.[60] "The acquisition of literacy transforms the human brain," as cognitive neuroscientist Stanislaus Dehaene et al. have shown, though the full extent of such transformation and its effects remain understudied.[61] The domestication of what Lévi-Strauss called "wild thought" through literacy can be seen occurring again and again in successive waves from the birth of tragedy to Romanticism and beyond.[62] Indeed, forced literacy may be a primary tool not only of civilization but of empire and slavery.[63] As Lévi-Strauss writes:

> The only phenomenon with which writing has always been concomitant is the creation of cities and empires, that is the integration of large numbers of individuals into a political system, and their grading into castes or classes. Such, at any rate, is the typical pattern of development to be observed from Egypt to China, at the time when writing first emerged: it seems to have favored the exploitation of human beings rather than their enlightenment.... My hypothesis, if correct, would oblige us to recognize the fact that the primary function of written communication is to facilitate slavery. The use of writing for disinterested purposes, and as a source of intellectual and aesthetic pleasure, is a secondary result, and more often than not it may even be turned into a means of strengthening, justifying, or concealing the other.[64]

The fundamentally oppressive function of written language may be resisted through literary *marronage*—through strategies of feral literacy such as surrealism, stochastic feedback, contradiction, aporia, or Glissantian errantry—but it cannot be escaped.[65] Indeed, there seems to be little hope that this bedrock inculcation in "thinking like a modern" can be undone so long as we continue to teach our children to read. Perhaps the churning mélange of images and emojis and TikToks at the heart of contemporary digital literacy creates new affordances for novel posthuman animisms at the same time as it opens new frontiers for cognitive and social colonization.[66] Perhaps new "flows of energy and matter" will give birth to new gods.[67] Perhaps literacy itself is a dead end. On such questions, who can say?

My point is merely that there is no going back. The worldview of the human in an oral culture is profoundly foreign to modern thought. The relationship humans have to language and concepts in oral cultures is radically different from our own. Walter Ong writes, "Fully literate persons can only with great difficulty imagine what a primary oral culture is like, that is, a culture with no knowledge whatsoever of writing or even of the possibility of writing."[68] In the words of David Abram, "It is exceedingly difficult for us literates to experience anything approaching the vividness and intensity with which surrounding nature spontaneously presents itself to the members of an indigenous oral community."[69] The way is blocked. There is no initiation into such mysteries. We cannot willfully become illiterate. We cannot undo decades or centuries of acculturation in literate thought, wish our way out of our minds, resurrect the teeming biodiversity of a lost world, chuck our jobs, and "become indigenous" in the sense of returning to premodern lifeways, not even those of us who identify as indigenous today.

Finally, to return to Fanon by way of Tuck and Yang, we must recognize that "decolonization is not a metaphor" and that the scale of disordering violence that genuine "decolonization" would demand outstrips anything advocates of indigenous epistemologies or uncivilization are likely to accept, since it would be indistinguishable from global genocide.[70] Indigenous ways of knowing and living have never in the history of the planet supported more than fifty million human beings at once; to envision humanity "becoming indigenous" in any real way would mean returning to primary oral societies with low global population density, lacking complex industrial technology, and relying primarily on human, animal, and plant life for energy. It would mean bringing down the entire system within which we now live—tearing down skyscrapers, blowing up gas stations, burning books, smashing screens, and dismantling complex machines, including harvesters and freezers and centrifuges—and letting more than 99% of Earth's current population die. As Fanon writes, "You do not disorganize a society . . . with such an agenda if you are not determined from the very start to smash every obstacle encountered."[71]

"Every obstacle" in the case of "becoming indigenous" means not only the intractable facts that we have never been modern, that indigeneity is always local, and that we cannot become illiterate. It means not only oil companies and corrupt politicians and the high priests of economics and industrialization. It means "not just our energy use . . . our modes of governance, ongoing racial injustice, and our understandings of ourselves as human"—not only the roots of plantation logic in forced literacy, centralized agriculture, and private property—not only the possibility that it may be "too late for indigenous climate justice," in the words of Kyle Whyte, because decarbonization and environmental justice are not the same and perhaps even contradictory—but much more.[72] "Every obstacle" means the material infrastructure of our lives; "every obstacle" means streets, houses, computers, cars, planes, medicine, cheap energy, cheap food, modern dentistry, refrigeration, central heating, electricity, synthetic-cotton blends, toothpaste, printing presses, PDFs, email, and university indigenous studies programs. Thus while pre-modern indigenous social formations are doubtlessly more ecologically sound than the ones offered by progressivist capitalism, the only path to reach them lies through the end of the world. And as much as we may be obliged to accept and even embrace such an inevitability, committing ourselves to bringing it about is another question entirely.

RADICAL HOPE

Like all those peoples whose worlds ended before our own was born, we cannot go back. Yet even if we abjure the nostalgic seductions of an apocalyptic return to pre-modern life, we might still productively study historical moments of collapse and transition as examples of how to navigate the disintegration of our world. Philosopher Jonathan Lear does just this in his book *Radical Hope: Ethics in the Face of Cultural Devastation*.[73] Lear considers the case of the great Crow (Apsáalooke) chief Plenty-Coups, who guided his people through the transition from being nomadic warriors and hunters to becoming peaceful sedentary ranchers and farmers. This transition involved a harrowing loss of meaning, yet Plenty-Coups was able to articulate for his people a meaningful and hopeful way forward by committing them to life in an unknowable future.

The Crow were a culture whose entire way of life focused on hunting buffalo, raiding other tribes, and showing courage in battle, but in the late nineteenth century they found this form of life becoming impracticable. Caught between Blackfeet to the north, Sioux to the east, and other tribes to the south, they were being squeezed out of their territory by white settlers coming across the plain. They couldn't move across their home range as they once had. They couldn't hunt buffalo because the buffalo were gone. The treaties they'd been forced to sign prohibited fighting with other tribes, and if they fought the whites, they knew they'd be destroyed: the US Army had better weapons, more troops, and a whole industrial society behind it. The Crow faced a desperate impasse, with no clear way forward.

So Plenty-Coups told his people about a dream he'd had as a boy.[74] In that dream, he met a buffalo who turned out to be a man in a buffalo robe. The man showed Plenty-Coups a hole in the ground where buffalo kept coming out, buffalo after buffalo until they blackened the plain, but after the last buffalo came out, they all disappeared. Then a different animal came out of the hole, like the buffalo but not as hairy, and spotted instead of brown all over. Now it was the white man's cattle filling up the plain.

The buffalo man showed Plenty-Coups an old man sitting under a tree and told him he was looking at himself. The sky turned black "with streaks of mad color going through it," and a great wind came from the north and the south, from the east and the west, a wild storm that twisted trees like blades of grass, tore down the hills, and scattered the people.[75] When the storm had passed, every single tree in the forest had fallen except one: the one old Plenty-Coups was seated under. Plenty-Coups asked the buffalo man what it meant, and the man told Plenty-Coups that the tree was Chickadee's lodge and that even though Chickadee was the least of all beings when it came to physical strength, he had the strongest mind of anyone, because he was always listening, always willing to work for wisdom and learn from others. When other people sat around bragging and complaining, Chickadee listened and learned from their mistakes.

When Plenty-Coups first told his elders his dream, as a boy, they had seen great wisdom in it. They saw how it foretold the end of the buffalo and the coming of the white man. But they also saw how it foretold their people's survival by following the way of the Chickadee. So when Plenty-Coups became a great chief, he told his story again. The Crow had been conquered, pent up on a reservation, and forced to watch their way of life be uprooted and destroyed like a tree in a storm, but they could survive, even if they no longer hunted buffalo or raided Blackfeet camps. Their world had come to an end, but *they* need not. They could start over in the new world they'd been forced to inhabit.

Lear begins his investigation by inquiring into an astonishing statement from Plenty-Coups, who says that after the passing of the buffalo and the coming of the white man, "nothing happened." Lear asks, "What is this possibility of things' ceasing to happen?"[76] This question, as Lear carefully demonstrates, is an ontological query into the conditions of the possibility of meaningful events. The way Lear interprets Plenty-Coups's statement is to understand him as saying that the Crow way of life had collapsed and actions and events following this collapse no longer took their meaning from within the rich web of shared signification, values, and goals that had been the Crow world. The Crow had survived, but they no longer lived as Crow.

Examining the practice of planting coup-sticks, where a Crow warrior plants a stick in the ground and defends it to the death, Lear shows how we can see such an action as facing failure in two ways, one meaningful and one not. A meaningful failure would be for an enemy like a Sioux to kill the Crow warrior and take his coup-stick. Thus the Crow would fail and the Crow border would have been breached. Yet in this case Crow reality would still exist, because it still had borders that could be breached. A meaningless failure would be if the Crow were forced onto a reservation and made to stop fighting, as happened with the treaty of May 7, 1868.[77] In this case, planting a coup-stick would no longer mean anything, since the Crow no longer possessed the autonomy to define their own borders, nor did they have the right to defend them. Lear argues that this kind of failure lay beyond Crow conceptual reality, which held to a logic of the excluded middle: "*Either our warriors will be able to plant their coup-sticks or they will fail.*" This other kind of failure, in which planting a coup stick was meaningless, was unthinkable. It didn't make sense. Yet through the difficult, visionary work of Plenty-Coups, the Crow were able to accept their loss and begin to articulate a meaningful existence beyond the end of their world.

Through reference to Kierkegaard, Lear compares Plenty-Coups's apophatic commitment to an unknown future to the binding of Isaac, the Akedah. Lear writes:

> Through his dream—and his fidelity to it—Plenty-Coups was able to transform the destruction of a *telos* into a teleological suspension of the ethical. A traditional way of life was being destroyed, and along with it came the destruction of its conception of the good life. The nature of human happiness became essentially unclear and problematic. In such conditions, the temptation to despair is all but overwhelming. And it was in just such a moment that Plenty-Coups's dream predicted that destruction and offered an image of salvation—and a route to it. The traditional forms of living a good life were going to be destroyed, but there was spiritual backing for the thought that new good forms of living would arise for the Crow, if only they would adhere to the virtues of the chickadee.[78]

We face a similar impasse today, forced on us by a changing planet. Catastrophic global warming and ecological disaster are bringing about the end of the world as we know it. Just as Plenty-Coups and his people could no longer live as Crow lived, we can no longer live as progressivist, capitalist, modern Humans. We thus face two challenges: first, whether we might successfully mitigate the worst possibilities of climate change, and second, whether we might successfully transition to a new way of life in a new world. Meeting the former challenge is a technocratic and political question about which one may justifiably be pessimistic. Meeting the latter challenge will require very different work: facing the losses we have already incurred, finding a realistic way forward, and committing to an idea of human flourishing beyond any hope of knowing what form that flourishing will take. "This is a daunting form of commitment," Lear writes, for it is a commitment "to a goodness in the world that transcends one's current ability to grasp what it is."[79]

Lear compares Plenty-Coups with Sitting Bull, the Sioux chief who was killed supporting the Ghost Dance movement in the weeks before the massacre at Wounded Knee. While Plenty-Coups sought a realistic compromise with the future, Sitting Bull turned to what Lear calls a wishful way of thinking: "It is a hallmark of the wishful that the world will be magically transformed—into conformity with how one would like it to be—without having to take any realistic practical steps to bring it about. The only activity in which one is destined to partake is ritual, in this case a dance."[80] The analogy to present-day climate marches, strikes, and even, tragically, the Dakota Access Pipeline protests is clear. Nevertheless, Sitting Bull's people, the Hunkpapa Lakota, survived and live on today, as do the Crow. Across the United States, Native American communities struggle with poverty, unemployment, poor housing, inadequate health care, high rates of youth suicide, disproportionate rates of incarceration, and educational failure.[81] At the same time, Native American poets, historians, singers, dancers, novelists, filmmakers, and thinkers work to keep their inherited cultures alive beyond mere survival. Danowski and de Castro describe the situation:

> If the Humans who invaded it represented the indigenous America of the sixteenth and seventeenth centuries as a *world without humans* . . . the surviving Indians, fully entitled Terrans from that New World, reciprocally found themselves as *humans without world*: castaways, refugees, precarious lodgers in a world in which they no longer belonged, because it could not belong to them. *And yet, it just so happens that many of them survived.* They carried on in *another world*, a world of others, their invaders and overlords. Some adapted and became "modernized," but in ways that bear little relation to what the Moderns understand by that word. Others still struggle to hold on to whatever little world is left to them, and hope that, in the meantime, the Whites will not manage to destroy their own White world, now become for all living beings the "common world."[82]

The point here is not to judge which Native American tribes adapted most successfully to conquest and genocide. The point is to ask what we might learn today from these people who survived the end of their world, not in the romantic sense of glamorizing the "ecological Indian" nor in a nostalgic longing for "becoming indigenous" but rather in the more mundane sense of learning from courageous and intelligent people who lived through extraordinary times.

For we, too, live in extraordinary times, an existential impasse defined by the accelerating breakdown of the postwar political order, the collapse of the biosphere, ecological overshoot, and planetary climate transition.

We don't face an Event coming tomorrow or in thirty years or in 2525, but ongoing dis-integration and decoherence, day by day, every day, that confounds our sense of meaning. Our challenge is to learn to live ethically within this impasse in such a way as to sustain a commitment to some as-yet-unimaginable form of future human flourishing, through the apophatic suspension of narrative Lear calls "radical hope."

It is not clear that we possess the psychological and spiritual resources to meet this challenge. The first step must be coming to terms with our situation, and even that has been the struggle of decades, the outcome of which remains obscure. We argue about messaging and "green growth" as the ground shifts under our feet, glaciers calve into the sea, and subarctic boreal forests catch fire. History offers no support for optimism: the long record of human accomplishment is little better than a march of folly, and the last thirty years of dealing with climate change have been a disgraceful farce. It's not clear that successfully coming to terms with our situation will even matter unless we see substantial immediate reductions in global carbon emissions. One alarming study suggests that at atmospheric carbon dioxide levels around 1200ppm, which we may hit within a hundred years, subtropical stratocumulus cloud decks may dissipate, radically decreasing overall planetary albedo and adding as much as 8ºC warming on top of the 3–4ºC warming already expected by that point.[83] It's hard to see how the human species could survive that much warming happening that quickly. Nevertheless, the fact that our impasse offers no way out does not free us of the obligation to deal with it. If we want to continue to lead meaningful lives, we must promise ourselves to the future, whatever form that future takes. Despite decades of failure, ongoing paralysis, a social order geared toward consumption and distraction, and the possibility that our great-grandchildren may be the last generation of humans to ever live on Earth, we must go on. We have no choice in that. The choice we have is how to live in the gap between our failing fictions and our unknowable future.

BETWEEN FICTION AND REALITY

Fictions are necessary, even when we know they are fictions, even when ultimately, in the face of death, they cannot help but fail. Kermode argues that we should thus make our fictions as complex and sophisticated as possible, so that they might more persuasively reflect reality and also offer in their complexity a transcendental experience of timelessness. The fictions "that continue to interest us move through time to an end," he writes, "an end we must sense even if we cannot know it; they live in change, until, which is never, *as* and *is* are one."⁸⁴ There is a touch of the sublime in this argument: sophisticated fictions are, for Kermode, compelling surrogates for the existential desiderata of cosmic coherence and meaning. But as you look up from the last page of your latest novel or the season finale of your latest show to face yet again the unraveling world in which you live, you may find the ideal of narrative sophistication wanting.

Despite Kermode's sophisticated arguments, that is, mere sophistication cannot make our "common project" cohere. When it comes to the "ways we make sense of our lives," aesthetic and intellectual complexity for its own sake cannot help but fail to justify our "ideas of order," not least because it lacks any serious ethical or political dimension. Indeed, the aestheticized politics one might extract from Kermode's approach leads to a kind of apocalyptic fascism, as Walter Benjamin foresaw. Let us not forget that Hitler was an artist.

> "*Fiat ars—pereat mundus*," says Fascism, and . . . expects war to supply the artistic gratification of a sense perception that has been changed by technology. This is evidently the consummation of "*l'art pour l'art.*" Mankind, which in Homer's time was an object of contemplation for the Olympian gods, now is one for itself. Its self-alienation has reached such a degree that it can experience its own destruction as an aesthetic pleasure of the first order. This is the situation of politics which Fascism is rendering aesthetic.⁸⁵

Kermode's attempt to save aestheticism from its consummation in fascism fails, and the best he can do is weave a beautiful fiction of detached complexity, the same pose of critical sophistication so many intellectuals adopt today. Meanwhile, party faithful on the right and left offer increasingly grotesque fantasies, expounded with a naïve force that only grows more vehement as our disintegration grows more obvious. Storytelling fails. Fiction fails. Every future timeline is horror show, farce, or both at once, not even dignified enough to be tragic. What fiction could possibly make this chaos bearable? What axis is left to us, "in the middest," to orient our lives along, as we face a future at once awful and absurd, catastrophic and unknowable?

Even scientists—*especially* scientists—are unsure about the twists and turns that lie before us. Evolutionary biologists Jon Bridle and Alexandra van Rensburg write, "Despite increasingly precise predictions of rises in average temperature and the frequency of extreme weather events, biologists still cannot predict how ecological communities will respond to these changes."[86] What happens when you dump billions of tons of carbon dioxide into the atmosphere? What happens when you fill the seas and skies with microplastics, toxic compounds, and radioactive waste? What happens when you wipe out nearly all large mammals except the ones humans eat? What happens when you cut down the forests, pave the wetlands, dam the rivers, plow the prairies, and poison the lakes and oceans? At what point does business as usual become an abrupt and unstoppable transition to a new state? No one knows. Scientists can only guess, estimating from previous abrupt transitions in the paleoclimate record, computer models, and observations of specific ecosystems. But as Bridle and van Rensburg soberly note, there is "a growing body of evidence for alarming, widespread losses of biodiversity and for rates of global change that now exceed the critical limits of ecosystem resilience."

This too is a fiction, in Kermode's sense, insofar as every human viewpoint is necessarily fictional, and it is apocalyptic as well, though this apocalyptic fiction is rather robustly backed up by evidence. And while it's notionally possible that the numerous researchers telling us in so many words that the world is ending are doing so in order to forge a concordance between their own lives and the majestic pattern of human existence on Earth, as a Kermodian reading might suggest, or fabricating a scary story out of self-interest, as climate change denialists insist, we have very good reasons for taking their observations as credible statements about physical reality. There is nevertheless something peculiar about this empirical fiction: like Kermode's fiction of sophisticated academic detachment, science's "fiction" is founded in a radical skepticism toward *all* fictions. Science learns what it knows only by learning what it doesn't—science grounds its claims on a recognition of the radical limits of scientific knowledge.[87] Thus the viewpoint articulated by Bridle, van Rensburg, and the scientific consensus on climate change at large critically undermines the credibility of our other contemporary fictions.

The truth is we cannot know the future. The best empirical knowledge we have about the present tells us we're on an accelerating trajectory toward unprecedented and catastrophic transformation, but we can't know in advance how quickly the planet will warm. We can't know when or whether it's too late to change. We can't know whether this civilization will survive the next century or even the next decade. We can't know how many species will eventually go extinct. We can't know how long Earth can sustain more than eight billion humans. We can't know how awful the next several decades might be. We can't know what kind of planet humans will be living on in 2150 or 2550. We can't know how or whether our grandchildren will survive. We can't know when our world will end, though we can trust someday it will. We can't know what world our descendants may find on the other side. We can't know who or what will live there.

As I have been arguing throughout this book, it is narratively, psychologically, philosophically, and politically difficult to come to terms with the paradoxical extremity of our situation. The rational mind quails before such an unveiling, such an ἀποκάλυψις. Our lives today are built around concepts and values that are undermined by climate change, ecological overshoot, and rapid simplification, and this cultural trauma exposes yet another way in which our impasse transcends our capacity to fashion a coherent fiction adequate to our circumstances. The emerging polycrisis, as some are calling it, fails to manifest as a meaningful event. Trapped within the conceptual framework of capitalist modernity and intellectually shackled to the Myth of Progress, we are like the inhabitants of Socrates's allegorical cave or the Crow trying to make sense of reservation life with coup-sticks: we cannot see climate change as a meaningful human occurrence. It cannot be quantified by markets, mastered by gadgetry, confronted by protests, or defeated by military mobilization. Something is happening, that much is clear, and the trajectory sketched by scientists is deeply alarming, but fundamental aspects of our situation remain beyond our grasp, and the solutions offered by our leaders are no more than shadows of puppets dancing on a wall.

Our fictions are failing, the future is unknowable, and the possibilities for meaningful action appear limited and compromised. We must somehow live through the end of our world, knowing we will never live to see the new one. And yet this existential destitution in no way frees us from the obligation to live and care for each other—and not only for each other but for the world that is passing and passed. Our poverty may free us from the delusions of infinite affluence, optimism, and progress, but it does not free us from the demands of ethics. As I will argue in the following chapters, suffering matters. You may not be obliged to save the planet, but you're still obliged to care.

I will spend the rest of this book arguing that a pessimistic recognition of our human limits, of the inevitability of human suffering, and of the interdependent character of human life entails a commitment to compassion and care. Such a recognition also offers a resilient, egalitarian, and empirically reliable framework for collective action. When we act like the humble chickadee and face our end with radical hope, our dogmatic faith in progress gives way to ethical pessimism.

PART TWO

The Leap

Five

GET HAPPY

The soul refuses all limits. It affirms in man always an optimism, never a pessimism.
—RALPH WALDO EMERSON, "COMPENSATION"

In 1979 two psychologists, Lauren Alloy and Lyn Abramson, conducted an experiment to see whether depressed people felt helpless.[1] Specifically, they were trying to find out whether, in accordance with Martin Seligman's theory of learned helplessness, depressed people tended to believe that they had less control over things than they actually did.[2] To test this, the psychologists had students sit, one at a time, in front of a machine with a green light and a button. Following the proctor's instructions, the student would either press the button or not. Sometimes the light flashed when the button was pushed, sometimes it didn't. Sometimes the light flashed on its own. Alloy and Abramson varied other factors as well. In one experiment, for instance, they gave students money when the light flashed. In another, they gave students money at the beginning, then took some back with each flash. The initial sample set was ninety-six students, half of whom identified as depressed.[3]

The results of the experiment were surprising. The evidence Alloy and Abramson found didn't confirm learned helplessness theory at all but rather suggested that well-adjusted students had an erroneously inflated sense of agency.[4] That is, nondepressed students thought they had more control over the green light than they actually did, especially when they didn't—for instance, when the light flashed independently of them pushing the button. When nondepressed students were emotionally invested (for instance, when money was involved), their judgment got even worse. They thought they had even more control when they were getting what they wanted and thought they had less when they weren't. Depressed students, on the other hand, retained greater clarity of judgment regardless of either frequency or incentive and were better able to judge how much control they had over the green light in general.

Alloy and Abramson's article describing their experiments, "Judgment of Contingency in Depressed and Nondepressed Students: Sadder but Wiser?," became the foundational study for the theory of depressive realism, which holds that

> depressed individuals tend to be more accurate or realistic than nondepressed persons in their judgments about themselves. Specifically, research suggests that nondepressed people are vulnerable to cognitive illusions, including unrealistic optimism, overestimation of themselves, and an exaggerated sense of their capacity to control events. This same research indicates that depressed people's judgments about themselves are often less biased.[5]

Depressive realism has been repeatedly confirmed, though the theory remains contentious and many people are resistant to the idea.[6] Depressive realism challenges cognitive-behavioral theories of therapeutic treatment, for example, since such treatment claims to be correcting erroneous thoughts, but if the theory of depressive realism is accurate, cognitive-behavioral therapy may actually reinforce self-deception. As Alloy and Abramson write, "the primary active ingredient in cognitive therapy may be the training of depressed clients to engage in . . . optimistic biases and illusions."[7] Perhaps most disturbingly, depressive realism "raises the possibility that a realistic and unbiased perception of oneself and one's relation to the world" might just be depressing, calling to mind T. S. Eliot's Nietzschean lines from *Four Quartets*: "Human kind / Cannot bear very much reality."[8]

There are, it must be granted, numerous thorny concerns and variables involved in such research, including the unreliability of self-reporting, problems of bias, delay in response, whether participants' judgments are made in public or in private, whether judgment is being tested about oneself or others, how data are interpreted, and how to define "reality." It's also notable that many studies suggest that depressive realism may only apply to judgments about oneself.[9] There's also the issue of severity. Studies suggest that while mildly depressed people may be more realistic, severely depressed people tend to see things as bleaker than they really are.[10] Overall, however, there seems to be good support for at least a weak version of the theory: we can say that mildly depressed people tend to see things a little more clearly.[11] Indeed, depressive realism may be a form of self-protection. Recent research on psychological responses to the COVID-19 pandemic, for instance, support the hypothesis that anxiety, depression-like symptoms, and fear are adaptive responses promoting defensive action against environmental threats.[12]

Yet what may be most striking about Alloy and Abramson's landmark experiment isn't what it shows us about the clear insights of the gloomy but what it reveals about the rose-tinted delusions of the well-adjusted. Leaving aside the complexities and implications of depressive realism, the key takeaway—conclusively established by a broad range of research—is that nondepressed, cognitively "normal" people are persistently, unrealistically, and irrationally optimistic. They believe they have more control over things than they do, they'll live longer than they will, and things will turn out better than they're likely to—they even remember things turning out better than they did.[13] Most people are deeply susceptible, that is, to optimism bias.[14]

THE OPTIMISM BIAS

In part 1 of this book, I considered some of the narrative, cognitive, political, and conceptual problems that we face in trying to understand, come to terms with, represent, and respond ethically to our impasse. I discussed the limits of progress, problems in climate change communication, political extremism, conceptual problems posed by climate change, and the end of the world, arriving at the conclusions that our situation does not seem to be comprehensible within progressive modernity and may not be comprehensible at all and further that our dependence on narrative, biases and heuristics, psychological dispositions, inherited political and cultural frameworks, and commitments to core social values may make our situation even harder to comprehend. These conclusions lead to a paradoxical and perilous ethical precipice: the only way to cross the gap between progressive carbon-capitalist modernity and whatever comes next is to make a leap of faith, committing ourselves to an unknowable future.

This is an awkward and risky situation, fraught with danger and temptation. Every crisis demands a response, without allowing us to ground that response in a stable existential framework. False prophets seduce us with happy endings and zealots get us drunk on hate, while the planet-devouring machinery of carbon-capitalism grinds on even as it unravels, offering the comforting illusion of an eternal present. How do we navigate such a time with integrity? How do we live ethically in a world of catastrophe?

In part 2, I sketch a framework for ethical decision-making under such conditions. I call that framework ethical pessimism, and in order to make the case for it, we must begin not with pessimism itself but with that characteristic of the human animal pessimism reacts against, a characteristic of great evolutionary advantage over the long term but that under conditions of global ecological catastrophe has proven highly maladaptive, a characteristic as widespread as it is dangerous: our bias for optimism.

Neuroscientist Tali Sharot writes, "Across many different methods and domains, studies consistently report that a large majority of the population (about 80% according to most estimates) display an optimism bias. Optimistic errors seem to be an integral part of human nature, observed across gender, race, nationality and age."[15] This human bias for optimism is not only pervasive but durable: since people tend to believe what they already believe and reject contradictory information, optimism is highly resistant to evidence.[16] In the words of neuroscientist Samuel Gershman, "Human beliefs have remarkable robustness in the face of disconfirmation."[17] As research from Sharot et al. shows, there even appears to be a neurological tendency to specifically ignore information that challenges positive assessments.[18] These biases afflict even those ostensibly trained to reject them, such as scientists and doctors.[19] *Homo optimismus* is neither a rationally self-interested profit-seeker nor a curious empiricist but "a charlatan trying make the data come out in a manner most advantageous to his or her already-held theories."[20]

Our disposition to optimism is probably an evolutionary adaptation and over the long term an effective one.[21] Psychologists generally agree that "unrealistically positive self-evaluations, exaggerated perceptions of control or mastery, and unrealistic optimism" are, for most individuals, positive and helpful.[22] Yet in a crowded world in the midst of ecological crisis, the costs of excessive optimism may outweigh the benefits, both for individuals and, more importantly, for society. Research from Roy Baumeister suggests there is an "optimal margin of illusion," where a smidgen of unrealistic optimism confers a benefit, but positive effects quickly drop off as self-serving delusion diverges from reality.[23]

The negative consequences of optimism run the gamut from the trivial to the catastrophic. Optimism can foster egotism, inflexibility, grandiosity, and megalomania.[24] Many people find unrealistically positive self-evaluation in others unpleasant: narcissism, bragging, frequently interrupting people, condescension, aggression, and demanding constant validation are just a few of the traits observed in people with overly positive self-evaluation.[25] Excessive optimism can be pathological and lead to "narcissistic, histrionic, and obsessive-compulsive personality traits."[26] It "can cause people to persist fruitlessly at unsolvable tasks" and may increase ideological extremism.[27] Unrealistic optimism is especially dangerous in scientists, journalists, and political leaders. Flawed, overly optimistic assessment can lead to poor planning, bad decisions, unrealistic deadlines, overconfidence, high-risk behaviors, complacency, information neglect, a lack of motivation to learn about or adopt preventative measures, incompetence, impeded adaptation, and a distorting perception of uniqueness.[28] Countless historical examples of disasters resulting at least in part from "unrealistically positive self-evaluations, exaggerated perceptions of control or mastery, and unrealistic optimism" are readily invoked, from the Trojan War to the Iraq War, including the failure of ruling elites to seriously address the threat of global warming.[29] Although erroneous, self-confirming optimism may have been adaptive under the harsh conditions of the Pleistocene, under modern conditions of human planetary dominance, with an array of powerful, poorly understood technologies at our disposal, such optimism may be suicidal.[30]

Considering the significant drawbacks of delusional optimism and keeping in mind Alloy and Abramson's theory of depressive realism, one can begin to see the value of pessimism, not as a counsel of despair but as a necessary corrective to a dangerous and pervasive bias. Indeed, when we look back over the horrors, suffering, and delusions that make up the balance of human history, from ritual cannibalism to modern genocide, we can understand why pre-modern religion and philosophy repeatedly emphasize moderation, harmony, balance, and restraint, offering their followers not the optimistic bromides so familiar from contemporary advertising and pop psychology but the kind of hard wisdom we today call pessimism. In contrast to delusional optimism, pessimism offers a prudent, ethical framework for existing in a world of suffering and uncertainty. This has always been the case, but under the pressure of global ecological crisis, pessimism's value increases. Facing an ambiguous situation with fraught and unclear choices, where the consequences may be not only disastrous but final, refusing to consider the worst is simply irresponsible.[31] Under such conditions, it is pessimism, not optimism or progress, that offers "the richest set of resources with which to confront the human condition."[32]

Such claims may seem outrageous, especially to Americans.[33] Compulsory positivity, variously identified by Barbara Ehrenreich as "brightsiding," Lauren Berlant as "cruel optimism," and Susan Cain as "toxic positivity," is central to American culture.[34] "There is no kind of problem or obstacle for which positive thinking or a positive attitude has not been proposed as a cure," writes Ehrenreich. "Like a perpetually flashing neon sign in the background, like an inescapable jingle, the injunction to be positive is so ubiquitous that it's impossible to identify a single source."[35] Optimism is "ingrained in the culture as an explicit, essential value," attests psychologist Edward Chang. "We're hit over the head with American freedom and liberty and rugged individualism so much so that explicit pessimism isn't actually tolerated that much in our society."[36]

SUFFERING AND TIME

In order to better appreciate the value of pessimism, we should first consider how philosophers and psychologists have defined optimism and pessimism, explore what empirical research has to say about them, look at the two terms' intertwined history, and consider the many forms pessimism has taken. It will also help to distinguish philosophical pessimism from other views with which it's often confused, such as nihilism, fatalism, and quietism.

What do we mean by *optimism* and *pessimism*? The terms were originally coined by eighteenth-century literary critics to describe contrasting philosophical views on reason and evil, as we'll discuss, but today they carry a range of meanings, both technical and commonplace, and are often used to signal other values. "Both concepts," writes philosopher Mara van der Lugt, "lend themselves to the kind of exaggeration that is at the same time a deflation, one that flattens these terms until they become almost trivial, denoting little more than a mental attitude, an outlook on day-to-day life."[37] In the most mundane sense, the terms are pejoratives: optimists are naïve Pollyannas who minimize human suffering in order to "always look on the bright side of life," while pessimists are black-pilled doomers "who get off on being wet blankets."[38] To identify as an optimist is to express a positive commitment to cosmic justice; a pessimist reckons it'll all go to hell. Optimists don't even need to identify as such, they only need to accept the story they're told every day in newspapers, advertisements, Hollywood films, and the latest hit show. In contrast, the pessimist is a misfit and pariah. As philosopher Eugene Thacker writes, "one always *admits* to being a pessimist," as if it were a personal failing.[39]

In a slightly more sophisticated register, the two terms describe psychological dispositions: those who see the glass half full against those who see the glass half empty, the cheerful versus the gloomy. This register connects with and bleeds into optimism and pessimism as descriptions of predictive stances, as expectations of good or poor outcomes. According to sociologist Joe Bailey, "Pessimism is the negative evaluation of the likely future," and optimism is likewise a positive evaluation of the same.[40] Once we're considering the scientific study of optimism and pessimism, however, things quickly get complicated. In psychological discourse, researchers distinguish between "explanatory styles" and "expectation": the story we tell about what happened versus what we expect to see in the future, which are related but not the same.[41] Psychologists also try to keep optimism and pessimism distinct from other cognitive and emotional phenomena. Edward Chang writes: "Optimism should be distinguished from other notable psychological variables such as internal control and self-esteem. Likewise, pessimism should not be confused with expressions of lack of control and self-effacement."[42]

Psychological research also suggests that optimism and pessimism are best understood not as opposed or antithetical extremes but as independent dimensions of personality.[43] "Some people are both strongly optimistic *and* strongly pessimistic," writes Julie Norem, "[while] others are neither particularly one nor particularly the other."[44] Psychologists have observed that humans paradoxically show both an optimism bias and a negativity bias: we tend to unrealistically expect the best while also overweighting bad experiences.[45] There are numerous possible explanations for this "positive-negative asymmetry," but the most persuasive is that it's a behavioral adaptation.[46] Specifically with regard to climate change, bioethicist Anders Nordgren argues that it is "problematic to speak about this pessimism and optimism in general terms. Pessimism and optimism may concern different aspects of climate change." For instance, he writes, "a person can be pessimist concerning climate change as an unmitigated or poorly mitigated process and optimist concerning mitigation of climate change, and be pessimist concerning one specific mitigation measure and optimist concerning another."[47]

Optimism and pessimism can also be cognitive strategies.[48] Psychologists Julie Norem and Nancy Cantor have identified both "strategic optimism" and "defensive pessimism" as methods for coping with uncertainty and anxiety.[49] Strategic optimists succeed by setting high expectations and refusing to dwell on failure: this is the classic optimistic approach that underlies such bromides as "think positive" and "dream big." Defensive pessimists succeed as well, but by setting lower expectations and rehearsing the ways things could go wrong.[50] Defensive pessimism is especially useful for anxious people coping with risk, but is potentially helpful to anyone making decisions in complex situations where the stakes are high, information is limited, and important factors may be out of one's control.[51]

Considering optimism and pessimism as cognitive strategies rather than as dispositions raises important questions: Can we choose? Can we train ourselves to "look on the bright side" and intentionally reject negative thinking? Conversely, can we train ourselves in tactical pessimism? Can we learn to check our biases toward unrealistic optimism and intentionally reflect on challenges, obstacles, and limits that we might not intuitively foresee? There are, after all, real benefits to both strategies. As Norem and Chang have shown, pessimism can help manage anxiety and mitigate risk and may in certain situations lead to better outcomes.[52] On the other hand, broad research supports the conclusion that generally expecting positive outcomes correlates with both emotional and physical well-being and may have health benefits.[53] "Across a variety of indicators, people who hold positive beliefs about (a) their personal attributes, (b) their ability to bring about desired outcomes, and (c) their future, are better off than those who are more realistic," write Jonathon Brown and Margaret Marshall.[54] Research on optimism, pessimism, and socioeconomic status shows another correlation: wealthier people tend to be more optimistic, while poorer people tend to be more pessimistic.[55] Relatedly, lower-class and lower-status people tend to show more accurate emotional cognition and may "display greater insight into others than their high-status counterparts," suggesting a status-based variety of depressive realism.[56]

Yet optimism and pessimism have a "chicken and egg" problem, since it's difficult to know whether optimistic people are healthier and wealthier, or if healthy and wealthy people tend to be more optimistic. While self-esteem is not the same as optimism, it is perhaps worth considering research from Baumeister et al. suggesting that self-esteem doesn't cause good performance but rather results from it.[57] Further research supports the hypothesis that optimism and pessimism are more effect than cause: the dispositions appear to be "sticky," resistant to external events and shifts in status. One longitudinal study shows that childhood poverty is associated with greater pessimism in adulthood, even in cases where people moved up in socio-economic status. "Childhood SES, as indexed by the parental educational level and occupational class, predicted lower optimism and greater pessimism in adulthood," writes Kati Heinonen et al., "even after controlling statistically for the adulthood educational level and occupational class."[58] Poverty may foster lifelong defensive pessimism: from a position of inadequate resources, all economic choices are losses, and every decision involves comparing negative scenarios.[59] Choosing *optimistically* in such circumstances would be unrealistic and self-destructive.[60] When faced with no-win scenarios, defensive pessimism—hedging losses—is the more prudent strategy.

How much freedom we actually have to choose between optimism and pessimism may remain an open question, entangled in deeper questions about free will, inequality, agency, and the limits of reason. I would like to believe that reflection and training can mitigate the dangers of instinctive optimism, institutional structures can be developed to support realistic assessment, and psychological insight can be deployed by well-intentioned agents to create socially beneficial framing, and I'm encouraged by the work of sociologist Karen Cerulo, which suggests that some specific communities do just that, particularly service-oriented groups with porous boundaries, a formal knowledge base, and high levels of individual autonomy, like doctors and IT professionals.[61] But hoping for widespread, socially validated pessimism would be unrealistically optimistic. The sad truth is that thinking is hard, collective decisions are no more rational than individual ones, and wisdom is rare.

Anyone who has worked in politics, served on a committee, led an organizational unit, or tried to influence the public and then reflected seriously on their experience will know how much sheer labor is required to come to a wise decision, convince others of its merits, implement it, and make it last, and how even in the best cases outcomes are often tenuous. Anyone who has struggled with addiction or mental health knows how hard it can be to change one's own thoughts and behaviors, never mind those of others. And as we saw with climate change communication, people tend to filter for evidence that confirms what they already believe, are motivated largely by social cues and inertia, and don't want to be told what to think or do. Humans can be difficult and irrational in the best of circumstances, and although behavioral economists have suggested we may be able to "nudge" people through "choice architecture," the evidence that nudging actually works is equivocal.[62] Ultimately, the very question of whether humans can collectively and intentionally adopt strategies to make our lives better, solve social problems, and increase general human well-being is the very question on which optimists and pessimists most fundamentally differ.

OPTIMISME ET PESSIMISME

To understand optimism and pessimism in their deeper historical and philosophical senses, we must understand where they came from. "The origins of pessimism and optimism lie in philosophy," writes Bailey, namely the early modern European philosophical movement we call the Enlightenment.[63] The earliest optimists were motivated by skepticism toward received wisdom and the rational search for empirical truth, while the earliest pessimists turned these optimists' skepticism against their own conclusions.

The word *optimism* (or *optimisme*, for it was in French) first appears in the February 1737 issue of the *Mémoires pour l'Histoire des Sciences et des Beaux-Arts*, an influential and controversial eighteenth-century French Jesuit academic journal, in a review by mathematician and physicist Louis-Bertrand Castel of Gottfried Wilhelm Leibniz's *Essays of Theodicy on the Goodness of God, the Freedom of Man and the Origin of Evil* (1710).[64] Castel coined the word *optimisme* to describe Leibniz's theodicy, which, as the title of his book indicates, sought to reconcile belief in a benevolent God with the fact of evil. "In terms of art," Castel writes, "he calls [his doctrine] the best reason, or even more skillfully, and Theologically as well as Geometrically, the Optimum system, or Optimism."[65]

Leibniz invented the term *theodicy*, meaning "justice of God" from the Greek, and penned his groundbreaking essay introducing that term as a riposte to the Calvinist thinker Pierre Bayle in a long-running fight between the two, which had its proximate origin in a dispute over the relationship between dogs' bodies and their souls.[66] The argument was about much more than canine metaphysics, however; it was about whether reason could make sense of suffering. The challenge that had initially provoked Leibniz came from Bayle's 1697 *Dictionnaire Historique et Critique*, specifically his entries on the Manicheans, the Paulitians, and the pre-Socratic philosopher Xenophanes.[67] In his *Dictionnaire* and other work, Bayle had scandalously asserted that reason was incapable of explaining free will or evil.[68] Leibniz, in response, sought nothing less than to demonstrate how free will could be reconciled with the idea of necessity and the fact of evil reconciled with the idea of a benevolent, omnipotent, and perfect God.[69] According to philosopher Mara van der Lugt, the "challenge of pessimism" Bayle posed, which inspired Leibniz to write his *Theodicy*, had three main components.

First, rejecting the standard scholastic interpretation of evil as a lack of goodness, Bayle argued that evil and suffering were as real as goodness and happiness and that "evils are thus ontologically symmetrical with goods."[70] Second, Bayle argued that evil and suffering offered "greater *experiential intensity*" than goodness and happiness, anticipating the theory of negativity bias.[71] Third, Bayle argued that most people would not choose to relive their lives, and that thus life was not, on the balance, worth living, but that even if this were the case for only one person in the entire world, it would still invalidate any rational appeal to God's plan. As van der Lugt puts it, "God's decision to create unhappy creatures (humans but perhaps also animals) cannot *rationally* be justified."[72]

The question of how to reconcile the persistence of human suffering with the idea of a benevolent God had long been a problem for the monotheistic traditions, receiving perhaps its most striking form in the Book of Job, but the problem took on new urgency in seventeenth- and eighteenth-century Europe, as thinkers attempted to reconcile the insights of Baconian natural philosophy, the rediscoveries of Renaissance humanism, the epochal expansion of their world through the Columbian Exchange, and their inherited Christian and scholastic traditions. The issue at stake was fundamentally the question Descartes opened in his *Meditations*: how to rationalize and justify the existence of God within the mechanistic, deterministic understanding offered by nonadaptive scientific empiricism. If the world was orderly and mechanistic, as the new science showed, then God must be responsible for laying down the natural laws that matter followed. Thus there must be some congruence between the moral order of Biblical theology and the physical order of God's universe. At the same time, the experimental method axiomatically rejected metaphysical explanations, meaning that neither God nor the Devil could be accepted as an immanent cause. Empiricism drove a wedge between the material and the divine, between physics and metaphysics.[73] Scientific determinism further threw into crisis the idea of free will, which is fundamental to the Christian doctrine of sin and redemption. Only a being endowed with the capacity to choose between good and evil can be judged for their choice. But if humans are subject to the same mechanistic laws as everything else, then we do not choose whether we sin; we merely do what we do—or what we have been programmed to do, so to speak.

Bayle's solution to this conundrum was a fideism so radical as to be indistinguishable from agnosticism. For Bayle, reason was utterly incapable of explaining ultimate questions because it was incapable of comprehending the divine. Leibniz, in opposition to Bayle, struggled to save reason for faith and vice versa. We might go so far as to see Leibniz as one of the last great medieval scholastics, in the tradition of Thomas Aquinas and Duns Scotus, working to unify reason and faith in a comprehensive, holistic worldview, aligning the intricacies of the visible cosmos in harmony with the eternal perfection of loving deity.[74] Key to Leibniz's efforts was the argument that among all possible worlds there was necessarily one that was the best, and we lived in it. Leibniz argued that the goodness of the world, supervenient on God's wisdom, was analogous to a numerical range, supervenient on the rationality of mathematics, and thus in order for there to be any world at all, there must be a best possible world, as well as presumably a worst, just as in a numerical range, there must by necessity be a maximum and minimum. He writes:

> As in mathematics, when there is no maximum nor minimum, in short nothing distinguished, everything is done equally, or when that is not possible nothing at all is done: so it may be said likewise in respect of perfect wisdom, which is no less orderly than mathematics, that if there were not the best (*optimum*) among all possible worlds, God would not have produced any.[75]

According to Leibniz, then, if God is the omnipotent, omniscient, benevolent creator of the world, it necessarily follows not only that this is the best possible world but that God could not have made it otherwise. For if our world were not the best possible world, God would have had to have been deficient in some respect: He didn't know how to make the best possible world, He didn't want this world to be the best, He wasn't capable of making it the best, or He had fallen prey to some other flaw incompatible with the idea of a perfect, loving, all-powerful deity. But none of this could be true, since such failures are by definition impossible for a being whose very existence is perfection itself.

Therefore every seeming deficiency in this world is but a misunderstood blessing, for all is part of God's perfect whole. In short, Leibniz concludes, "the greatest felicity here on earth lies in the hope of future happiness, and thus it may be said that to the wicked nothing happens save what is of service for correction or chastisement, and to the good nothing save what ministers to their greater good."[76] Recast in the taut rhymes of Alexander Pope, who popularized Leibniz's ideas in his *Essay on Man* (1733–1734):

> All Nature is but Art, unknown to thee;
> All Chance, Direction, which thou canst not see;
> All Discord, Harmony, not understood;
> All partial Evil, universal Good:
> And, spite of Pride, in erring Reason's spite,
> One truth is clear, "Whatever IS, is RIGHT."[77]

The lesson Reason imparts, according to Pope and Leibniz, is that everything has its proper place: "All are but parts of one stupendous whole, / Whose body Nature is, and God the soul."[78] Yet however much such a theodicy may have appealed to reactionary aristocrats and intellectuals anxious to preserve the idea of an orderly moral universe justifying their own positions of privilege, it proved inadequate against the corrosive spread of scientific rationality. It also provoked a rather pointed response from the sharp-witted writer François-Marie Arouet, better known as Voltaire.

Two decades after Castel coined the term *optimism*, Voltaire satirically adopted the word for the title of his novel *Candide, or Optimism*, in which he lampoons Leibniz in the character of Doctor Pangloss, professor of "metaphysico-theologo-cosmolonigology," who famously argues that

> things cannot be otherwise, for, everything being made for an end, everything is necessarily for the best end. Note that noses were made to wear spectacles, and so we have spectacles. Legs are visibly instituted to be breeched, and we have breeches. Stones were formed to be cut and to make into castles; so my lord has a very handsome castle . . . and, pigs being made to be eaten, we eat pork all year round.[79]

When we meet Pangloss, he is tutor to the Baron Thunder-ten-Tronckh's daughter, Lady Cunégonde, and young Candide, the baron's bastard nephew. We then follow Candide, upon his eviction from the baron's castle for the crime of kissing Cunégonde, through a series of misadventures, from being dragooned into the Bulgarian army to witnessing the Lisbon earthquake, from Buenos Aires to Eldorado to Venice and finally to a small farm outside Constantinople, where he retires to "cultivate his garden."

Along the way, Candide finds his old teacher Pangloss, who had been reduced to pauperdom by syphilis and war, then hanged for heresy, miraculously alive and enslaved on a Venetian galley.

> "My dear Pangloss," Candide asks him, "when you had been hanged, dissected, racked with blows, and rowing in the galleys, did you still think that all was for the very best?"
>
> "I am still of my first opinion," the optimist answers, "for after all I am a philosopher, and it is not fitting for me to recant."[80]

Candide rescues Pangloss, and although his old tutor remains an optimist, Candide himself learns to mitigate his optimism, in part through dialogue with the Baylean philosopher Martin, a companion he'd picked up in Surinam. In the end, settled on Candide's farm, Pangloss still insists everything worked out for the best, in spite of all their suffering along the way:

> All events are linked together in the best of all possible worlds; for after all, if you had not been expelled from a fine castle with great kicks in the backside for love of Mademoiselle Cunégonde, if you had not been subjected to the Inquisition, if you had not traveled about America on foot, if you had not given the Baron a great blow with your sword, if you had not lost all your sheep from the good country of Eldorado, you would not be here eating candied citrons and pistachios.[81]

Candide is an incomparable distillation of style and thought, a founding text of philosophical pessimism, and a turning point in the life of an intellectual provocateur: like the novel's eponymous hero, Voltaire retreated from a decade of tumult and strife to "cultivate his garden." Indeed, even if, as Ira Wade argues, the "broad movement from an optimistic to an increasingly pessimistic period is far too general to accord with the facts" of Voltaire's life, the decade preceding the writing of *Candide* might have strained the morale of the greatest optimist.[82] In 1749, Voltaire's lover, interlocutor, and friend of nearly two decades, the philosopher Émilie du Châtelet, died, leaving him grief-stricken. Once he began to recover, he left France for Berlin, where he served ambivalently as chamberlain to Frederick of Prussia, miring himself in financial scandal by illegally importing cut-rate Prussian bonds.[83] Voltaire compounded his difficulties by picking a fight with another French philosophe at Frederick's court, Pierre Louis Maupertuis, head of the Berlin Academy, which soured his relationship with Frederick. Voltaire left Sanssouci in March 1753, carrying off a book of

Frederick's private poetry, provoking the Prussian king to send his agents to track Voltaire down and arrest him, holding him for almost a month until he returned the poems.

Finally cut loose from Frederick and banned from Paris by Louis XV, Voltaire settled in Geneva, where the republic's approach to censorship and an indiscreet article in d'Alembert's *Enyclopédie* made things difficult. He stayed, however, and it was there in Geneva that Voltaire heard about the catastrophe that had struck Lisbon: a massive earthquake and tsunami that had killed thousands, destroyed much of the city, and sent shockwaves through Europe's intellectual community. Voltaire was inspired to write a poem about the event, sketching in draft the pessimism that would suffuse *Candide*:

> Sad is the present if no future state,
> No blissful retribution mortals wait,
> If fate decrees the thinking being doomed
> To lose existence in the silent tomb.
> All may be well; that hope can man sustain,
> All now is well; 'tis an illusion vain.[84]

Numerous scholars have discussed the impact of the Lisbon disaster on European intellectual culture, including philosopher Susan Neiman, who sees the event as the beginning of modern philosophy. For Neiman, philosophy since Lisbon is concerned essentially with the problem of evil, because it "is fundamentally a problem about the intelligibility of the world as a whole."[85] Although Neiman doesn't frame her account in this way, what she offers in her book *Evil in Modern Thought* is a history of modern philosophy understood through the opposition between optimism and pessimism—that is, between two opposing views on whether reason can overcome suffering.

Tracing the history of philosophy from Lisbon to Auschwitz, from Bayle, Voltaire, and Leibniz to Arendt, Adorno, and Rawls, Neiman identifies two broad approaches to the problem of whether or not reality is ultimately susceptible or even intelligible to human reason: an optimistic view holding that reality is controllable or at least orderly and a pessimistic view holding that it is neither.[86] Both sides agree that "appearance gives us a world of misery," but where they split is on the question of "what to take more seriously: the stark and painful awareness that we have . . . when confronted with any form of evil; or the ideas and explanations that allow us to transcend it."[87] Neiman's argument offers a provocative and revealing counter-history of modern philosophy, dissenting from the more conventional, deflationary story of philosophy from Descartes to David Chalmers as a tradition split between epistemological questions increasingly constrained by the rise of scientific rationality on the one hand and ethical questions increasingly ungrounded by metaphysical materialism on the other, to help us see just how central optimism, pessimism, and the Myth of Progress are to understanding how we make sense of reality in the modern era.

But as Neiman herself makes clear, the Lisbon earthquake wasn't an event that changed the course of European philosophy as much as an objective correlative for emerging conceptions of suffering, meaning, and time. The earthquake didn't inspire the birth of philosophical pessimism; rather, philosophical pessimism found in the earthquake a thought-image that gave it narrative and conceptual shape. If even the pious city of Lisbon could be destroyed by God, what hope might we have that Reason could say why?

Theodor Adorno writes that "the earthquake of Lisbon sufficed to cure Voltaire of the theodicy of Leibniz," but it's not clear that Voltaire ever subscribed to Leibniz's rationalization in the first place.[88] The development of Voltaire's pessimism was not sudden but gradual, the result of numerous events as well as philosophical reflection and empirical study. Voltaire's *Essai sur les Mœurs et l'Esprit des Nations*, for instance, which he'd been working on by that point for twenty years, had already judged history "a continuous succession of crimes and disasters."[89] Meanwhile, in Paris, the great *Encyclopédie* of d'Alembert and Diderot, the grandest intellectual effort of the age, foundered on the rocks of state censorship. Simmering colonial conflict between France and England in North America spilled over into Europe in 1756 as French forces took the British island of Minorca, and King Frederick took the opportunity to "invade Saxony under pretext of friendship, and make war upon the Empress Queen of Hungary," as Voltaire writes in his memoirs, "[and] singly changed the whole system of Europe."[90] In March of 1758, Prussian-Hanoverian forces crossed the Rhine and defeated the French army at the Battle of Krefeld. Grief, earthquakes, religious persecution, deceit, failure, intolerance, and war: Voltaire had all the evils of *Candide* within arm's reach.

He finished the book over the summer in Geneva. By the time it was published, in January 1759, he had moved back to France, taking up residence a few miles from the Swiss border, where he would spend his last decades, writing and "tending his garden." *Candide* was immediately suppressed by authorities in Paris and Geneva, yet was so popular that it went through more than twenty editions in 1759 alone and sold between 20,000 and 30,000 copies just that year.[91] The first recorded use of the word *pessimism* (or *pessimisme*, again in French) followed hard upon. In a review of *Candide* in the journal *Année Littéraire*, also published in 1759, literary critic Elie Fréron writes:

> We give the name Optimism, from the adjective Optimum which means the best, to the doctrine of the Philosophers who called themselves Optimists, who maintain with Leibniz and Father Malebranche, that God made the right things according to the perfection of his ideas, that is to say, the best he could, and that, if he could have done better in the creation, he would have. This opinion is reversed in *Candide*; we replace it with an absolute contrary, and the horrible consequence that we draw from this writing is that this world is the worst of worlds possible; in a nutshell, instead of Optimism, it is Pessimism, if I may put it that way.[92]

Thus was christened optimism's antithesis. Yet Fréron's description is not quite accurate: Voltaire's novel presents a wretched world but makes no claim that it couldn't be worse. After all, the novel ends with Candide reunited with his beloved Cunégonde, self-sufficient, and happily settled, and although the trials and tribulations that brought him to that point were certainly awful, any one of them could have easily left him crippled or dead.[93] Candide, like Voltaire, managed to endure the vicissitudes of life with his spirit intact, if chastened.[94]

Voltaire is thus a rather ambiguous founding figure when it comes to philosophical pessimism. Despite his keen satiric wit and gloomy view of history, Voltaire is generally regarded as a meliorist, who believed in the potential of human reason to decrease suffering.[95] Voltaire may have been "much too intelligent to overwork the idea of progress," as Karl Löwith writes, but he "believed in moderate progress, interrupted by periods of regression and subject to chance."[96] Voltaire was, in a word, a prototypical Enlightenment thinker, and his resistance to Leibniz wasn't a resistance to the idea that things could be good, but rather a rejection of the idea implied in Leibniz's thought that since we already lived in the best of all possible worlds, we could hardly hope for them to be better. Whether Voltaire's critique of Leibniz was accurate is a question I must leave to specialists, since Leibniz himself seems at times to be attempting to reconcile the static, monadic perfection of his metaphysics with the idea of temporal progress, particularly in *On the Ultimate Origination of Things* (1697), and his optimism may finally be more complex than either Voltaire's satire or Pope's popularization grant.[97] Nevertheless, it's worth taking Voltaire's critique on its own terms to note how different the conceptions of optimism and pessimism were, as originally staked out, from what we take them to be today.

Though Castel's review of Leibniz's *Theodicy* gave us *"optimisme"* in 1737 and Fréron's review of *Candide* introduced *"pessimisme"* in 1759, both concepts grew from seeds planted a century or more earlier, in Pico Della Mirandola's *Oration on the Dignity of Man* (1486), Bacon's *Novum Organum* (1620), and Descartes's *Discourse on Method* (1627).[98] Leibnizian optimism used reason to defend faith, arguing for the unchanging perfection of a world created by a perfect and unchanging God. In contrast, Voltairean pessimism used reason to attack faith, arguing that the world as it existed was deeply imperfect, but we might have hope that it could be made better through human action, which is precisely what Voltaire saw himself doing, playing the Socratic gadfly, "arousing and persuading and reproaching."[99]

Forty years later, the debate between optimism and pessimism took a dramatic new turn, as it was violently brought to life in the explosive upheavals of the French Revolution. Leibnizian holism and Voltairean meliorism merged in the idea of Progress, articulated most powerfully by the French aristocrat the Marquis de Condorcet and the English essayist and novelist William Godwin. From its inception, the idea of Progress inspired fervent faith and equally fervent critique from a range of thinkers, including the clergyman Thomas Robert Malthus, to whom we now turn.

Six

A MELANCHOLY HUE

Thomas Malthus is one of those notable writers who enjoy the ambivalent distinction of having been abstracted into an adjective: like *Orwellian*, *Kafkaesque*, and *Marxist*, *Malthusian* serves as a placeholder for a range of associations that vary depending on a particular speaker's needs, though in this case the associations are generally pejorative—typically callousness toward the poor, wrongheaded doomism, and being bad at math. One might wonder how such a paragon of insipidity managed to claim the attention of thinkers from Darwin and Marx down to John Maynard Keynes and Herman Daly, were it not for the fact that common conceptions of Malthus and his signature work, *An Essay on the Principle of Population* (1798–1826), tend to be superficial at best, where they are not completely wrongheaded.

Properly understanding Malthus and his *Essay* requires some sense of his project's goals and context and, most importantly, a clear view of his main target, *An Enquiry Concerning Political Justice* (1793), by British journalist, novelist, and political philosopher William Godwin. On publication, Godwin's *Enquiry* had taken the reading public of Britain by storm. As William Hazlitt later wrote, "No work in our time gave such a blow to the philosophical mind of the country."[1] And Godwin's *Enquiry* is indeed an original and visionary work, inspired by the tradition of English religious dissent, the French Enlightenment, and the French Revolution, combining a radical commitment to reason with a heady faith in progress. According to Godwin, all the evils that beset human life are due to inequality, property, and government and thus they can all be resolved by the proper application of reason.

For Godwin, ends and means are the same: the open, equal association of free-thinking individuals. Although the *Enquiry* is rightly regarded as the founding text of philosophical anarchism and argues for abolition of marriage, equalization of property, and dissolution of government, Godwin himself was no revolutionary. He vociferously rejected coercion, trusting entirely in "the diffusion of knowledge through the medium of discussion," private judgment, the perfectibility of moral reason, the indestructibility of knowledge thanks to print, and the inevitable progress of human society.[2] "There is no characteristic of man," Godwin writes, "which seems at present at least so eminently to distinguish him, or to be of so much importance in every branch of moral science, as his perfectibility."[3]

Reason ruled all. "To a rational being there can be but one rule of conduct, justice, and one mode of ascertaining that rule, the exercise of understanding."[4] The human mind, as Godwin saw it, was at once a blank slate, a purely deterministic mechanism bending ineluctably toward truth, and an omnipotent force: "We bring into the world with us no innate principles," he writes, and while "the existence of physical causes cannot be controverted ... their efficacy is swallowed up in the superior importance of reflection and science."[5] He opposed national education, believing it fostered fixed opinions rather than rational inquiry and had an "obvious alliance with national government."[6] He argued that improvement in moral reason and the practice of cheerfulness could eventually lead to immortality, and he envisioned humankind filling the Earth to the point where "they will cease to propagate, for they will no longer have any motive, either of error or duty, to induce them." In Godwin's vision:

> The whole will be a people of men, and not of children. Generation will not succeed generation, nor truth have in a certain degree to recommence her career at the end of every thirty years. There will be no war, no crimes, no administration of justice as it is called, and no government. These latter articles are at no great distance; and it is not impossible that some of the present race of men may live to see them in part accomplished. But beside

this, there will be no disease, no anguish, no melancholy and no resentment. Every man will seek with ineffable ardour the good of all. Mind will be active and eager, yet not disappointed.[7]

Astonishingly, this utopia was to be achieved solely by the inevitable spread of truth, "at first so slow as for the most part to elude the observation of mankind," eventually "rapid and decisive."[8]

To Thomas Malthus, then a young parson in the rural southeast of England, Godwin's utopianism seemed doubtful. Despite having been raised in an "intellectually radical environment" by an acolyte of Jean-Jacques Rousseau and educated by the controversialist dissenter and abolitionist Gilbert Wakefield, Malthus was skeptical of the radical progressivism then circulating in British parlors and coffeehouses.[9] The picture painted by Godwin and Condorcet didn't seem to match what he knew of history or what he saw around him. While a student at Cambridge, Malthus had focused his studies on stadial history (including Gibbon's *Decline and Fall of the Roman Empire*), geography, and applied mathematics and had overcome a cleft palate to excel as a student, secure a preaching parish, and earn election as a fellow of Jesus College. Malthus's parish posting, Okewood Chapel, was located in what historian Robert J. Mayhew describes as a dreary, isolated wasteland, scarred by mining and ordnance factories, characterized by "agrarian poverty . . . high birth rates, and short life spans."[10]

Over the decade he spent preaching at Okewood before writing the first edition of his *Essay on the Principle of Population*, Malthus would have observed several other phenomena that must have seemed at odds with Godwin's progressivist narrative. British population rose significantly over the final decades of the eighteenth century (from 5.9 million to 8.7 million, an increase of around 55%), probably due to earlier marriages causing larger families, while agricultural production over the same period lagged, causing the very kind of demographic tension Malthus theorizes in his essay.[11] This tension between human reproduction and grain shortage led to food riots in 1727, 1756–57, 1766–68, and 1773–74, and came to a crisis in 1794–1796, after bad weather and poor harvests caused two years of famine that provoked widespread rioting, including several standoffs between housewives and armed soldiers.[12] In 1795, a London mob surrounded the king's coach, shouting for bread.[13] Meanwhile, waves of unrest swept England in the form of labor riots and machine breakages, war in Europe threatened to spill across the Channel, and British imperial power was staggered by a series of failures and insurrections overseas, culminating in the Irish Rebellion of 1798.[14] Such was the context in which Malthus "sat down with an intention of merely stating his thoughts" on "the general question of the future improvement of society," which question was so vast that he reckoned it "would be much beyond the power of an individual" to canvas "all the causes that have hitherto influenced human improvement."[15] The thoughts he began to sketch sparked new ideas that would inspire further research, raise difficult questions, and provoke heated debate for more than two centuries, particularly as they took form in his "principle of population": the theory that population tends to exceed its subsistence wherever it is not held in check by natural and social constraints.[16]

Malthus's aim was to puncture Godwin's grandiloquently confident progressivism, but as Alison Bashford and Joyce Chaplin write, "it would be simplistic to see the two Englishmen as revolutionary versus reactionary," since "both men utilized arguments about humanity meant to better the human condition . . . and each insisted, with consistency and conviction, that he intended for the best."[17] Close attention to the *Essay* shows what sympathetic readers have attested, that Malthus was neither arguing that nothing could be done for the poor nor that attempts to improve society were necessarily doomed but rather that there were real limits to society's ability to transcend the laws of nature. "Necessity," he writes, "that impervious all-pervading law of nature, restrains [the germs of existence] within the prescribed bounds. The race of plants and the race of animals shrink under this great restrictive law; and the race of man cannot by any efforts escape from it."[18] In the words of literary scholar Maureen McLane, "Malthus's essay is nothing if not the work of sensitive husbandry—husbandry in all its senses: agricultural, matrimonial, economical, theological."[19]

Although we cannot escape nature's laws, Malthus agreed we might bend them, with the caveat that "a careful distinction should be made between an unlimited progress, and a progress where the limit is merely undefined."[20] His essay was an attempt to better ascertain where some of those limits might lie, so that public policy might respond more humanely by encouraging moral restraint rather than relying on misery and vice. Similarly, his criticisms of the English Poor Laws weren't attacks on social policy in principle but critiques of what he saw as a failed policy that exacerbated precisely what it was supposed to ameliorate. By increasing food prices and encouraging procreation, Malthus argued, the Poor Laws actually drove more people into poverty. He argued that the only way to actually improve the conditions of the poor would be to "raise the relative proportion between the price of labour and the price of provisions"—that is, to raise real wages or make food cheaper.[21] Furthermore, Malthus's several revisions of his *Essay* and substantial further research, particularly his greatly expanded second edition of 1803, show not reactionary zealotry but a sedulous commitment to empirical evidence.

There is much more to be said about Malthus, even given the voluminous existing literature, but for our purposes it's enough to note that the parson of Okewood is a key and underappreciated pessimist thinker, an important early figure in political economy, and a major Romantic writer. His work inspired Darwin, Marx, and Keynes, among others, and his *Essay* is a key precursor to environmental economics. It's true that Malthus held ideas many readers today would consider questionable, such as his theological opposition to birth control (which was not the reliable, mass produced suite of technologies available today) and the belief that suffering motivates human industry, or as he put it, "Evil exists in the world, not to create despair, but activity."[22] Yet he was also a Whig reformer who advocated for national education, the abolition of slavery, and free medical care for the poor, among other social programs.[23] Whether or not Malthus was right about the Poor Laws, whether or not his intentionally reductive and illustrative comparison of geometric versus arithmetic growth rates was misguided, and whether or not he failed to foresee the development of carbon-fueled capitalism and various technological revolutions in agriculture (for which he can hardly be faulted), he remains important and discussed and vilified today because his pessimistic account of society so clearly articulates an insight that refuses to be repressed, despite inexhaustible attempts to deny it: the laws of nature apply to humans.[24] This is a hard pill to swallow, even for people who should know better, such as science journalist Michael Shermer, whose 2016 article for *Scientific American*, "Why Malthus Is Still Wrong," blames Malthus for eugenics and the Nazis, then offers this piercing refutation: "Humans are thinking animals. We find solutions."[25] Indeed we do, and Malthus wouldn't have argued against that. What he did argue against was the complacent optimism of progress, echoing down the centuries from William Godwin to Steven Pinker, assuring us that our solutions are perfectly adequate. What Malthus offered instead was a view of society that tried to account for the data he'd collected, the history he'd surveyed, and the reality he saw around him. "The view which he has given of human life has a melancholy hue," Malthus wrote of himself in the third person, "but he feels conscious, that he has drawn these dark tints, from a conviction that they are really in the picture."[26]

ECOPESSIMISM

Malthus is not only the most important philosophical pessimist between Voltaire and Schopenhauer, but he is also the founding father of what we might call ecopessimism, the philosophical view that there are real ecological limits to human flourishing, exemplified in more contemporary works such as *Limits to Growth* and *The Sixth Extinction*.[27] Indeed, in its weaker form—as the empirical recognition that humans are biological entities dependent on and entangled with their environment—ecopessimism is indistinguishable from mainstream ecology. A weak ecopessimism is even compatible with ideas like sustainable capitalism and zero-growth economics. In its more severe forms, ecopessimism verges into antihumanism, with some ecopessimists arguing that humans are constitutionally incapable of living within our natural limits. In this vision, progressivist human exceptionalism is inverted, and humanity is seen as uniquely destructive.

Perhaps the fullest account of the ecopessimist tradition thus far has been put forward by two of its opponents, Pierre Desrochers and Joanna Szurmak, in their 2018 book *Population Bombed! Exploding the Link Between Overpopulation and Climate Change*. Desrochers is a Canadian geographer at the University of Toronto with links to various free-market institutions such as the American Institute for Economic Research, the Fraser Institute, and the Property and Environment Research Center. Szurmak is a librarian at the University of Toronto Mississauga and, at the time of this writing, a PhD candidate at the Graduate Program in Science and Technology Studies at York University. Their book was published by the disingenuously named Global Warming Policy Foundation, a think tank that promotes climate denial. Like Steven Pinker's *Enlightenment Now*, Desrochers and Szurmak's

Population Bombed! is a tendentious and poorly argued brief for the wonders of technology and human ingenuity; in contrast to Pinker, however, Desrochers and Szurmak actually offer useful scholarship.

Desrochers and Szurmak trace ecopessimist thought from what they identify as its deep roots in *The Epic of Gilgamesh*, Confucius, Plato, Tertullian, Saint Jerome, and Machiavelli through its fuller emergence in the work of Giovanni Botero and Giammara Ortes, whom they call "population catastrophists," to its archetypal form in Malthus's *Essay*.[28] From Malthus they turn to George Perkins Marsh's 1864 classic *Man and Nature*, then leap to the twentieth century, following the thread through Edward Murray East's *Mankind at the Crossroads* (1923), Graham Vernon Jacks and Robert Orr Whyte's *The Rape of the Earth* (1939), William Vogt's *Road to Survival* (1948), Henry Fairfield Osborne Jr.'s *Our Plundered Planet* (1948), Harrison Brown's *The Challenge of Man's Future* (1954), Rachel Carson's *Silent Spring* (1962), the work of Indian sociologist and Minister of Health and Family Planning Sripati Chandrasekhar, Garrett Hardin's seminal essay "The Tragedy of the Commons" (1968), Paul and Anne Ehrlich's *The Population Bomb* (1968), and the Club of Rome's *Limits to Growth* (1972), all the way up to contemporary warnings about the dangers of climate change and ecological collapse from such hysterical doom-mongers as David Attenborough and Pope Francis. Desrochers and Szurmak might well have added Charles Darwin, Alfred Russell Wallace, and the German zoologist Ernst Haeckel, who coined the term *ecology*, not to mention the science of ecology itself, which by taking a systemic empirical view of human-environmental interactions accepts natural limits as unavoidable, as well as much of the American nature-writing tradition, including Thoreau, Melville, Aldo Leopold, Edward Abbey, Annie Dillard, and Bill McKibben, plus cultural discourses around local food production, indigenous traditions, and the idea of sustainability. Indeed, one of the ambivalent features of Desrochers and Szurmak's broad-brush approach is that it paints anyone who questions the most ardent growth-oriented techno-optimism in ecopessimist black and green.

Their argument, of course, is that such thinking is wrong and bad. In their own words, their book "is an attempt to present a relatively concise case for the environmental benefits of economic development, population growth and the use of carbon fuels."[29] Like Dienstag, they see "two main perspectives on the relationship between humans and nature," which they label optimism and pessimism, one arguing that we "should reshape the natural world to our benefit," the other that "humanity should live within natural limits."[30] They elsewhere identify "eco-pessimism" as a worldview that coheres around the following arguments:

> 1) in a finite world, continued demographic and economic expansion is impossible; 2) everything else being equal, a reduced population will enjoy a higher standard of living; 3) in a world where resources are finite, economic growth will become increasingly expensive and environmentally damaging over time; 4) the risks inherent to new technologies make it preferable to restrict population growth and to live within limits than to rely on human ingenuity.[31]

These are neither well-thought-out definitions nor applicable even to their own examples. A better definition of ecopessimism accounting for their own evidence would be a viewpoint holding that the resources of the planet Earth are finite, human flourishing ultimately depends on living within natural limits, and while technological innovation may expand or loosen natural limits, it cannot eliminate them.

In any case, according to Szurmak and Desrochers, it's all sheer balderdash. They assert that "population growth and the development and growing use of carbon fuels liberated human labor and brains from subsistence agriculture while simultaneously reducing pressures on both wild flora and fauna," and they see no reason to worry that our bountiful harvest will end.[32] Like Pinker and Lomborg, they take the last two hundred years of industrial development and the last eighty years of industrial acceleration as only the beginning of humanity's ever-increasing prosperity. In Desrochers and Szurmak's view, ecopessimism has disgraced itself mainly by being so obviously wrong. Elaborating the classic argument against Malthus ad nauseam, they insist that those who assert there are limits to human growth must be in error because we haven't hit any yet. We keep growing, therefore we can. And according to their logic, we won't hit any limits as long as we keep burning fossil fuels, growing our economy, breeding, and letting market capitalism run wild.[33] "Provided human creativity is allowed to flourish in the context of market economics," they write, "future energy transitions will deliver both economic and environmental benefits."[34]

What's more, they argue, ecopessimists are not only mistaken but immoral. They're authoritarian, elitist, undemocratic, self-interested, self-righteous, ignorant, and vengeful—"blind, blinkered, and bought."[35] Desrochers and Szurmak could have made their moral case against ecopessimism even stronger by discussing antihumanist strains like those articulated by Peter Zapffe, H. P. Lovecraft, Thomas Ligotti, Michel Houllebecq, and David Benatar, never mind the historical connections between ecology, environmentalism, eugenics, and fascism from Ernst Haeckel and Madison Grant to Ludwig Klages and Martin Heidegger, from green Nazism to the internal conflict about immigration that almost destroyed the Sierra Club in the 1990s. Yet in the final analysis, what may be most striking is that while Desrochers and Szurmak are basically advocating mainstream cultural optimism, they sound like unhinged Randians. Trust in capitalism, faith in progress, and an implicit optimism toward the future are core tenets of modern society. Ecological thought is marginal at best, where it hasn't been co-opted and corrupted into ESG and "green growth," and ecopessimism is generally regarded with contempt. Annual global GDP is estimated at more than $110 trillion and our lives are structured from birth to death by the commodification of human activity, yet Desrochers and Szurmak somehow manage to make it sound like capitalism is an embattled victim of ideological persecution.

A judicious and comprehensive history of ecopessimism remains to be written. Ecopessimism, like sociobiology and evolutionary psychology, challenges widely held modern beliefs about human exceptionalism, the importance of free will, and the primacy of human culture over nonhuman nature—indeed, it challenges the very idea, going back to Aristotle, that culture and nature are separate domains. Seeing humans first and foremost as biological entities, either in terms of their ecological dependence on their habitat or in terms of their social organization and evolution, seems to undermine the idea of moral choice and perpetually risks devolving into social Darwinism. Ecopessimism is also tainted by historical association with biological racism. Yet none of these objections resolve the challenge that understanding human beings as organisms within ecological systems poses to our elaborate, self-justifying cultural narratives, nor do they obviate the need to reckon with how the planet we live on is finite, how human society is dependent on the long-term health of our habitat, and how exceeding the boundary conditions for human flourishing will have dire consequences. Ecopessimism—like pessimism more broadly—may challenge our optimistic prejudices and raise intractable moral dilemmas, and it may have repugnant historical associations, but none of that makes it wrong or inaccurate. The planet we live on has limits, as do human societies. Pessimism demands we reckon with what that means.

FROM BABYLON TO POSTHUMANISM

While philosophical pessimism is a distinctly modern tradition, that tradition has proto-modern and pre-modern roots in Montaigne, La Rochefoucauld, Cervantes, the Stoics, and the pre-Socratics, as well as in various wisdom literatures from the Bible to the Diamond Sutra, from the Bhagavad Gita to ancient Babylonian parables.[36] "Not to be born is, beyond all estimation, best," wrote Sophocles in the fifth century BCE, "but when a man has seen the light of day, this is next best by far, that with utmost speed he should go back from where he came."[37] The Hebrew Bible has strong pessimistic currents, in Ecclesiastes for instance: "And I declared that the dead, who had already died, are happier than the living, who are still alive. But better than both is the one who has never been born, who has not seen the evil that is done under the sun."[38] The betrayal, trial, and crucifixion of Jesus is a deeply pessimistic story as well, even redeemed by the miracle of resurrection. Afropessimist philosopher Jared Sexton goes so far as to argue that pessimism "is as old as philosophy itself and profoundly shapes the latter's coeval development with a whole range of literary forms."[39] Nevertheless, and with all due respect to Sexton, such pessimisms *avant la lettre* should be distinguished from the modern tradition. The key difference is that before the modern era, negative assessments were simply part of how most people viewed life, especially as they grew older: what we call pessimism, they called wisdom. The only hope for redeeming our lot came if at all from the heavens. The Myth of Progress displaced not only revealed religion but common sense, promising heaven on earth if we only "follow the science."

Tracing through Western philosophy and culture over the three centuries between Leibniz and ourselves a transgenerational effort to redescribe Christian metaphysics in science-friendly terms, we can see Christianity and Scientific Rationalism absorb each other, as the Holy Spirit transmutes into the spirit of Reason in History, the problems of sin and evil become problems of ignorance and savagery, and the "true and only heaven" of New Jerusalem descends from metaphysical eternity to become a utopian future here on Earth, where technology shall finally answer every human need. Leibnizian holism and Voltairean meliorism merge in the Myth of Progress, the foundational tenets of which are that human knowledge is cumulative, human life is getting better and will continue to do so, perennial problems like prejudice and conflict will eventually be solved, and we can not only improve the conditions of life but make ourselves better, smarter, and happier.[40]

For modern progressivists, Leibniz not only won his argument with Bayle but out-maneuvered Voltaire from the grave. Like Godwin asserts, progress is as undeniable as it is inevitable. After all, as modern-day Pangloss Steven Pinker insists, just look around you. We were meant to spend our days ruining our posture and mental health clicking buttons on tiny screens to get serotonin hits, and thus we invented smartphones. We were meant to heat the planet to levels not seen since the Paleocene-Eocene Thermal Maximum, and thus we invented the internal combustion engine. We were meant to live under the constant threat of global annihilation, and thus we invented nuclear missiles. Not accepting the manifest truth of progress is obviously a form of self-indulgent nihilism that leads only to despair. For its adherents, belief in progress is the only meaningful belief, and all others are either trivial or invalid.

Pessimism today remains scandalous and annoying, as it was for the French authorities who banned *Candide*, but instead of targeting theological certainties, it now mostly targets human ones. Optimism, meanwhile, has been turned inside out: no longer a faith that we live in the best of all possible worlds, it has become a faith in the world's continual improvement through progress. Thus we can understand optimism not only as a pervasive cognitive bias that fosters delusional overconfidence, erroneous risk assessment, and the persistent rejection of contradictory evidence but as the core creed of modern industrial civilization. Barbara Ehrenreich got it right: we're bright-siding ourselves to death.

The fundamental tenet of philosophical optimism and the intellectual justification for the ancestral Myth of Progress is the claim that human reason can explain suffering and improve the human condition. Pessimism, by contrast, holds that reason can do neither.[41] Both are attempts to make sense of the world after the delegitimization of theological authority.[42] As Mara van der Lugt, Susan Neiman, and Joshua Foa Dienstag all argue in their different ways, optimism and pessimism are antithetical modern frames for making sense of suffering and time.[43] In Dienstag's words:

> The optimistic account of the human condition is both linear and progressive. Liberalism, socialism, and pragmatism may all be termed optimistic in the sense that they are all premised on the idea that the application of reason to human social and political conditions will ultimately result in the melioration of these conditions. Pessimism . . . denies this premise, or (more cautiously) finds no evidence for it and asks us to philosophize in its absence.[44]

Dienstag holds that pessimism entails a belief in linear time, though I would argue it's not so simple, and that some pessimists or varieties of pessimism see time more ambiguously (Nietzsche, for instance). Further, it's not clear that a stance on temporality as such is definitive. Mara van der Lugt distinguishes Dienstag's "forward-oriented" conception of pessimism, in which the central question is whether life can get better, from her own "value-oriented" conception of pessimism, in which the central question is whether life is worth living, which she sees as more important.[45] Either way, the key question is whether reason can justify suffering. For the pessimist, the answer is no.

Philosophical pessimism is thus best understood as a critique made by reason against reason, arguing that the capacity for human beings to change things for the better is doubtful at best, and that historical transformation must be explained without presuming the metaphysical conceit of a teleological end or final form. Indeed, pessimism could be seen as empirical reason's self-lacerating reductio ad absurdum: a truly empirical view of human existence leads, as it were, beyond good and evil. "Philosophy at first refutes errors. But if it is not stopped at that point, it goes on to attack truths," writes Bayle. "And when it is left on its own, it goes so far that it no longer knows where it is and can find no stopping place."[46] Scientifically speaking, the idea of human progress is nonsense, since there is no goal for the species to progress toward. There is only energy, evolution, adaptation, and entropy. As Karl Löwith writes, "The problem of history as a whole is unanswerable within its own perspective. Historical processes as such do not bear the least evidence of a comprehensive and ultimate meaning. History as such has no outcome."⁴⁷ Pessimism thus rejects both the consolation of theodicy and the promise of progress, the one saying all our problems will be solved by God, the other by future generations. Pessimism further insists that some of our most important problems, such as suffering, evil, and death, can never be "solved": they are ineradicable, perhaps even definitive aspects of the human condition. This may be why Eugene Thacker calls pessimism "a lyrical failure of philosophical thinking": at its root, pessimism rejects any salvific potential in reason at all.⁴⁸

Beginning in skepticism, philosophical pessimism develops not as a doctrine or school but rather as a non-systematic or even anti-systematic counter-tradition of affiliated thinkers. Dienstag identifies a cluster of propositions that characterize this counter-tradition—namely, "that time is a burden; that the course of history is in some sense ironic; that freedom and happiness are incompatible; and that human existence is absurd," though fundamentally the only core commitment is the rejection of any optimistic claims for human reason.⁴⁹ After Bayle, Voltaire, and Malthus, the next great pessimist in Western philosophy is Arthur Schopenhauer, whose work inspired a veritable efflorescence of pessimism in nineteenth-century Germany (including the anti-pessimistic pessimism of Friedrich Nietzsche) and later inspired Wittgenstein, Horkheimer, and Adorno.⁵⁰ The pessimist counter-tradition might also include Giacomo Leopardi, David Hume, Julien Offray de la Mettrie, the Marquis de Sade, Agnes Marie Constanze von Hartmann, Max Weber, William James, Oliver Wendell Holmes, Miguel de Unamuno, George Santayana, W. E. B. Du Bois, Peter Wessel Zapffe, Jose Ortega y Gasset, Sigmund Freud, Albert Camus, Ernst Jünger, Lewis Mumford, Michel Foucault, Emil Cioran, Jean Améry, James Baldwin, Sarah Kofman, Thomas Ligotti, and David Benatar, along with "close associates in writers like Sartre, Arendt, Benjamin, Wittgenstein, and Weil," as well as ambiguous affiliates like Ralph Waldo Emerson and Jean-Jacques Rousseau.⁵¹

This counter-tradition of philosophical pessimism bears a strong resemblance to Richard Rorty's idea of the "edifying" strain in modern philosophy, that train of "figures who . . . resemble each other in their distrust of the notion that man's essence is to be a knower of essences" and who "are often dubious about progress."⁵² Rorty's "edifying philosophers"—

including Dewey, Wittgenstein, Heidegger, Kierkegaard, and Nietzsche—are "intentionally peripheral" discontents who "destroy for the sake of their own generation," speak in "satires, parodies, aphorisms," and are "reactive" against the main stream of optimistic philosophy.[53]

> They refuse to present themselves as having found out any objective truth (about, say, what philosophy is). They present themselves as doing something different from, and more important than, offering accurate representations of how things are. It is more important because, they say, the notion of "accurate representation" itself is not the proper way to think about what philosophy does.... Whereas less pretentious revolutionaries can afford to have views on lots of things... edifying philosophers have to decry the very notion of having a view, while avoiding having a view about having views. This is an awkward, if not impossible, position.[54]

As awkward and impossible as this kind of philosophy might be, Rorty argues that it nevertheless serves a valuable purpose. Likewise, Rorty's deflationary vision of philosophy may seem hopelessly inadequate to our impasse, which of course it is, but as we struggle with the awkward and impossible task of making sense of our own end, "keeping the conversation going," as Rorty would put it, is not only worthwhile but necessary.

The roll call of pessimist thinkers "would grow considerably longer," Dienstag adds, "if one included poets and fiction writers," for instance, Herman Melville, Mark Twain, Edith Wharton, Henry Adams, Gertrude Stein, Samuel Beckett, Ralph Ellison, Ingeborg Bachman, and William Burroughs, and even longer if one included other artists.[55] The philosopher Raihan Kadri suggestively argues that "the Surrealist movement consciously set itself up as . . . involved with the intellectual current of philosophical pessimism."[56] Philosophical pessimism thus presents a rich and vital tradition with robust contemporary strains, including ecopessimism (discussed previously), liberal pessimism, conservative pessimism, cultural pessimism, posthumanism, and Afropessimism.[57]

Liberal pessimism may seem a contradiction. Liberalism is, after all, deeply and optimistically committed to the Myth of Progress. In receiving the nomination for the US presidency in 1960 from the Liberal Party of New York, then-Senator John F. Kennedy defined liberalism as "a faith in man's ability through the experiences of his reason and judgment to increase for himself and his fellow men the amount of Justice and freedom and brotherhood which all human life deserves."[58] In the contemporary American political imaginary, liberalism falls somewhere in the optimistic middle-left, neither a revolutionary call for utopian rupture nor reactionary resistance to change but a melioristic progressivism characterized by a willingness to work within the system to improve the human condition. Adherents of liberalism believe in the perfectibility of humankind, yet as literary scholar Amanda Anderson argues:

> Liberalism's self-correcting movements do not merely reflect a progressivist confidence. . . . On the contrary, the self-critical and transformative nature of liberalism throughout its history, its responsiveness to ethical, philosophical, and historical challenges, brings to the fore an entirely different facet of the liberal character that has been present since the beginning—a pessimism or bleakness of attitude that derives from awareness of all those forces and conditions that threaten the realization of liberal ambitions.[59]

Anderson illuminates this "bleak liberalism" in the novels of Dickens and Eliot and the work of John Stuart Mill, but writes, "This constellation of concerns is perhaps nowhere more evident than in the liberalisms of the twentieth century, particularly what often goes under the name cold war liberalism."[60] For Anderson, writers such as Lionel Trilling, Doris Lessing, and Ralph Ellison exemplify such pessimistic optimism. The clear-eyed, complex work of John Hersey would as well.[61] Political scientist Dillon Tatum similarly identifies a mid-twentieth century tradition of "pessimistic liberalism." He writes, "Isaiah Berlin, Jacob Talmon, Raymond Aron, and, to some extent, Karl Popper represent a break from previous theorizing about liberal world order, a new way of painting a bleak picture onto the canvas of a liberalizing world."[62] Sara Marcus's study of leftist political failure and persistence, *Political Disappointment: A Cultural History from Reconstruction to the AIDS Crisis*, could also be considered in dialogue with liberal pessimism.[63]

The work of mid-century theologian Reinhold Niebuhr, whose powerful intellectual influence on American thought remains underappreciated (Arthur Schlesinger called him "the most influential American theologian of the 20th century"), could also be categorized as liberal pessimism—or conservative pessimism, depending on how we define the terms.⁶⁴ Niebuhr's politics were distinctly liberal, with socialist tendencies, though he was a committed Christian, decidedly anti-Communist, and conservative on segregation; the intellectual milieu he inhabited is difficult to imagine today. Niebuhr's "Christian realism," grounded in Augustine's *De Civitas Dei* and the concept of original sin, offers a tragic, pessimistic worldview that denies the fundamental tenets of optimism and the Myth of Progress. "Modern man does not regard life as tragic," Niebuhr writes. "He thinks that history is the record of the progressive triumph of good over evil. He does not recognize the simple but profound truth that man's life remains self-contradictory in its sin, no matter how high human culture rises."⁶⁵

To those for whom liberal pessimism seems self-contradictory, the relationship between pessimism and conservatism may appear obvious: pessimism is conservative and conservatives are pessimists, since both reject the idea of progress. As historian Andrew Bacevich points out, thinkers in the modern American conservative tradition "have typically viewed modernity as a threat, responding to it with a mixture of apprehension, alarm, and horror."⁶⁶ In his mission statement for the *National Review*, William F. Buckley famously defined a conservative as one who "stands athwart history, yelling Stop, at a time when no one is inclined to do so, or to have much patience with those who so urge it."⁶⁷ Conservative and reactionary thinkers often present themselves as hard-minded realists who are opposed to liberals' dewy-eyed visions of socially engineered utopias and sometimes identify outright as pessimists. Thus, for adherents of liberal progressivism, pessimism and conservatism are effectively indistinguishable, a single irrational and reactionary complex of "irritable mental gestures which seem to resemble ideas," fundamentally "a matter of temperament, not of reason."⁶⁸

Yet such a prejudicial view is badly mistaken: it flattens the complexities of political conservatism as a tradition deeply entangled with both liberalism and radical leftism and it ignores the fact that conservatism typically finds redemptive potential in cultural tradition, individual liberty, the market, and religion.⁶⁹ Conservatism has pessimistic strains and affinities, to be sure, and we can identify a conservative or at least conservatively oriented pessimism in the work of Edmund Burke, Randolph Bourne, Joan Didion, Christopher Lasch, Wendell Berry, John Gray, Roger Scruton, and

Andrew Bacevich, but the nostalgia, religiosity, and faith in capitalism that characterize much American conservatism are alien to pessimism's core rejection of optimism.[70] Conservative nostalgia, after all, is as utopian as liberal progressivism, only oriented toward the past rather than the future, and the idea that capitalism might be a just and effective means for organizing society is deeply optimistic in all the worst ways. Indeed, many on the far right and far left see conservatism and progressivism as "right" and "left" flavors of liberalism—that is, as two strains of the same capitalist ideology. And while pessimism is compatible with religion, it is not compatible with religious fervor: any truly pessimistic faith would necessarily be limited, even apophatic, more like Bayle's fideism or Hume's skepticism than doctrinaire theological commitment.

Conservative pessimism sometimes emerges as declinism, or cultural pessimism, which Oliver Bennett defines as "the conviction that the culture of a nation, a civilization, or of humanity itself is in an irreversible process of decline."[71] Distinct from the empirical analysis of societal collapse, cultural pessimism projects value judgment along a temporal axis: things were once better, now they're getting worse. While this may be "pessimistic" in its rejection of progress, it stops short of a full commitment to the core truth of pessimism, which is our inability to solve suffering. The declinist believes that our problems could be solved if only we returned to the old ways, making it a form of the Myth of Renewal. But philosophical pessimism not only rejects progressivist narratives about utopian futures, it also scorns idealized versions of the past; philosophical pessimism as it emerged from the work of Voltaire and Bayle began by criticizing inherited theological authority. Intellectually robust forms of conservative thought may overlap with a nuanced philosophical pessimism, but declinist cultural pessimism and reactionary nostalgia are specious forms of pessimism at best, since they depend on a belief that things were once better and could be again.[72]

Posthumanism offers another contemporary strain of pessimist thought. Thinkers at the intersection of literary criticism and European post-structuralist philosophy, including Jane Bennett, Lee Edelman, N. Katharine Hayles, Donna Haraway, Timothy Morton, Eugene Thacker, and Cary Wolfe, have taken up questions about the value of life and the idea of the human in relation to scientific advances in robotics, cybernetics, and bio-technology in ways that are often recognizably pessimistic.[73]

Philosopher Francesca Ferrando writes: "'Posthuman' has become a key term to cope with an urgency for the integral redefinition of the notion of the human following the onto-epistemological as well as scientific and biotechnological developments of the twentieth and twenty-first centuries."[74]

Posthumanism encompasses a broad range of questions and approaches critical of human exceptionalism and anthropocentrism, too various to address in detail here, comprising or at least engaging with transhumanism, new materialism, object-oriented ontology, antihumanism, anti-natalism, media theory, and the digital humanities. Literary theorist Cary Wolfe argues that posthumanism should be understood as a project thinking beyond the human as well as beyond humanism, particularly the inherited tradition of positivistic Enlightenment thought. "When we talk about posthumanism," he writes, "we are not just talking about a thematics of the decentering of the human in relation to other evolutionary, ecological, or technological coordinates . . . we are also talking about *how* thinking confronts that thematics, what thought has to become in the face of those challenges."[75]

Posthumanist thought intersects with philosophical pessimism through questions of human extinction, the critique of progress, and historicizing "the Human." At the same time, many posthumanist thinkers remain committed to some notion of progress and see the end of "the Human" as an evolutionary and liberatory process, as in Donna Haraway's classic "Cyborg Manifesto," Ray Kurzweil's transhumanism, and other forms of accelerationist thinking. Yet even in these cases, the progress in question is apocalyptic, disjunctive, and distinct from incrementalistic ideas about improving the human condition; posthumanist ideas of progress envision humans radically superseded, either by AIs, cyborgs, or transhumans, in such a way that traditional human values are no longer relevant.

AFROPESSIMISM

One of the most robust strains of pessimist thought to emerge in recent years, Afropessimism, developed in dialogue with posthumanist theory, critical race studies, and postcolonial theory. Defined by one of its main proponents as a "critique without redemption or a vision of redress except 'the end of the world,'" Afropessimism argues that a binary racial ontology is both foundational and essential to modern civilization.[76] It presents itself as "a lens of interpretation that accounts for civil society's dependence on antiblack violence," rejects "globalization, cultural optimism, [and] optimism in culture's emancipatory potential," and argues "that the black (or slave) is an unspoken and/or unthought sentience for whom the transformative powers of discursive capacity are foreclosed *ab initio*."[77] Afropessimists argue that the idea of progress depends upon a "structural antagonism between humans and blacks" and that the only way to imagine the end of racism would be to imagine the end of the world as we know it. This austere and challenging position takes us to the limits of philosophical pessimism and raises important questions about what it means to be human, what it means to live in a world, and what it might mean for the world to end.

There are multiple Afropessimisms, of course, as there are various Afropessimists, but the two figures most responsible for developing this line of thought are Jared Sexton, professor of African American Studies and Film and Media Studies, and Frank B. Wilderson, III, professor of African American Studies and Drama, both at the University of California, Irvine. Sexton and Wilderson have, over the past twenty years, engaged various interlocutors and fellow travelers, apprenticed and inspired numerous students and readers both directly and indirectly, and claimed for themselves a capacious intellectual lineage that is often disputed and more often misapprehended. Although numerous critical assessments of Afropessimism have emerged, a full intellectual history of the phenomenon, locating its emergence in the context of the Global War on Terror, accelerating climate change, Hurricane Katrina, the Obama and Trump presidencies, and the rise of the #BlackLivesMatter movement, tracing the full range of its antecedents, and putting it into dialogue with contemporaneous related work, remains to be written.[78]

The central idea in Afropessimism is that the conceptual and narrative structure of contemporary modernity is predicated on an ontological division between Human being and Black being, which division emerged historically through the development of chattel slavery in the Americas and the trans-Atlantic slave trade, and because this division is foundational to the modern world in an existential or phenomenological sense, there can be no possible assimilation, recuperation, or resolution of Black being into Humanity. In Wilderson's words: "Human life is dependent on

Black death for its existence and coherence. Blackness and Slaveness are inextricably bound in such a way that whereas Slaveness can be separated from Blackness, Blackness cannot exist as other than Slaveness. There is no world without Blacks, yet there are no Blacks who are in the world."[79] This argument, complex and contentious on its face, entails numerous ramifications, one of the most controversial being the claim that Blackness and "the grammar of Black suffering" occupy an exceptional and singular cultural space that cannot be analogized or compared with other kinds of racism, colonial oppression, or social injustice, not even genocide.

The first clear articulations of Afropessimism can be found in a 2003 issue of *Qui Parle*, in a special section titled "Dossier on History, Representation, and the Impossible Subject of Race," edited by Jared Sexton and Huey Copeland.[80] The dossier includes an introduction by Sexton and Copeland; articles on race, media, literature, and culture by Kara Keeling, David Marriott, and Neferti X. M. Tadiar; a series of photographs from the artist Hank Willis Thomas with an accompanying essay on the series by Copeland; and, most notably, an interview with Saidiya Hartman, conducted by Frank Wilderson, titled "The Position of the Unthought." The impetus of the dossier was, in Sexton and Copeland's words, to take up "the notion derived from Fanon, of the impossibility of representing race, either for the slave or the master, outside of an entrenched visual schema predicated on the fungibility of the black body"; Wilderson's rich dialogue with Hartman explores this question in sharp and precise terms.[81] Indeed, though Sexton and Copeland identify Fanon as the source of their inquiry, the more immediate interlocutor with which their dossier engages is Hartman herself.

In important ways, Afropessimism should be seen as building on the argument put forth in Hartman's groundbreaking 1997 book, *Scenes of Subjection: Terror, Slavery, and Self-Making in Nineteenth-Century America*. That argument begins from an observation that might seem obvious: the black body under chattel slavery was objectified not only in material, physical, and legal senses in that the slave was a commodity to be bought and sold, with no legal right to autonomous decision-making and hence no personhood under the law as a rights holder or citizen, but also and perhaps more importantly in representational, performative, and psychological senses. That is, the black body was objectified first in slavery, then again in white fantasies of suffering, violence, and redemption, fantasies ritualized and narrativized across nineteenth-century American culture in minstrelsy, the spectacle of the coffle procession, and mass lynchings.

As Hartman carefully shows, these "scenes of subjection" were not restricted to the overtly repressive practices of chattel slavery. They were also central to the abolitionist imagination, the slave narrative, and—Hartman's most troubling point—the narratives and practices of emancipation. Hartman's painstaking readings of *Uncle Tom's Cabin*, *Slave Life in Georgia*, juba songs, *State of Missouri v. Celia*, Harriet A. Jacobs's *Incidents in the Life of a Slave Girl*, abolitionist John Rankin's anti-slavery epistle to his brother, and handbooks written for emancipated former slaves such as *Advice to Freedmen*, among numerous other nineteenth-century texts, show how Black subjectivity is foreclosed in the very sites and moments of its supposed emergence into self-consciousness, and argue that—calling to mind Spike Lee's *Bamboozled*—the slave is *always* dancing for the master's pleasure, most of all in those moments when she is ostensibly performing her freedom. In a phrase, we might gloss Hartman's analysis as showing that, in the case of the emancipated slave, subjectification is objectification.

The conclusions this argument leads to are devastating for liberal narratives of racial progress and for narratives of progress in general. The first conclusion is that it's difficult if not impossible to talk about autonomous slave subjectivity. We might see Hartman as adapting Gayatri Spivak's question "can the subaltern speak?" and offering a similarly complex deflationary answer.[82] Second, the historical unfolding of emancipation did not offer a genuinely revolutionary, post-slavery subjectivity but rather granted no more than the *promise* of subjectivity as the ever-receding reward for submission to newer and subtler forms of domination: freed slaves became Blacks, their political emancipation taking form within a racial hierarchy of social being as the lowest form of the Human (if Human at all), their self-representations of *Bildung*, self-expression, and overcoming reinscribed as ritualized social fantasies of suffering and redemption in which the Black is forever striving to achieve the impossible goal of becoming White. The logical consequences of Hartman's work are recognizably pessimistic: while slavery is historically distinct from its afterlife, there is a continuity between the two in their shared foreclosure of Black subjectivity and shared ontological structure of racialized being. As Hartman demonstrates, looking closely at emancipation shows us how racialized domination is articulated through the spectacular performance of Black emancipation, Black freedom, Black fugitivity, and Black joy.

Although in her later essays and books such as "Venus in Two Acts," *Lose Your Mother*, and *Wayward Lives, Beautiful Experiments*, Hartman has worked against the conclusions suggested by *Scenes of Subjection*, Afropessimist thinkers radicalize Hartman's argument with three key moves: first, rhetorically positioning her argument as an uncompromising provocation; second, connecting it to the black radical tradition's critique of Western civilization, specifically the work of Frantz Fanon, CLR James, Cedric Robinson, and Sylvia Wynter (among others); and third, merging it with Orlando Patterson's concept of "social death," or institutionalized marginality.[83] Thus, according to Afropessimism, the modern world is constituted by a structural antithesis between Human and Black that cannot be resolved without undoing that world, since the very idea of the Human depends on its excluded antithesis, the Black. As Calvin Warren puts it, in the American context,

> the American dream . . . is realized through black suffering. It is the humiliated, incarcerated, mutilated, and terrorized black body that serves as the vestibule for the Democracy that is to come. In fact, it almost becomes impossible to think the Political without black suffering. . . . Progress and perfection are worked through the pained black body and **any recourse to the Political and its discourse of hope will ultimately reproduce the very metaphysical structures of violence that pulverizes black being**.[84]

In Afropessimism, Hartman's Foucauldian historical analysis of American slavery becomes a Nietzschean transvaluation of all values: an apocalyptic call for the end of *this* world, so that a new world might be born.

Afropessimism has garnered attention from large-audience liberal platforms such as the *New Yorker*, while also generating significant pushback from scholars in Black and African American Studies. It has, as Sexton writes, both "captured the imagination of certain black radical formations" and "struck a nerve among others, all along the color line, who fear that open-minded engagement involves forsaking some of the most hard-earned lessons of the last generation."[85] Writer and activist Annie Olaloku-Teriba, for instance, accuses Wilderson and Sexton of a "project of mystification" that conflates historically distinct processes of identity formation, undermines solidarity, and offers at best "a fight without a purpose."[86] British writer Kevin Ochieng Okoth sees Sexton and Wilderson's lifting "Afropessimism" from the Africanist framework in which it first emerged as a "bizarre use of [a] historically loaded term." What's more, according to Okoth, the proponents of Afropessimism misread Fanon, advance an "ontologically flat conception of Blackness," offer a shallow critique of Marxism, and "frequently erase or distort beyond recognition, the various Black liberation movements that fought against racism, colonialism, and imperialism through the Global South."[87] African American literature scholar Greg Thomas similarly writes: "Reading what some are today calling 'Afro-pessimism' invites no small amount of amnesia, myopia, as well as illiteracy. There is little if any Africa to this discourse at all, its nominal Afro-hyphenation notwithstanding."[88]

One of the more generous critiques of Afropessimism comes from Jesse McCarthy, professor of English and of African and African American Studies at Harvard University, in an essay first published in the *Los Angeles Review of Books* and reprinted in his book *Who Will Pay Reparations on My Soul?* In this essay, McCarthy situates Afropessimism in the black radical tradition, and offers a compelling reading of Wilderson's memoir *Afropessimism* as both Black revolutionary narrative and slave narrative: "There is a complicated sense," McCarthy writes, "in which Wilderson has written a slave narrative *in the present*, insofar as it is his contention that he actually is, under the terms of his philosophy, a slave right now."[89] McCarthy carefully identifies and explains some of the important theoretical foundations underpinning Sexton's and Wilderson's work, such as "racial exceptionalism, political immutability, 'antiblackness' as structural antagonism, and abjection in the form of 'social death.'"[90] In the end, however, McCarthy concludes that Afropessimism is an ahistorical and even nihilistic academic reaction to a very real problem, the total social failure to adequately frame "the question of how to realize the fullest aspirations of a racially integrated democracy." McCarthy summarizes his charge:

> The fact that the main current of Afropessimist thinking runs counter to all of black political history and tradition thus far; the fact that the foundational thinker for this perspective, Frantz Fanon, came to completely opposing conclusions with respect to the nature of politics and solidarity in struggle; the fact that the theory often appears to evade scrutiny or con-

testation by proclaiming itself "meta-theoretical" or "ontological"; the fact that it asserts a "mandate" for which no empirical evidence is provided and in the face of overwhelming evidence that it constitutes at best a minoritarian and class-specific position—all of this has to be reckoned with by those who want to take Afropessimism to heart.[91]

McCarthy is a thoughtful reader, and he levels serious criticisms against Wilderson, in particular regarding the contradictions within his book *Afropessimism*, his use of hyperbole, his evacuation of "the entire material history of antebellum slavery," and his transformation of slavery "into a portable and fundamentally *psychological* relation unrelated to any historical memory."[92]

These are valid points, yet one thing McCarthy and other readers seem to miss in their criticisms of Wilderson's hyperbole, irony, exaggeration, temporal disjunction, and catachresis is how he uses just these tools to expose what he sees as irresolvable contradictions within contemporary American culture. What McCarthy misses and in some sense dismisses are the existential, rhetorical, and *comic* aspects of Wilderson's work.[93] In addition to the genres of the black revolutionary narrative and the slave narrative, that is, Wilderson is also working in the tradition of African American satire.[94] Satire, writes Jonathan Greenberg, "combines, inhabits, or transforms other genres. . . . [and] mixes subject matter, linguistic registers, and literary traditions."[95] In writing a "slave narrative in the present," Wilderson may be satirizing the slave narrative itself. Indeed, Darryl Dickson-Carr's summary of George S. Schuyler's 1931 satirical novel *Black No More* seems to describe Wilderson's memoir *avant la lettre* in the way it "repeatedly installs, subverts, then reinstalls racism as the agent of ideological and political irrationality and chaos, ending with a pessimism that suggests the permanency of racism in the absence of a transformation of the American body politic."[96] Reading *Afropessimism* as satire may seem a stretch: the book is a theoretically dense polemic on one of the most fraught topics in contemporary American intellectual culture. But satire isn't always funny. Sometimes it is deadly serious.

Satire, like obscenity, is difficult to define, but Northrop Frye's definition of satire as "militant irony" neatly describes Wilderson's approach.[97] From *Afropessimism*'s opening pages, Wilderson walks a line between bathos and tragedy, his over-the-top prose delivered with a straight-faced intensity that all but dares the reader to laugh. "A psychotic episode is no picnic," he begins, "especially if you know you can't call it madness because madness assumes a change in the weather, a season of sanity."[98] It's difficult to know how to read Wilderson's self-mocking litotes in the opening clause, at once extreme and dismissive, his phrasing torqued with chiastic assonance, "is no" repeating the vowel sounds of "episode" but lacking the structuring plosives, "picnic" offering a muted echo of "psychotic," or how the rest of the sentence undermines and complicates the opening clause, asserting that the "psychotic episode" invoked is no episode at all but merely business as usual: what Wilderson's narrator promises to describe is not "madness" (a word practically poetic in its anachronism, in our era of "mental health") but the norm. Wilderson's use of alliteration mimics the stymied frustration of a stutter, the *p*'s of the opening clause giving way to the gratuitous repetition of "madness," which slides hissing into "assumes," "season," and "sanity." With his first sentence, Wilderson plunges us into a noisy, kairotic, and contradictory negativity, where psychosis can be compared to a picnic, temporality is unpredictable, an episode is not an episode, madness is not madness, and the norms of everyday life are in suspension.

This isn't the jagged formal disarrangement mimetic of psychological extremity you get from Antonin Artaud or LeRoi Jones. What distinguishes Wilderson's prose, even in hyperbole, is its control, and it is this controlled indeterminacy, this methodological negation flirting consistently with satire, black humor, and bathos, that characterizes Wilderson's *Afropessimism* as a masterfully ironic work of literary art. The proper key is struck in Wilderson's comparison between madness and weather, which evokes Hamlet's sly lines from act 2, scene 2, "I am but mad north-northwest. / When the wind is southerly, I know a hawk from a hand saw."[99] With these lines, Hamlet tries (and fails) to establish solidarity with his university friends, Rosencrantz and Guildenstern, clueing them in to his strategic use of performative "madness." So, too, Wilderson makes a bid to us, the reader, sheerly through literary style, to follow him into the labyrinth of Blackness.[100]

From this opening, Wilderson's narrative dives abruptly into the scene of his mental break—"I was moaning. Sobbing."[101]—while sustaining the same disorienting disjunction between style and substance, cool control against complete disruption, and continuing to walk the line between satire and nonsense: "The hinges of my jaws made moans or howls but not words. *How funny is that?* I answered him in the words of a bird as its throat is slit." Over the next few pages, we are introduced to the narrator, "a middle-aged graduate student," incapable of speech and wanting to bark like a dog, alternating between laughter and tears, a raw-nerved trainwreck of a man, thrown off the tracks by language: "I knew it had started in the mirror. I was washing my face when a stanza of poetry came to me."[102] Wilderson recounts that break, then brings us back into the scene's present of sensory and psychological overload, pure trauma in the classic sense, where hawks are indistinguishable from hand saws, only to use it as a portal collapsing into the past: "Now, alone in the clinic, trombones of light blistered my eyes and the room grew cold. But if I closed them a string of past lives skidded down my skull like a train that had jumped the rails above a ravine. Each cascading car was a carriage of time."[103] The phrase "trombones of light blistered my eyes" stands out as exemplary: visually evocative, grotesquely surreal, the explosive fusion of image and sound pushes language beyond the edge of meaning.

And by this point, we have gone beyond ourselves. Bracketing the broader contextual and paraliterary framing that inevitably shapes a reader's experience of beginning a book, we are implicated in a riddle in the chapter's title: "For Halloween I Washed My Face." Halloween, of course, is a cheerful, campy holiday when children dress up and go trick or treating for candy, as well as an atavistic ritual appeasing malicious supernatural forces—namely, the spirits of the dead. Children become ghosts or goblins demanding satisfaction on threat of violence: give us a treat or you'll get tricked. In the poem that closes the chapter, we see the role-play reversed, though the dynamic remains the same:

> *for Halloween I washed my*
> *face and wore my*
> *school clothes went door to*
> *door as a nightmare*[104]

Wilderson dramatizes himself as a Black child performing the role of avenging spirit by showing up dressed for school, interpellated as a non-human subject within the progressive social framework of public education. By posing his Black childhood as one in which Halloween involves washing his face and wearing his school clothes, Wilderson condenses Orlando Patterson's concept of social death into a concrete image while also presenting the reader with a complicated and implicating bargain. Reading *Afropessimism*, Wilderson seems to suggest, will be more ambiguous than a Dantesque journey through hell, and its narrator will be more ambivalent than some kind, old Virgil. In closing his tumultuous opening chapter with an image of himself as a child on Halloween, Wilderson presents himself not as a trustworthy figure of reason but as an unappeasable spirit. What treat, after all, could anyone offer a Black child doomed to ontological exile? Snickers?

You see the problem. The rest of *Afropessimism* modulates tone, rhythm, and intensity, weaving and warping temporality as Wilderson shifts smoothly between narrative recollection, high theory, and sharply drawn scenes of racial conflict, repeatedly pushing language to the breaking point, verging at moments into bitter, Juvenalian satire. As Frye writes, "A slight shift of perspective, a different tinge in the emotional coloring, and the solid earth becomes an intolerable horror."[105] In chapter 2, young Frank accidentally almost kills one of his white friends by cracking his fontanelle with "a plastic bottle of emerald-colored Palmolive," provoking a grotesque and absurd moment of racial tension.[106] In chapter 3, perhaps the most fervid and disorienting part of the book, Frank and his older lover Stella, a radical Black activist, slide into a paranoid nightmare, believing they are being followed, targeted by the government, deliberately poisoned with radioactive material by their upstairs neighbor, and subjected to a hostile conspiracy of forces that might be COINTELPRO or might simply be racism. Chapter 4 offers cutting academic satire, and chapter 7 jarringly juxtaposes the petty microaggressions of workplace racism against the political violence of South African apartheid. Throughout, Wilderson never lets the disjunction tip into comedy. Wilderson's satire keeps a straight face, always refusing the redemptive gesture of a punchline.

There is much more to be said in favor of reading *Afropessimism* as a complex, layered, intentionally ambiguous provocation within a tradition of postmodern Black satire, alongside works like Ishmael Reed's *Mumbo Jumbo*, Percival Everett's *Erasure*, Paul Beatty's *The Sellout*, the stand-up comedy of Richard Pryor and Dave Chappelle, Jordan Peele's horror films, and Donald Glover's *Atlanta*. In addition, Wilderson's lampooning of contemporary progressivist pieties cannot help but call to mind Voltaire's *Candide*. Among the many insights Wilderson's project offers is the inkling of a deep, underexplored connection between philosophical pessimism and literary satire.

Toward the end of his essay on Afropessimism, McCarthy writes that African Americans "are a historic people with a world-historical destiny that understands our suffering as endowing us with both the right and responsibility of *civilizing* the United States in such a way that it reflects the values that our historical experience bring to it, the freedoms, equalities, and cultural pluralisms that we have made vital and central to its identity."[107] For McCarthy as well as for Wilderson, that is, the Black is the negative determination of the Human. The difference between them is that while McCarthy sees that negation recuperated within a teleological narrative of social progress in which African Americans "are a historic people with a world-historical destiny," Wilderson sees that negation as ontologically grounding modernity as such. For McCarthy negation in the present subtends the promise of a better future, while for Wilderson negation in the present is the very condition of our present's possibility. In one case the Black is "the unfulfilled"; in the other, "the unthought." Which is to say that while McCarthy's criticisms are unanswerable, they are also beside the point since McCarthy takes as given the very thing Wilderson denies.

One might even say that McCarthy takes Wilderson's bait by arguing that "we lift each other and rise together with the spirit of history at our backs."[108] McCarthy clearly means to show that Afropessimism lies "on the wrong side of history," but he does so (necessarily, Wilderson might argue) by reprising a racialized progressivism that makes Black bodies the symbolic bearers of the future's suffering, hope, and eventual redemption. Yet while McCarthy seems to optimistically believe things can get actually better within the world that exists and thereby redeem the utopian potential of Black suffering, Wilderson does not.

Afropessimism is not interested in reconciling us to the world that is but rather in breaking our attachment to this world and lifting our eyes to the possibility of another. Afropessimism may thus be understood as the theoretical successor to a narrative and performative lineage of pessimistic African American satire in which jokes are played not for laughs but as a form of resistance to ongoing and irresolvable racial degradation. As Wilderson put it in an interview with Patrick Farnsworth:

> There cannot be a World as we know it with Black people who are free. And when this World as we know it ends, there will be sentient beings like you and I on the other side. But they will not be Black and they will not be Human. They will be another category. And that's not gonna come about through negotiation or consciousness raising.[109]

What marks the *pessimism* in Afropessimism is its refusal of any reconciliation—its refusal to play the role of performative abjection—its refusal to go for the laugh in order to restore the unity of a world defined by white supremacy. Instead, the Afropessimist subverts that role and insists on the seriousness of his hyperbole, turning away from the unjust world that actually exists to gesture apophatically toward an unknown world we cannot even yet begin to imagine.

Seven
OK, DOOMER

Parades wind through streets bursting with cheer. Elders, children, and workers dance and sing. They have every reason to celebrate, since they have no king, no president, no police, no army, no slaves, no racism, no bombs, no guns, no advertisements, no stock market. They live lives of plenty, indulging in whatever luxuries they like, gratifying themselves with orgies, beer, nonaddictive and sweet-smelling *drooz*, pizza, pastries, music, everything pleasurable, nothing ruinous. They have public transit, battery-powered scooters, and renewable power. They have no need for cars. They have no carbon footprint, no pollution, no guilt. They live in a land of plenty, without coercion or want. They live in a city called Omelas.

Omelas means satisfaction, and the people who live there are happy but not simple. "These were not . . . dulcet shepherds, noble savages, bland utopians. . . . They were mature, intelligent, passionate adults whose lives were not wretched."[1] Their architecture is noble, their music poignant, their literature profound. They are compassionate and brave. The only problem is that their city is cursed. "In a basement under one of the beautiful public buildings of Omelas, or perhaps in the cellar of one of its spacious private homes, there is a room. It has one locked door, and no window."[2] In that dark and lonely room huddles a child reeking of urine, her hair a filthy tangle, her rough smock stained with feces. She is always hungry. She gets one half-bowl of cornmeal and grease a day. Her body is covered with untreated sores. She is the happiness of Omelas.

Maybe it was a wizard who cursed their city, maybe it was a god. No one really remembers, but the terms of the curse demand that the child must suffer. No one is required to actively torture her, but she must remain locked in the cellar alone, ignored, unwashed, and underfed. "The terms are strict and absolute; there may not even be a kind word spoken to the child." In return, the citizens of Omelas enjoy wealth, prosperity, and ease. It is no secret:

> They all know it is there, all the people of Omelas. . . . Some of them understand why, and some do not, but they all understand that their happiness, the beauty of their city, the tenderness of their friendships, the health of their children, the wisdom of their scholars, the skill of their makers, even the abundance of their harvest and the kindly weathers of their skies, depend wholly on this child's abominable misery.[3]

At some point, of course, parents have to explain the curse to their children. It's never pleasant. The kids don't always take it well. They ask questions the parents can't answer. They try to come up with solutions, imagining there's some way to negotiate the terms and ease the child's suffering. Sometimes the older ones organize protests. But eventually most of them accept their life in Omelas. No wonder: it's a good life.

Omelas is the creation of Ursula K. Le Guin, first introduced in 1973, in a short story published in the science fiction anthology *New Dimensions 3*, the year after Peter Singer published his essay posing the problem of the drowning child, a year before Garrett Hardin formulated his lifeboat ethics.[4] Whereas Singer and Hardin present their parables as realistic moral dilemmas that make ethical claims on the reader, Le Guin makes no effort to hide the factitiousness of her fiction. She fudges details, accentuates the fairy-tale quality of her narrative, and admits that some of her choices are arbitrary. Maybe they have a cure for the common cold, maybe not; it doesn't matter. What matters is the kernel, which presents an unacceptable yet irrefutable truth, transformed just enough to make it bearable: the ease and pleasure of some depend upon the misery of others.

We live in a world of lifeboats, where every day, Princeton philosophers walk past drowning children and a vast global hierarchy of wealth rises like a ziggurat from an even vaster foundation of misery. The situation in Omelas is stark but not nearly as bad as our own. For Omelas to prosper, only a single child must suffer, whereas the prosperity of our world depends on the suffering of millions. According to World Bank estimates, approximately 3.6 billion people around the world today live below the poverty line, meaning they live on less than $6.85 a day, and more than 719 million people live in extreme poverty, meaning they live on less than $2.15 a day.[5] In the United States alone, around 37.9 million people live in poverty, more than 11% of the population.[6] If the US were Omelas, it would be a place where every tenth child was locked in a basement; if the world were Omelas, if would be closer to half of all children. Admittedly, Omelas is simple and egalitarian, whereas our world is complicated and highly differential. In our world even millionaires suffer, not least from their envy of billionaires. Like the specific details of Le Guin's utopian fiction, however, none of that really matters. What matters is that we accept the terms.

Or not. Le Guin's parable offers one more twist, implicit in its title: "The Ones Who Walk Away from Omelas."

> At times one of the adolescent girls or boys who go see the child does not go home to weep or rage, does not, in fact, go home at all. Sometimes also a man or a woman much older falls silent for a day or two, then leaves home. These people go out into the street, and walk down the street alone. They keep walking, and walk straight out of the city of Omelas, through the beautiful gates. They keep walking across the farmlands of Omelas. Each one goes alone.[7]

Nobody knows where they go, and Le Guin admits that whatever their destination, it's even less imaginable than Omelas itself. "I cannot describe it at all," she writes. "It is possible that it does not exist. But they seem to know where they are going, the ones who walk away from Omelas."[8]

It's tempting to read this final turn as the moral to Le Guin's fable: it seems to offer the promise of liberation, a way to break free from the curse, if not to break the curse itself. Those who walk away offer, with their stoic if futile gesture of resistance, the happy ending we crave. They make their own bargain, of course, leaving the child in its misery, but in leaving Omelas at least they are no longer complicit. They no longer profit personally from the child's suffering. In giving up the curse, they also give up the blessing. Yet as Le Guin makes clear, wherever it is they think they're going may not even exist; it may be truly utopian, a "no-place." The darkness they walk into may swallow them.

In offering this turn at the end of her story, Le Guin seems to be inviting the reader to pick sides. The obvious choice is to decide that the people of Omelas are evil and the ones who walk away are good. A contrarian position would hold the opposite. Philosopher Paul Firenze, for example, argues that "those who remain in Omelas are, in important ways, morally 'superior' to those who walk away."[9] But like Borges's "Funes, His Memory," Le Guin's story is less a moral fable than a funhouse mirror reflecting back to us our own desire and imagination. Which is to say that when we pick sides, we miss the point. The ones who walk away pay a terrible cost for their gesture: they cut themselves off from their society, give up joy for the sake of purity, and do nothing to alleviate the child's suffering. The ones who stay, on the other hand, may be haunted and corrupt, but they're also forced to reckon with their own complicity and to recognize how their own happiness depends on other human beings. As Firenze writes:

> The existence of the child requires the people of Omelas to locate the source of their own happiness outside themselves and their own subjective experience. Walking away is an attempt to break this link. It says, "My happiness is my own affair, my own doing." For those who remain, the child serves as an anchor, a constant living reminder that our happiness requires others, and that others' happiness requires us.[10]

"The Ones Who Walk Away from Omelas" offers an even more complex and pessimistic view than at first it seems, a view unspeakable in polite society unless it's wrapped in black humor: not only is the happiness of some dependent on the suffering of others, but to live at all means to suffer and cause harm. We cannot exist without harming other beings, just as we cannot help but be vulnerable to the harm they inflict on us. These are the terms of existence. This is our curse. Yet it's also the source of our compassion and the grounding of our wisdom. Only by recognizing our vulnerability and our responsibility toward others can we begin to understand how much we depend on each other, and only in recognizing our interdependence can we begin to fathom the depth of our obligations for care, repair, and prudence.

Only through pessimism, that is, can we begin to comprehend what being human really demands.

On learning of our curse, some seek escape or revolution. Others blame a scapegoat. Some simply refuse to see, optimistically assuring themselves that everything will turn out okay. Where optimism doesn't reject the terms outright, it rationalizes. Leibnizian theodicy claims that manifest suffering is only the worldly appearance of a rational, beneficent order: we see but the tiniest corner of the tapestry, the whole of which is beauty and justice itself. Progressivism projects the best of all possible worlds into the future: today's suffering will be redeemed by tomorrow's fulfillment. Mostly we put the curse out of our minds, and if our minds are working properly, it's easy. Optimism is our adaptive default, complacency rationalized, the business of business as usual.

Pessimism, in contrast, is a parti-colored scandal. The pessimist doesn't flee, but she doesn't celebrate either. Pessimism may be extreme or mild; it may be specific or broad; it may be cultural, metaphysical, or existential. At the heart of its many variations, pessimism is an attempt to live ethically with suffering. Pessimism is characterized by the rejection of optimistic rationalization, disdain for escapism, contempt for scapegoating, and an insistence on human limits, which all emerge from the core recognition that to live is to suffer and cause suffering. The pessimist neither justifies the curse nor attempts to escape it, but confronts it, accepts her responsibility, and commits herself to living with it.

PESSIMISM AND ITS DISCONTENTS

Pessimism begins with suffering, and paying attention to suffering is itself painful, which may be one reason why so many people reject it out of hand. Pessimism, critics say, is nothing but fatalism, a counsel of despair, the most craven kind of quietism, and a self-fulfilling prophecy to boot. "One of the oldest objections against pessimism," writes philosopher Frederick Beiser, "is that it is immoral because it saps the motivation to fight evil and to alleviate suffering."[11] According to its critics, pessimism is variously misanthropic, racist, sexist, privileged, classist, and anti-democratic. And of course, most importantly, it's wrong. Things are getting better. People are basically decent. Human ingenuity and goodwill are winning. Just look around you! Against the manifest evidence of human progress, critics assert, pessimism reveals itself as nothing more than willful nihilism, a cheap pose struck to incite scandal and spite the hard work of well-meaning people. To reject the Myth of Progress is to reject all that is beautiful, to *choose* death and chaos and noise. Pessimists wallow in the suffering they melodramatically claim is the inescapable fact of the human condition.

None of these criticisms hold water. To begin with, pessimism is not nihilism: a pessimist believes in suffering and death because a pessimist believes in reality, and suffering matters to pessimists because there are things in life worth valuing. As philosopher Paul Prescott argues, pessimism is incompatible with nihilism, both what he calls axiological nihilism and practical nihilism.[12] Axiological nihilism holds that because there are no objective values, nothing is worth valuing; practical nihilism holds that nothing of value can be realized. But as Prescott points out, pessimism holds that things *do* matter, *are* worth valuing, and *can* be realized, which are the very reasons why death, failure, and suffering are problems. And while a pessimistic recognition of death, failure, and suffering is often confused with the nihilist claim that existence is meaningless, it is not pessimists who insist on conflating the two, but their critics. Joshua Foa Dienstag writes:

> To say that our lives are always on the way to death is not at all to say that they are pointless, but simply to set out the parameters of the possibility for our existence. Pessimism may warn us to acknowledge our limitations—but it does not urge us to collapse in the face of them. Death is merely the ultimate reminder that we do not control the conditions of our existence and are not ever likely to.[13]

Likewise, it is critics of pessimism who assert that the recognition of suffering and death must lead to quietism and resignation, not pessimists themselves. Beiser points out that while many nineteenth-century German philosophers involved in the Pessimism Controversy "argued that pessimism leads to quietism," others "protested that pessimism had no such quietistic implications; they insisted that it gave people every reason to strive to make the world a better place because that alone would diminish evil and suffering."[14] German pessimist philosopher Eduard Hartmann, for instance, argued that pessimism was at odds not with morality but with egoism, and hence it undermined *selfish* reasons for acting morally but not compassionate or ethical ones.[15] Similarly, Buddhist pessimism emphasizes compassionate action as an ethical response to the three marks of existence: impermanence, suffering, and non-self (anicca, dukkha, and anattā).

In contrast to fatalism and cynicism, "pessimism per se does not rule out human agency, or commit one to a particular view of human nature."[16] Thinking pessimistically does not mean believing we have no control over the future. It means accepting that the amount of control we have is radically limited. It means playing the hand we're dealt, not imagining we're responsible for which cards wind up in our hand. As psychologist Julie Norem's research on defensive pessimism shows, pessimism can be an active strategy for managing uncertainty.[17] "Defensive pessimism helps people harness their anxiety and direct it toward action to prevent the negative outcomes they envision," writes Norem, "and research shows that using defensive pessimism is associated with better outcomes than other strategies people often use."[18] Further, being realistic about our limitations doesn't mean giving up on doing better. "Contra both the cynic and the fatalist alike," writes Prescott, "one can be pessimistic—even about the human condition—while affirming both the possibility of human agency and the potential for good in human beings."[19]

Nor is pessimism synonymous with despair.[20] Despair, according to the *Oxford English Dictionary*, is "the action or condition of . . . losing hope; a state of mind in which there is entire want of hope; hopelessness," and while pessimism could certainly lead to hopelessness, it might also lead to a more realistic assessment of what to hope for.[21] In any case, hope and hopelessness are not the same kind of mental states as optimism and pessimism. Hope is a desire or will, while optimism and pessimism are better understood as frameworks, stances, or heuristics aligned by narrative and expectation. Optimists do not merely *hope* things will get better, they *believe* it. Similarly, pessimists who believe human suffering is ineradicable and unavoidable may still hope to avoid it.

Philosophical pessimism does raise the question of whether life is worth living.[22] Indeed, Camus's existential pessimism famously reduces all philosophy to the question of whether to commit suicide.[23] But ethical pessimists cannot kill themselves simply because they suffer: while suicide may end one's own pain, it adds to the pain of others.[24] If I have friends, family, colleagues, and community, I must persist even if I suffer. Life may not be worth living for me, in itself, or at all, but that doesn't free me from my obligations. On the other hand, pessimism in no way rules out martyrdom, as sacrifice is one of the fundamental ways humans create meaning.[25]

Making meaning depends on culture, community, and a commitment to a collective future, and it's true that pessimism can put that commitment into question. In his book *Death and the Afterlife*, philosopher Samuel Scheffler argues that the meaningfulness of our actions today depends on a robust faith in the near-term persistence of human life after our deaths.[26] Although no one is terribly bothered by the idea that the increasing luminosity of the sun will evaporate the oceans in a billion years, Scheffler argues that most of us would be devastated if we learned that all human life was going to end thirty years after our own. The reason for this, Scheffler argues, is that our projects, investments, commitments, goals, and values are to a significant degree dependent on an implicit belief in the continuation of human life. Without a belief in future generations who might build on our work, benefit from it, celebrate it, or remember it, the work itself loses much of its meaning. Scheffler uses the example of cancer research, but other examples abound: Why have a child? Why write books? Why work for labor rights or racial justice? Why do anything?

Scheffler's point is not that we *should* resort to nihilism in the face of extinction, only that many people *would* because of the rarely considered observation that most of our actions today make sense only if we believe there will be humans tomorrow, to the point that we are, on some level, more committed to the idea of collective human survival than we are to our individual lives. And while we face a much more uncertain future than the one in Scheffler's thought experiment, we must still come to terms with the possibility of near-term human extinction, the end of civilization as we know it, and the brute fact of ongoing environmental disaster. Our heuristics and biases make these realities hard to see and harder to internalize, but for many, the dim awareness of humanity's existential precarity is increasingly a source of anxiety, fear, and even despair. Adopting a stance of pessimism may seem to be "giving in" to these feelings, and in a way it is: pessimism means surrendering the unrealistic demands of the ego to the humbling constraints of reality. It means accepting our anxiety, fear, and despair as legitimate responses to a frightening and depressing situation. It means accepting that our suffering is real and to some extent unavoidable, and that our fate is not entirely in our hands.

Viewed strictly from a progressivist standpoint, of course, pessimism *is* unethical, because it rejects the fundamental tenet of progress, namely that we have the power to permanently alter the collective conditions of human existence for the better. But while pessimism rejects the axiomatic truth of progress and is thus unethical in progressivist terms, it is not unethical *as such*. An ethical system may be understood as a framework of values and meaning that guides and explains behavior. "The Greeks called the character ηθος (êthos), and its expressions, i.e., morals, ηθη (ĕthē)," writes Schopenhauer. "But the word comes from εθος (éthos), custom; they chose it in order to express metaphorically constancy of character through constancy of custom."[27] Christian ethics differ from progressivist ethics, both of which differ from Jewish, Muslim, Hellenic, Stoic, and military ethics. By this understanding of ethics, pessimism is one ethical approach among many, and while pessimism may be unethical in progressivist terms, the converse holds true as well: according to a pessimist ethos, progressivism is a dangerous and immoral delusion.

In a broader sense, however, ethics has come to mean something like due recognition toward other sentient beings. As Aldo Leopold puts it, "An ethic, ecologically, is a limitation on freedom of action in the struggle for existence. An ethic, philosophically, is a differentiation of social from anti-social conduct. These are two definitions of one thing."[28] In this view, different ethical approaches may be compared on the basis of how they balance individual action against social and ecological welfare. Seen this way, pessimism is not only highly ethical but significantly more ethical than progressivism. Pessimism's core concern, after all, is to give due weight to suffering, and from this concern arises pessimism's recognition of the limits we face as interdependent beings in a complex world. Progressivist ethics, on the other hand, can justify any amount of suffering in the present for the sake of a better future.

Finally, as to the question of whether or not pessimism is accurate, the pessimist in me is tempted to throw up my hands. What more can be said of the misguided efforts by dogmatic optimists to shore up their dangerous and erroneous delusions? Things may be going well for some humans at the moment, given the vast stock of cheap energy we've plundered, but as I've argued, there are good reasons to believe we're in serious trouble. Even if we choose to believe that people are well-meaning and sometimes altruistic, they are also fallible and limited. Indeed, believing in people's essential goodness should lead to an even deeper pessimism than believing in their selfishness, since the persistent gap between our good intentions and the suffering we cause makes a mockery of our attempts to make the world a better place. Finally, "better" is both subjective and relative, and material plenty does not entail existential satisfaction. Even if some things are temporarily or locally "better," the empirical observations behind philosophical pessimism hold: we are inescapably subject to mortality, error, impermanence, and suffering.

Pessimism may not be the only justifiable view of life, but it is justifiable. It may not be the only ethical view, but it is ethical. And it may not be the only empirical view, but it is quite a bit more empirical than optimism: a pessimist must by definition believe in a reality that exists independently of human cognition and desire, since reality is precisely what stymies us. As William James writes, "healthy-mindedness is inadequate as a philosophical doctrine, because the evil facts which it refuses positively to account for are a genuine portion of reality." What's more, James argues, not only are the "evil facts" that stymie our will a "genuine portion of reality," but "they may after all be the best keys to life's significance, and possibly the only openers of our eyes to the deepest levels of truth."[29]

EVERYBODY DIES

I would hope that any reader who has come this far with me won't be surprised to hear me say that while facts may be one kind of thing and values another, I'm skeptical that we can really disentangle the two. Our "facts" about the world are value-laden and indeed comprehensible only within the fictional frameworks that give them meaning. Conversely, our social norms and values depend upon and derive from our claims about reality, which are sometimes empirical, sometimes metaphysical. In any case, the position I'm putting forth is more an orientation than a program, since one of my core claims is that human reason cannot solve our deepest problems. What pessimism has to offer is not a full-scale ethical system but more of a stance: a view of reality that entails an obligation to other beings and other realities—as Kermode put it, a "common project, truth in poverty," to meet our "common need, solidarity of plight in diversity of state."[30] This stance is grounded in the following empirical claims.

1. Human life is beset by suffering, which is never more than temporarily relieved.
2. Consciousness itself is a source of suffering.
3. The unpredictability of complex events threatens our projects in ways that can never be comprehensively anticipated.
4. Human agents are fallible and limited.
5. Conflict is inevitable.
6. History has no inherent, intelligible meaning or trajectory.
7. Suffering matters.
8. Everybody dies.

It may seem crude to lay out such trivial claims so baldly, since they ought to be obvious to every mature adult human being, but history and the daily news suggest that they are not. We believe we can eradicate suffering. We believe we can avoid conflict or impose unanimity without having to compromise. We build castles in the air that take no account of our ignorance and limits. We believe reason can save us. We believe in progress. We repress and deny the foreknowledge of our inevitable deaths.

This being the case, it's worthwhile to consider these claims in turn. First, **human life is beset by suffering, which is never more than temporarily relieved.** Put more simply, suffering exists. We don't even need to refer to physical pain, hunger, illness, grief, sorrow, fatigue, and despair, though these are all too common. Boredom, anxiety, and envy—the mainstays of daily life in consumer capitalism—are enough to prove the point.

Many argue that suffering is caused by lack. Buddhism, for instance, holds that desire itself is suffering, and that the satisfaction of desire doesn't last or leads to boredom, which is to say, more suffering. Schopenhauer writes: "So long as our consciousness is filled by our will, so long as we are given up to the throng of desires with their constant hopes and fears, so long as we are the subject of willing, we can never have lasting happiness nor peace."[31] Hunger and sexual desire are the most salient examples here, but grief, loneliness, envy, and despair are also examples of privative or negative suffering.

Yet not all suffering comes from lack: as Bayle argues (building on Malebranche), pain is more than the absence of pleasure, illness more than the absence of health.[32] Pain is a positive force, an action of the nerves. Illness is more like possession than evacuation, as Virginia Woolf so insightfully illustrates in her essay "On Being Ill." To be ill is to be occupied, to feel your body and soul become the battlefield of invading forces, to be subject to forces beyond your control.[33] What's more, mental and physical anguish are not merely the consequences of illness, but inevitable features of consciousness. Monthly menstrual cramps and the pains of pregnancy are not the privation of health, as Aquinas might have put it, but its fullness. The pain that comes with growth is not a symptom of emptiness but of increase. And the pain of love is not merely the pain of desire, as Lacan argues, but the raw vulnerability of being entangled with another living being. As Elizabeth Stone has said about being a parent, "Making the decision to have a child—it's momentous. It is to decide forever to have your heart go walking around outside your body."[34] Such common, unrelenting, everyday pain is not insufficiency but vitality itself.

Pain is an unavoidable fact of life. How we choose to make sense of this fact is an ethical decision. As Ernst Jünger writes: "Pain is one of the keys to unlock man's innermost being as well as the world.... Tell me your relation to pain, and I will tell you who you are!"[35] Yet perhaps the most salient fact of suffering is how easily we forget what it's like once it goes away. Mostly we repress the actual experience and remember only that it happened. Were this not the case, the human species would surely have died out long ago, since few women would willingly suffer the trauma of giving birth a second time. Powerful chemicals and neurological affordances warp and occlude our memory of pain, contributing to *Homo sapiens*' tenacious and irrational optimism. We can survive torture, desolation, dismemberment, childbirth, a child's death, the loss of our partner, homelessness, and starvation. People do it every day.

Taking the fact of suffering seriously, one must reckon with Sophocles's hard wisdom that it is best not to be born at all. This position is often labeled "anti-natalism," which makes it sound like a campaign against babies, though it is nothing of the kind. The claim that, in David Benatar's words, "coming into existence is always a harm" is based on a recognition of life's tragic character.[36] It is the paradox of Omelas. And it is a challenge to our sense of ethics, indeed the very reason ethics exists, for if we could live without harm, then there would be no need to weigh conflicting courses of action: there would be only one morality. We need ethics precisely because we cannot live without causing harm and being harmed. Being born damages our mothers' bodies. Eating harms the beings we eat. We cannot grow or learn without pain: the physical and emotional pain of failure—the pain of coming to terms with death and loss—the pain of growing up, becoming different from who we were, losing our innocence—the pain of puberty—the pain of old age—the pain of love.

Benatar builds his case for anti-natalism on the asymmetry of pleasure and pain: pain is bad and pleasure good, but while the absence of pain is recognizably good, the absence of pleasure is not inherently bad. We might give these qualities the values of -1 for pain, +1 for pleasure, +1 for the absence of pain, and 0 for the absence of pleasure. Adding pleasure and pain together gives you a net value of 0, while adding the absence of pleasure to the absence of pain gives you a net value of +1. According to this reductive logic, life is a null value, but avoiding life a net positive. Thus not existing is better than existing, and conversely, existing is always a harm.[37]

Quantifying the worth of existence in this simplistic way may seem both counterintuitive and tasteless, but Benatar is only giving abstract form to the bitter wisdom of the ancients, as well as to established empirical phenomena like the negativity bias, negativity dominance, loss aversion, and the endowment effect.[38] As discussed previously, negativity bias is the principle that "in most situations, negative events are more salient, potent, dominant in combinations, and generally efficacious than positive events."[39] Loss aversion describes our psychological bias against loss, often characterized by the shorthand "losses loom larger than gains."[40] The endowment effect describes how we value emotional attachment over objectively measured costs.[41] Considering loss aversion and the endowment effect together supports Benatar's contention that coming into existence is always a harm, since never having had a life is a lesser loss than losing the life you have—which loss is, of course, inevitable.

In general, negative events, feelings, and memories are more salient than positive ones, despite our optimistic bias toward the future, and we tend to weigh losses more heavily than gains. Thus suffering looms larger than happiness overall, *even if pain and pleasure are objectively equal*. From this perspective, the pessimistic conclusion that Sophocles, the author of Ecclesiastes, and David Benatar reach seems not only reasonable but conclusive. "The fact that one enjoys one's life does not make one's existence better than non-existence," Benatar writes, "because if one had not come into existence there would have been nobody to miss the joy of leading that life and thus the absence of joy would not be bad."[42]

But *my* life isn't so bad, you might say, and even if it is, surely all those people posting videos on TikTok must be having a great time, right? In truth, even the best life is worse than you think.[43] Optimism bias and duration neglect make self-reporting unreliable. Self-delusion, social constraints, heuristics, biases, storytelling—all our human illusions lead us to believe our lot is better than it actually is.

Disillusionment is painful. Learning is also painful, in part because learning something new means finding out that you were wrong, which is humiliating and embarrassing. But learning is also painful because **consciousness itself is a source of suffering**. This troubling insight is powerfully dramatized by the ancient narrative of the Garden and the Fall. According to the Biblical tradition, suffering is the consequence of self-consciousness caused by eating from the Tree of the Knowledge of Good and Evil.

We might ask what consciousness even is. Such a question is difficult to answer—consciousness may not exist—and too complex for me to get into here.[44] But whatever consciousness might be, it's a pain: consciousness involves self-awareness, and self-awareness means awareness of one's limits, one's failings, and the painful gaps between ourselves and others, our "self" and the world, what we have and what we want. Learning that reality doesn't bend to our will hurts—just watch a frustrated infant cry. Becoming aware of our physical limits, mental limits, and mortality also hurts. Consciousness of death is painful. Consciousness of others' pain is painful. Mere consciousness of existence can wound, as George Eliot illustrates in *Middlemarch*: "If we had a keen vision and feeling of all ordinary human life, it would be like hearing the grass grow and the squirrel's heart beat, and we should die of that roar which lies on the other side of silence."[45]

Even sitting still can hurt. According to one study, most men would rather give themselves electric shocks than be left alone with their thoughts for fifteen minutes.[46] This astonishing observation might help explain the role of meditation in the Buddhist tradition: the practice of sitting is not a practice of relieving suffering by ending consciousness but of learning to pacify the mind by accepting the suffering it cannot hope to avoid. This lesson is violently dramatized in a koan from *The Gateless Gate*:

> Bodhidharma sat facing the wall.
> The Second Ancestor [Dazu Huike] stood in the snow.
> He cut off his arm and said, "My mind has no peace as yet! I beg you, Master, please put it to rest!"
> Bodhidharma said, "Bring your mind here and I will pacify it for you."
> The Second Ancestor said, "I have searched for my mind, and I cannot take hold of it."
> Bodhidharma said, "Now your mind is pacified."[47]

The restless struggle of consciousness in response to pain only increases our pain and restlessness. Fixating on one's misery makes it worse. The only freedom from this endless cycle comes from recognizing the mind itself as a primary source of suffering.

One of the more lyrical if misanthropic formulations of the relationship between consciousness and suffering comes from Peter Wessel Zapffe's essay "The Last Messiah" (1933).[48] Zapffe's work was deeply influenced by Nietzsche and Schopenhauer, and he was lifelong friends with the deep ecologist Arne Naess. Zapffe in turn influenced the weird fiction writer Thomas Ligotti, who offers an extended meditation on Zapffe's anti-natalism in *A Conspiracy Against the Human Race*.[49] In "The Last Messiah," Zapffe argues that human consciousness is a tragic maladaption that alienates man from nature and hence from his own being: "A break in the very unity of life, a biological paradox, a monstrosity, an absurdity, a hypertrophy of the most catastrophic kind."[50] For Zapffe, excessive consciousness leads eventually to a deterministic understanding of the universe's ultimate meaninglessness, which provokes a "cosmic panic": "Soon he sees mechanics behind everything, even behind that which he used to hold dear, his beloved's smile."[51] Fortunately for us, according to Zapffe, "most people manage to save themselves by artificially paring down their consciousness."[52] Zapffe makes a strong depressive realist claim that neurosis and depression are actually symptomatic of clear perception, "indications of a deeper, more immediate experience of what life is all about, bitter fruits of the genius of the mind or emotion."[53] Such cosmic pessimism might, as Kermode would argue, be no more than a projection of our own gloom onto the biological narrative of the species, yet the chastening truth behind it remains.

Not only is consciousness a source of suffering, but its benefits are limited. Foresight is dim at best, and **the unpredictability of complex events threatens our projects in ways that can never be comprehensively anticipated.** To put it more pungently, shit happens. To have a genuinely accurate predictive model of reality would mean building a map as big as the territory it represents: it would mean re-creating the universe itself. In making our way, we face not only "the immense complexity of the societal and biospheric totality of effects, which defies all calculation," as Hans Jonas writes, but also "the essentially unfathomable nature of man, which always springs surprises on us; and the impossibility to predict future inventions, which would amount to preinventing them."[54]

Then there's the fact that **human agents are fallible and limited**. People make mistakes. One could claim with good evidence that people are inherently tribal and self-interested, yet even if we reject such a view and assume that humans are, as a rule, well-intentioned and compassionate, one must nevertheless grant that people are still flawed. Even the best of us screw up. Psychology and behavioral economics have shown conclusively that human cognition and perception are inherently biased. People are not "rational actors." This fact on its own is sufficient justification for pessimism, even if we still hoped that social engineering might ameliorate our worst excesses. We must accept human fallibility and the limitations of human knowledge as real constraints.

Given that the unpredictability of complex events threatens our projects in ways that can never be comprehensively anticipated, that human agents are fallible, and that different people have different goals, we may conclude that **conflict is inevitable**. Since different people want different things, sometimes those desires will come into opposition. Sometimes different people want the same thing, which is also a source of conflict. One might hope that the inevitable competition, jealousy, envy, and aggression one sees acted out in any kindergarten could be smoothed away by a combination of rules, care, and justice, but any utopian program must contend not only with such vices but with simple errors in judgment, inherent biases, external shocks, and the grating bumps and confusions that perpetually threaten to blossom into grudges and feuds. Conflict may be mitigated or managed, but never wholly eliminated.

Thus the progressivist faith that history itself will lead eventually to a better tomorrow cannot be supported. Ultimately, we must face the realization that **history has no inherent, intelligible meaning or trajectory**. There is change, and the past two hundred and fifty years have seen an astonishing growth in social complexity and technological sophistication, which we easily confuse with moral or social progress, but such a conflation is fallacious. Change does not equal progress. There is no universal history, not even, in Adorno's famous phrase, "one leading from the slingshot to the megaton bomb."[55] Even if it seems today that human existence is getting better, as Steven Pinker's whiggish teleology proposes, we cannot claim to know how it's going to turn out "in the end." A kayaker going down a river may be having the best day of her life until she goes over a waterfall. Recall the Buddhist joke about the monk who fell off a cliff: "Halfway down... so far so good." The last seventy years of the Great Acceleration, the last two hundred and fifty years of industrial society, and the last seven thousand years of agriculture, literacy, and empire are statistically insignificant against the hundreds of thousands of years of *Homo sapiens*' existence, never mind the millions of years in which our hominid ancestors were walking upright and using tools. And even if the planet burns up, civilization collapses, and we face a millennial dark age of chaos and suffering, that still won't fulfill any final goal or meaning for human history, for as long as humans go on existing they will go on existing "in the middest," weaving collective life in the present from stories of the past and visions of the future.

Questions about whether life is getting better or what history means offer no objective or empirical resolution, since they are on the one hand concerned with purely subjective judgments—a well-fed twenty-first-century octogenarian living in Phoenix may *feel* their life is miserable and meaningless, just as an illiterate pre-colonial Inuit hunter scraping a hazardous subsistence from Arctic ice floes may *feel* their life is a rich tapestry of connection and signification, or vice versa—and on the other hand concerned with our collective human trajectory into the future, a trajectory that can only be judged from the far side of its conclusion and that would only make sense in terms of a universal human subject who, despite the earnest efforts of theologians, philosophers, and intellectuals, does not exist.

What does the Roman Republic mean? The Ming dynasty? The Mughal empire? The Enlightenment? Who can say? As Karl Löwith writes, "To ask earnestly the question of the ultimate meaning of history takes one's breath away; it transports us into a vacuum which only hope and faith can fill."[56] The question of what human history would mean could finally be answered only by someone outside human history, which means it doesn't really matter what it means, since human meaning is *human*, and any posthuman or exobiological meaning-making entity concerned with the question of what happened to our precocious Terran ape would necessarily approach it from their own epistemological framework. The position from which one could judge the entirety of human history is one that can never be reached by any human historian, since being alive means being entangled in an open-ended trajectory, and the only kind of being for whom it would be possible to presume to judge the entirety of the human saga could emerge only after that saga's end, which is to say after our extinction. The famous line from *Oedipus Rex*, "count no man happy till he dies," applies with mordant force to the lives of empires, nations, peoples, cities, even our species. Of course, Leibniz may be right: God may have a plan. But we have no justified hope of ever knowing what that plan is, nor of understanding it were we granted such a revelation. How could a single human mind comprehend God's plan for the entire universe? The mere suggestion is absurd.

Yet although history may have no inherent, intelligible meaning, that does not mean that our lives are meaningless. For the pessimist, even more than for the existentialist, existence precedes essence, and the key fact of existence is suffering, which is to say that **suffering matters**. Suffering matters because it's how we learn that reality exists independently of our desires and fantasies. Suffering matters because it's how we learn about danger, consequence, and risk. Suffering matters because it teaches us that our happiness depends on other people. Suffering matters because it's one of the most important ways that we learn anything at all. And suffering matters because it's the one thing we all share with every other living being: the recognition of suffering is the root of compassion, and the recognition of shared suffering the source of our solidarity.

Finally, **everybody dies**. Yes, we live on in others. Yes, our names may resound through the ages. Yes, our children carry on our genes. And yes, we live longer today than ever, if we're wealthy and lucky. But there is no solution for death. Change, change, everything is change, and every living being must die.

THE VIRTUES OF PESSIMISM

We can now turn to a brief discussion of what I take to be pessimism's virtues. By virtue I mean both an ethical ideal to which pessimism is committed and a moral benefit that pessimism offers. If one adheres to the virtues of pessimism, that is, one will become a better pessimist and a more ethical person. My argument is distinct from other arguments for approaching the problem of climate change through virtue ethics, such as those offered by Dale Jamieson, J. Baird Callicott, Jeremy Bendik-Keymer, Ronald Sandler, and Allen Thompson, particularly in that I'm not making an argument for *environmental ethics*, nor am I arguing that we should adopt environmentally oriented virtues in order to protect nonhuman nature.[57] In one sense my argument is more radical, in another more conventional: I am simply arguing that pessimism is virtuous. By this I mean that pessimism fosters specific virtues already valued in contemporary society—namely, empirical accuracy, resilience, compassion, egalitarianism, inclusion, and responsibility. Pessimism may lead to more environmentally sustainable behavior or not, but that's not the point. Pessimism isn't about conservation, nor is it about "sustainability." Rather, in the face of inescapable human suffering, radical uncertainty, and an unprecedented transformation of the human lifeworld, pessimism offers an ethic of survival.

Pessimism begins from an acceptance of suffering. It is concerned first of all with *what is*: with seeing reality as clearly as possible and understanding what's really happening beyond our optimistic biases, hopes, fears, desires, and delusions. This empirical insistence on facing reality tends to make pessimism more accurate than optimism. "What is a pessimist anyway?" asks Clarence Darrow. "It is a man or a woman who looks at life as life is."[58] This quality of empirical attention to external reality is part of the reason why pessimism isn't quietist or fatalist, as its critics insist, but rather an effective tool for collective action. As philosopher Kathryn Norlock writes:

> The heavy knowledge any student of evil eventually acquires ... warrants pessimism but also clarifies appropriate goals, and shifts attention to different sorts of hopes as well as other attitudes. We are better off with the heavy knowledge that evils recur than we are with idealizations of progress, perfection, and completeness, and if we cultivate an appropriate ethic for living with such heavy knowledge, it should not prevent us from doing our best to resist evils, improve the lives of victims, and enjoy ourselves.[59]

Buddhist pessimism brings us to right action and the Eightfold Path. The austere pessimism of Ecclesiastes ends with the call to rejoice and "remove anger from your heart, and take evil away from your flesh."[60] Defensive pessimism helps us manage anxiety and reach our goals. Stoic pessimism helps us act with humility, discipline, and self-control. Ecopessimism helps us recognize our dependence on nonhuman nature. Afropessimism helps us understand how deeply the idea of progress depends upon ontological presuppositions of racial inequality. And existential pessimism challenges us to take control of our own lives, ask hard questions about who we are and what matters, and turn away from the distractions and illusions that foster complacency.

Pessimism helps us be resilient because it helps distinguish between realistic and unrealistic goals. Everyone knows the Serenity Prayer, which may be one of the most concise formulations of pessimism available. In what is probably its best-known phrasing, it goes:

> God grant me the serenity to accept the things I cannot change,
> Courage to change the things I can,
> and Wisdom to know the difference.[61]

The wisdom of pessimism is in recognizing just how little we can actually change, both in the world and in ourselves. Pessimistic adages such as Murphy's Law ("If anything can go wrong, it will") help us plan for unforeseen obstacles.[62] The precautionary principle, which broadly argues that innovations or policies developed out of ignorance of their possible consequences should be avoided and specifically states (as formulated by Nassim Taleb) "that if an action or policy has a suspected risk of causing severe harm to the public domain ... the action should not be undertaken in the absence of scientific near-certainty about its safety," applies pessimism as a guideline for wise leadership.[63] Likewise, German philosopher Hans Jonas's heuristics of fear, the rule that responsible consideration of decisions which may have existential consequences must give significant weight to the worst possible outcomes, applies pessimism as a useful tool for long-term decision-making.[64]

Not only is pessimism more accurate and resilient than optimism, it is more compassionate. One of the core criticisms pessimists have of optimists is that they fail to take full account of suffering and evil, for if they did, they could not be optimists. Expecting less from the world and the people in it helps one cultivate sympathy, patience, and generosity. Everyone is fallible, even (especially) oneself. Situations are always more complex than we think. Outcomes are unpredictable. Pessimism favors humility and forgiveness, since even the best intentions are often frustrated. Pessimism vitiates egoism and arrogance, as it makes people less likely to blame others for their situation, more inclined to reject inherited hierarchies, and more likely to see commonalities while also appreciating differences. Schopenhauer writes:

> *Pardon's the word to all!* Whatever folly men commit, be their shortcomings or their vices what they may, let us exercise forbearance; remembering that when these faults appear in others, it is our follies and vices that we behold. They are the shortcomings of humanity, to which we belong; whose faults, one and all we share; yes, even those very faults at which we now wax so indignant, merely because they have not yet appeared in ourselves.[65]

The core truth of pessimism, after all, is that suffering is inevitable and universal, yet always particular. As van der Lugt argues, "At its best and deepest, what this kind of dark thinking, which is also a *fragile thinking*, achieves, is neither desperate nor passive nor fatalist: it is to open up new horizons of compassion and consolation."[66]

Thus while optimism slides easily into complacency, obliviousness, and narcissism, pessimism's combination of empiricism and compassion leads to the realization that reflexive self-determination—guided by more-or-less accurate knowledge of obdurate reality—is the only ethical way to make collective decisions. In a word, pessimism is egalitarian. Whereas optimists may believe they know what's best and thus feel justified in adopting whatever means necessary to achieve their ends, pessimists understand that we're all just muddling through, so we may as well muddle through together.

Pessimism's criticality and compassion also mean that it's more inclusive than optimism. There's less likelihood of a pessimist imposing false or oppressive rationalities on others, because what would be the point? The pessimist is skeptical toward *all* ideological justification, *all* rationalization, since existence as such is inherently unjust and reason offers no recourse. Finally, against the criticism that pessimism leads to apathy and despair, one may point out that pessimism actually puts *more* responsibility on individuals to act together, since we can neither trust institutions to take care of us nor depend on those in authority to fix our problems.[67] In the words of journalist and editor Rosie Warren, "Pessimistic is just another word to describe those who fear we might be doomed but are fighting anyway . . . who have no certainty about the way forward except that it cannot be this."[68]

These are, as I see it, the virtues of pessimism: empirical accuracy, resilience, compassion, egalitarianism, inclusion, and responsibility. I think these are good virtues, and I think they're especially useful now, facing cascading systemic failure, widespread suffering, unpredictable chaos, and the end of the world as we know it. As I argued in part 1, our impasse poses intractable challenges that we have little hope of overcoming. Indeed, our need for narrative structure, our adaptive heuristics and biases, and our contemporary cultural paradigms all make the problem harder, if not impossible, to comprehend. As I've argued throughout, our troubles are compounded by a pervasive human bias for optimism, which bias is unfortunately strengthened by our civilizational faith in the Myth of Progress. In part 2, I've argued that pessimism is an accurate, ethical, and appropriate response to this impasse. I showed how progressivist optimism in the early Enlightenment period emerged in dialogue with a counter-tradition of philosophical pessimism, sketched the history of that counter-tradition, discussed some contemporary varieties of pessimism, demonstrated that the main philosophical criticisms typically levied against pessimism do not hold, and concluded by making an argument for ethical pessimism, outlining both its empirical foundations and its virtues. My argument may not be conclusive, but on its own terms no argument for pessimism could be, since pessimism fundamentally rejects the conceit that reason can make reality cohere.

I might have begun where I now end, except that doing so would have misrepresented my argument. My point isn't that philosophical pessimism is a valid stance that can be applied to climate change, but rather that the existential impasse caused by climate change demands ethical pessimism. If we are to have any hope at all for the future, that is, it must be a hope that rejects optimism, abjures progress, and accepts the end of the world. "If this sounds almost nonsensical," writes Dienstag, "perhaps it is because we have been told for so long that progress is the rational thing to hope for."[69] Yet as Georg Henrik von Wright paradoxically puts it, "to debunk" the "false mythologies" of progress may be "the greatest service intellectuals of our time can do to the cause of progress."[70] Pessimism is an accurate, appropriate, and above all ethical response to our situation—indeed it may be the only ethical response available. Pessimism is fundamentally about recognizing and living within natural human limits. It's about recognizing that suffering is inevitable but not unbearable. It's about learning to die and learning to live with death. And finally, it's about committing to a radical and paradoxical hope: the hope that life might be worth living after the end of the world.

Afterword
THE CHILDREN OF RUIN

Oh, plenty of hope, infinite hope—only not for us.
—FRANZ KAFKA

The beginning is easy, nothing more American. A simple white Dutch chapel, founded in 1767, First Reformed—neither a ruin nor a monument but a humble working church. We see a hand writing, hear a voice speak. A man of the cloth transcribes his thoughts: "I have decided to keep a journal not in a Word program or in a digital file, but in longhand, writing every word out so that every inflection of penmanship is recorded, every word chosen, scratched out, revised, to set down all my thoughts and the simple events of my day factually and without hiding anything."[1]

Cut to daylight, interior, a weary face washed in winter light: Dominie Ernst Toller (Ethan Hawke) stands at the pulpit clad in black, finishing his homily.[2] After communion, a congregant named Mary (Amanda Seyfried) asks the Reverend to speak with her husband Michael (Philip Ettinger), an environmental activist recently released from jail. Mary is pregnant, and Michael wants her to have an abortion.

Toller and Michael talk the next day in what seems to be Michael's office.

A poster on the wall shows the nine boundary conditions for human life on earth, four of which have been crossed (by now, six). A characteristic hockey-stick graph dominates the darkness behind the desk. The screen saver plays a NASA graphic in which the Earth goes from green to yellow to orange to red, again and again. Photos of environmental destruction decorate one wall, photos of murdered environmental activists another. Michael is a rag twisted hard and thrown into a chair, like something out of Dostoyevsky. His eyes leak anguish. His scraggly beard seems grown to be wrenched. He wears a gray fleece, gray socks, and gray-green pants.

The conversation begins politely enough, but things turn when Michael asks Toller his age. Toller says he's forty-six. "Thirty-three," Michael responds. "That's how old our child will be in 2050.... Do you know what the world will be like in 2050?"

"Hard to imagine," Toller responds with a chuckle.

"Yeah, you think?" Michael asks, bitter as wormwood. The question hangs in the air for a moment before he unleashes the deluge: ecological devastation, rising temperatures, "severe, widespread, and irreversible impacts ... and when scientists say stuff like that ..."

"He went on like that for some time," Toller's voice cuts in. "By 2050 sea levels two feet higher on the East Coast. Low-lying areas underwater across the world. Bangladesh, twenty percent loss of land mass. Central Africa, fifty percent reduction in crops due to drought. The western reservoirs dried up. Climate change refugees. Epidemics. Extreme weather." It is a familiar litany.

Then Michael comes to the real question: "How can you sanction bringing a girl ..." he asks, "for argument's sake let's say my child is a girl ... a child full of hope and naïve belief into a world ..." His voice fades as he ponders the idea, then comes back raw: "When that little girl grows to be a young woman and looks you in the eyes and says, 'You knew all along, didn't you?' What do you say then?"

Toller takes the question seriously, but as he and Michael struggle, he guides their discussion away from climate change and toward the deeper existential question of how to live with the knowledge of evil, the pain of loss, and uncertainty about the future. Toller admits of despair himself, recounting his own tale of woe: once an army chaplain like his father, Toller convinced his son to serve, then had to bury him when he was killed in Iraq. "Now Michael," Toller says, "I promise you that whatever despair you feel about bringing a child into this world cannot equal the despair of taking a child from it." Toller's eyes shine with pain. "Courage is the solution to despair," he tells him. "Reason provides no answers."

I've taught this scene from Paul Schrader's 2018 film *First Reformed* several times since I first saw it in a theater in San Francisco, near where I was staying in a cheap hotel on a street crowded with heroin addicts, four blocks from Twitter headquarters. I had been invited to speak at the Commonwealth Club, "the nation's oldest and largest public affairs forum," where I was scheduled to appear in dialogue with Climate One founder Greg Dalton and Episcopalian priest Matthew Fox, a former Dominican friar expelled for disobedience and heresy.[3] We would be discussing climate change, faith, and my then-most-recent book, a collection of essays titled *We're Doomed. Now What?* Like Toller, I was a veteran. Like his son, I'd been to Iraq. Like Michael, I was haunted by visions of catastrophe.

"I felt like I was Jacob," Toller reflects on his conversation with Michael, "wrestling all night long with the angel. Fighting in the grasp. Every sentence, every question, every response a mortal struggle. It was exhilarating."

I felt similarly exhilarated the first time I watched my thoughts play out on screen. I too had felt the thrill of grappling with the philosophical and spiritual implications of catastrophic ecological transformation, the desperate battle against despair, the "fighting in the grasp" to make sense of senseless suffering, failure, and the end of life as we know it. My first true intimation of what climate change meant came in the summer of 2013, and, like Toller, I found myself transformed by the encounter.[4]

Toller and Michael wrestle to a stalemate: a victory for the minister, but fleeting. Michael commits suicide a few days later. After that, we see Toller take up Michael's cause, growing more fervent as he struggles with stomach cancer and his growing yet ambivalent intimacy with Mary. We watch him don an explosive vest Mary had found among Michael's things.

There are many ways *First Reformed* resembles writer and director Paul Schrader's best-known screenplay, *Taxi Driver*, not least in its basic structure: the story of an isolated hero plunging toward existential violence. Yet whereas *Taxi Driver* ends in a sequence of redemptive carnage, *First Reformed* swerves in its last moments into a scene so ostentatiously at odds with everything coming before that accepting it at face value would seem to mean denying the film up to that point had anything serious to say.

The setting is the church's 250th re-consecration. Community luminaries have gathered, the church's repaired organ resounds through the sanctuary, and the local megachurch's choir director Esther (Victoria Hill) sings the 1887 gospel hymn "Leaning on the Everlasting Arms." In his rectory, Toller has donned and armed Michael's explosive vest, then covered it with a black Genevan gown and white stole, preparing to sacrifice himself, the attendant luminaries, and First Reformed itself in an act of environmental protest. He writes a last reflection in his notebook, puts his pen down next to an untouched tumbler of scotch, then gets up to go in to the church.

As Toller's hand reaches to open the rectory door, he sees Mary going up the steps to the chapel. The sight shocks him out of his reverie. Stricken, distraught, screaming into his fist, he disarms the bomb, tears off the vest, and scourges his naked torso with rusted barbed wire before donning a snow-white alb, which then blossoms with red flowers. Toller returns to his desk, dumps the scotch from its tumbler, and fills the glass with Drano.

He is about to drink when Mary bursts in. He turns, drops the glass, and walks toward her. She calls him by his first name, "Ernst," and the two clasp, eyes shut, and kiss greedily, while the camera spins and spins and spins and—nothing. Roll credits.

In his book-length study, *Transcendental Style in Film,* Paul Schrader argues that for a film in the transcendental style, form supersedes content. He quotes Bresson: "The subject of a film is only a pretext. Form much more than content touches a viewer and elevates him."[5] He writes, "In transcendental style the form *must* be the operative element, and for a very simple reason: form is the universal element whereas the subject matter is necessarily parochial, having been determined by the particular culture from which it springs."[6] According to Schrader, "Transcendental style is simply this: a general representative filmic form which expresses the Transcendent."[7] By *Transcendent,* Schrader means that which "is beyond normal sense experience." He writes:

> Transcendental style seeks to maximize the mystery of existence; it eschews all conventional interpretations of reality: realism, naturalism, psychologism, romanticism, expressionism, impressionism, and, finally, rationalism. . . . To the transcendental artist these conventional interpretations of reality are emotional and rational constructs devised by man to dilute or explain away the transcendental.[8]

The transcendental is, for Schrader, expressible only through contradiction: neither the implausible redemption at the end of *First Reformed* nor the relentlessly pessimistic two hours preceding it but the very juxtaposition of these incommensurable ontological frames.

Film, like writing, is ineluctably linear, an experience cast in time. Yet the transcendent *ekstasis* Schrader's art calls for is, by definition, beyond time: if there is any single characteristic by which we may characterize the divine, it is that it is not subject to human temporality. Representing the transcendent is therefore, Schrader argues, impossible. The best art can do is to create formal conditions that might inspire a genuine sense of mystery, not as emotion but as experience, not as catharsis but as kenosis, achieved not through simultaneous identification with contradictory ethical worlds, as in the Hegelian understanding of tragedy, but through confronting an even deeper contradiction: the incommensurability of a world of meaningless suffering and a world redeemed by love.

Neither can touch the other. They cannot be synthesized or sublated. Each makes its contrary unreal, even absurd. Mary and Toller's communion at the film's end is one truth, the two hours preceding it another, and the aesthetic experience offered by *First Reformed* is not a linear progression from the latter to the former, nor the supersession of the latter by the former, but rather the challenge to hold both truths in our mind simultaneously. As Toller himself tells Michael, early in the film, "Wisdom is holding two contradictory truths in our mind. . . . Hope and despair. A life without despair is a life without hope. Holding these two ideas in our head is life itself." *First Reformed* only *seems* linear, only *seems* narrative, only *seems* concerned with climate change. Ultimately, the subject is a pretext, and the film paradoxically presents in linear narrative an aesthetic experience of *ek-stasis*, achieved through the breaking of its narrative frame.

According to the best empirical evidence we have, observed present trends indicate that the future we face is one of accelerating planetary transformation, with catastrophic consequences for human society. At the same time, this future is radically unknowable. The massive, rapid perturbations in the carbon cycle and global climate we're currently experiencing are unprecedented in the geological record of the planet, and we cannot predict how they will unfold or how human society will respond. Neither the Myth of Progress nor human reason can meaningfully make sense of what's happening. Nevertheless this book is an attempt to do just that. I've approached the problem through years of research and reflection, as well as critical discussion, painstakingly concerned with evidence and argument, and presented my work as a hybrid of practical philosophy and environmental humanities scholarship, as if I could make the incoherence of our world cohere. There is a "complexity and a sense of failure" behind this intentionally anti-systematic meditation on a stubbornly intractable set of problems, which ultimately offers no more than the pretense of a solution for what is in truth an irresolvable impasse.

Fiction is inescapable. Myth inescapable. We cannot live without imposing order on the chaos of experience. And while some fictions may be more accurate than others, none offer unmediated access to reality. The best we can hope for is to better understand the dynamic between reality and the fictions that give it shape, with due respect for the perspectival, aporetic, contradictory, and paradoxical character of the attempt: to grow a little wiser in our failure. We may never experience the transcendent totality of being, though we may sense it. We may never close the gap between consciousness and the universe, but by the multiplication of perspectives, we might be able to triangulate our position and direction a little more accurately. Climate change may be transforming reality in ways we can neither avoid nor comprehend, but we might yet find ways to live through it.

"The chief danger to philosophy," writes Alfred North Whitehead, "is narrowness in the selection of evidence. Philosophy may not neglect the multifariousness of the world—the fairies dance, and Christ is nailed to the cross."[9] The ultimate virtue in criticism, then, may be just what Toller suggests, simultaneously holding multiple contradictory truths in our mind, or what Keats called "negative capability": the willing embrace of existence lived "in uncertainties, Mysteries, doubts, without any irritable reaching after fact and reason."[10] As T. S. Eliot wrote in *Four Quartets*, "The only wisdom we can hope to acquire / Is the wisdom of humility."[11] Thus the question of whether to commit to the Myth of Progress, the Myth of Renewal, the Myth of Marxism, the Myth of Managed Simplification, the Myth of the Multiverse, the Myth of the Anthropocene, or whatever myth you choose, may not be the right question at all.

The situation we face is unprecedented, but the truly revelatory content of our apocalyptic fictions is that the world has always been ending. The question we face is how to embrace our impasse: how to abjure the imposition of narrative on a rapidly changing reality, acknowledge the transience of the present, and see in the death of *what is* the birth of *what shall be*. Plotinus called the path of unknowing *apophasis*; Zen Buddhists call it *kenshō* or *satori*; Jonathan Lear calls it "radical hope." Schrader's transcendental aesthetic, Kermode's "sense of an ending," and Wilderson's Afropessimism may be versions as well. In our case, facing a global ecological transformation beyond the scope of anything human civilization has ever confronted, I call it ethical pessimism: a commitment to future existence that by definition cannot be imagined. Ethical pessimism recognizes that we cannot know how climate change and ecological catastrophe are going to transform our world, how human societies will respond, how human beings will adapt, or who we will become in consequence. Ethical pessimism rejects the claim that we can talk or think our way out of the situation. Yet ethical pessimism remains committed to some future human existence, *no matter what form that existence takes*, no matter *who* that human is.[12] Whoever they may be, they will suffer. Whoever they may be, they will love. Whoever they may be, they will live as we do, in the middest, making it up as we go.

What do we make of our end? "The question is no longer to know how to live life while awaiting it," writes Achille Mbembe. "Instead it is to know how living will be possible the day after."[13] Yet this question is precisely the one we cannot answer. The world of tomorrow will be unrecognizable to those of us alive today, just as the world we live in today would have been unrecognizable to our ancestors. We have no more power to decide what the present means to the future than our ancestors had over what they mean to us. The past is an ambiguous inheritance, unconditioned by the benefactor. What our wondrous, astonishingly destructive civilization ultimately means is not something to be decided by those who caused it or even those doomed to live through its end, but only by those who come after: the children of our ruin.

ACKNOWLEDGMENTS

I'm thankful for the many conversations with friends and colleagues, support from various institutions, and efforts of other scholars and researchers that have made this work possible, as well as for the careful attention from the two anonymous reviewers who responded to this book's proposal and the three reviewers who commented on the final manuscript. I'm also deeply indebted to the labor of those at Stanford University Press without which this book would not exist, especially my peerless copyeditor, Barbara Armentrout; production editor Emily Smith; Adam Schnitzer and the marketing team; and Erica Wetter, whose faith in and commitment to this project helped me carry it over the finish line. Warm thanks are also due to Bonny McLaughlin, who completed the index. While a final accounting of my thanks would go on at some length, such a list would tax any reader's patience, and in the event perhaps seem more gratuitous than grateful, so I will try to keep my acknowledgments brief.

I am grateful to friends, fellow writers, and scholars Omar El Akkad, Meehan Crist, Amitav Ghosh, Lawrence Jackson, Dale Jamieson, Jesse McCarthy, China Miéville, Naomi Oreskes, Hilary Plum, Paul K. Saint-Amour, Jake Siegel, Joseph Earl Thomas, and Martin Woessner for their conversation, fellowship, support, and feedback over the years, as well as to the members of the climate crisis reading group (Bathsheba Demuth, Lacy M. Johnson, Emily Raboteau, Elizabeth Rush, and Meera Subramanian, plus Meehan and occasional others), for their companionable ongoing discussion of topics central to this project. I'm also grateful to have phenomenal colleagues across the disciplines at Notre Dame, whose generous engagement on environmental issues, social justice, and ethics forms the matrix out of which this book emerged. Specifically, I want to thank Ellis Adams, Xavier Navarro Aquino, Tobias Boes, Diogo Bolster, Paul Cunningham, Anna Geltzer, Donna Glowacki, Robert Goulding, Dionne Irving, Debra Javeline, Lionel Jensen, Paul Kollman, Drew Marcantonio, Kate Marshall, Jason McLachlan, Rahul Oka, María Rosa Olivera-Williams, Joyelle McSweeney, Dan Miller, Emma Planinc, Francisco Robles, Joshua Specht, Meghan Sullivan, Jen Tank, and Julia Adeney Thomas, among others, as well as John Sitter and Laura Dassow Walls, both now retired. I also want to thank the many graduate and undergraduate students whose earnest grappling with climate change is a continual inspiration, particularly the students in my Postcolonial Anthropocene graduate seminar (2021 and 2024) and my undergraduate course Witnessing Climate Change (2022 and 2023), as well as those involved in the Environmental Humanities Initiative, only a few of whom I can name here: Makella Brems, Arman Chowdhury, Tim Fab-Eme, William Kakenmaster, Eduardo Febres Muñoz, Kristen Sieranski, Austyn Wohlers, and Tianle Zhang. Thanks are also due to my graduate research assistant, Oliver Ortega, who helped clean up later drafts of this work. Any remaining mistakes, of course, are my own. In addition, I am thankful for the support shown by Dean Sarah Mustillo, Associate Dean Ernest Morrell, Diogo Bolster, Robert Goulding, Meghan Sullivan, and Jen Tank for my efforts in building the environmental humanities at Notre Dame.

I also want to take a moment to thank everyone who participated in the 2022 conference I organized at Notre Dame, "1000 Years of Ice and Fire: Ecological Collapse and Migration from Vinland to the Anthropocene," which was formative for this project, especially Dipesh Chakrabarty, Claire Colebrook, Sam White, Robert Markley, Lucinda Cole, Kellie Robertson, and Malcolm Sen, as well as everyone who participated in the Kellogg Institute's 2024 manuscript workshop for this book, including several people already named, as well as Miguel Centeno, Joshua Foa Dienstag, Zach Schrank, Jacob Swisher, Linus Recht, and Luke Kemp. I also want to thank Therese

Hanlon, Karen Clay, Denise Wright, and the rest of the team at the Kellogg Institute for their logistical support.

I've been lucky and honored to have been the beneficiary of intellectual and financial support in ways large and small from a range of institutions, including the John R. Simon Guggenheim Foundation, the V. Kann Rasmussen Foundation, the University of Southern California, and, specifically at the University of Notre Dame, the Institute for Scholarship in the Liberal Arts, the Notre Dame Institute for Advanced Studies, the Kellogg Institute, the College of Arts and Letters, and the Reilly Center for Science, Technology, and Values. I am especially grateful to the Notre Dame Institute for Advanced Studies for allowing me to be their inaugural Teaching Lab Fellow in 2021–2022 and giving me the opportunity to develop my course Witnessing Climate Change, which fed into this project, and also for the support provided by Institute staff Maria Di Pasquale, Kristian Olsen, and J'Nese Williams.

Versions of parts of this book were delivered as talks at the Center for Environmental Futures at the University of Oregon, Central Washington University, the Environmental Change Initiative at the University of Notre Dame, the 2019 Göteborg Film Festival, Harvard University, the Humanities Research Institute at the University of Illinois Champaign-Urbana, the Penn Program in Environmental Humanities at the University of Pennsylvania, the Planetary Limits Action Network, the 2023 Society for US Intellectual History Conference, the University of South Carolina, and Xavier University. Among numerous thoughtful interlocutors at these various events and institutions, some already thanked above, I want to especially single out Bethany Wiggin and Stephanie Le Menager for their generous engagement.

Earlier versions of parts of this book were published in different form in the following publications. I'm grateful to the editors and publishers for the opportunity to write for them and for their close critical attention to my work.

> "Memories of My Green Machine: Posthumanism at War." *Theory & Event* 13:1, 2010.
>
> "Children of Thanos." *The Yale Review*, January 2019.
>
> "Learning to Live in an Apocalypse." *MIT Technology Review* 122:3, May/June 2019.

"We Broke the World: Facing the Fact of Extinction." *The Baffler* 47, September/October 2019.

"Climate Change Is Not World War." *New York Times*, September 18, 2019.

"Narrative in the Anthropocene Is the Enemy." *Literary Hub*, September 18, 2019.

"Beginning with the End." *Emergence*, April 30, 2020.

"Is Ecohumanism Possible?" Roundtable on *The Ecocentrists*, Society for US Intellectual History Blog, August 11, 2020, https://s-usih.org/2020/08/is-ecohumanism-possible-roundtable-on-the-ecocentrists-pt-1/.

Finally, this book wouldn't exist without Sara and Rosalind, who have traveled with me from long before we hit Chaco and who make the work both possible and meaningful. I owe them everything. Since I have already dedicated a previous book to Sara, it is to Rosalind I dedicate this one, with unbounded love: may it be of use in the troubles to come, and may it help you practice joy and compassion in even the worst of times. As my Jewish namesake Baruch Spinoza wrote, "Happiness is a virtue, not its reward." May it prove to be one of the virtues of pessimism.

NOTES

Preface

1. On evidence of cannibalism at Chaco Canyon, see Tim D. White, "Once Were Cannibals," *Scientific American* 285, no. 2 (2001): 58–65; Richard A. Marlar et al., "Biochemical Evidence of Cannibalism at a Prehistoric Puebloan Site in Southwestern Colorado," *Nature* 407, no. 6800 (2000): 74–78; and Christy G. Turner and Jacqueline A. Turner, *Man Corn: Cannibalism and Violence in the Prehistoric American Southwest* (Salt Lake City: University of Utah Press, 1999).

2. Washington Matthews, "Noqoìlpi, the Gambler: A Navajo Myth," *The Journal of American Folklore* 2, no. 5 (1889): 89–94. See Steven Lekson, *The Chaco Meridian: One Thousand Years of Religious Power in the Ancient Southwest*, 2nd ed. (Lanham, MD: Rowman and Littlefield, 2015), 148–149, for further citations. The Anasazi are today typically referred to as Ancient Puebloans, as the etymology and meaning of the Diné word *Anasazi* are controversial. According to the *Oxford English Dictionary*, the word *Öanaasází*, meaning "enemy ancestors" or "alien ancestors," is derived from the conjunction of *ÖanaaÖ*, meaning "war," and *Öazázi*, meaning "ancestor." *Oxford English Dictionary*, s.v. "Anasazi (n. & adj.), Etymology," July 2023, https://doi.org/10.1093/OED/4058701570. Leon Wall and William Morgan, *Navajo-English Dictionary* (New York: Hippocrene Books, 1994), define *'anaasázi* as "ancient people; enemy ancestors"; and Alyse Neundorf, *A Navajo/English Bilingual Dictionary: ÁŁchíNí Bi Naaltsoostsoh* (Albuquerque: University of New Mexico Press, 1983), defines *anaasází* as "cliff dwellers, ancient people; enemy." See also Harry Walters and Hugh C. Rogers, "Anasazi and 'Anaasází: Two Words, Two Cultures," *Kiva* 66, no. 3 (2001): 317–326, http://www.jstor.org/stable/30246362.

3. Immanuel Kant, *The Critique of Practical Reason*, trans. Mary Gregor (Cambridge: Cambridge University Press, 1997), 5:162. The original reads: "Zwei Dinge erfüllen das Gemüt mit immer neuer und zunehmender Bewunderung und Ehrfurcht, je öfter und anhaltender sich das Nachdenken damit beschäftigt: der bestirnte Himmel über mir und das moralische Gesetz in mir."

4. Dacher Keltner, "Why Do We Feel Awe?" *Greater Good Magazine*, May 10, 2016, https://greatergood.berkeley.edu/article/item/why_do_we_feel_awe. See also Paul K. Piff et al., "Awe, the Small Self, and Prosocial Behavior," *Journal of Personality and Social Psychology* 108, no. 6 (2015): 883–899, https://doi.org/10.1037/pspi0000018.

5. Edgar L. Hewett, *Ancient Life in the American Southwest: With an Introduction on the General History of the American Race* (Indianapolis: Bobbs-Merrill, 1930), 299.

6. Lekson, *The Chaco Meridian*, 14, 35–37.

7. See, for instance, John A. Ware, *A Pueblo Social History: Kinship, Solidarity, and Community in the Northern Southwest* (Santa Fe: School for Advanced Research Press, 2014), ch. 5.

8. While Hewett and earlier archaeologists estimated that the permanent population of Chaco Canyon may have passed 25,000 at its peak, later estimates have been more modest, ranging between 4,000 and 6,000, or even as low as 2,000. Kendrick Frazier, *People of Chaco: A Canyon and Its Culture*, rev. ed. (New York: W. W. Norton, 2005), 155–157.

9. Joseph A. Tainter, *The Collapse of Complex Societies* (Cambridge: Cambridge University Press, 1988), 184.

10. Barbara J. Mills, "What's New in Chaco Research?," *Antiquity* 92, no. 364 (2018): 855–869, doi: 10.15184/aqy.2018.132.

11. Kim McLean, *Chetro Ketl* (Tucson: Western National Parks Association, 1980).

12. Oregon State Parks, "Fort Stevens State Park," n.d., https://stateparks.oregon.gov/index.cfm?do=main.loadFile&load=_siteFiles%2Fpublications%2F%2F46556_Fort_Stevens_Historic_Guide_%28Web%29011632.pdf.

13. Paul Virilio, "Bunker Archaeology," from *Architecture Principe 1966 and 1996*, trans. George Collins (Besançon: Les Éditions de l'Imprimeur, 1997), reprinted in *The Paul Virilio Reader*, ed. Steve Redhead (Edinburgh: Edinburgh University Press, 2004), 12–13. See also Paul Virilio, *Bunker Archeology: Texts and Photos*, trans. George Collins (Princeton: Princeton Architectural Press, 1994).

14. François-René Chateaubriand, *The Genius of Christianity, Or, The Spirit and Beauty of the Christian Religion*, trans. Charles White (Baltimore: John Murphy & Co., 1856), 466–467.

15. Susan Stewart, *The Ruins Lesson: Meaning and Material in Western Culture* (Chicago: University of Chicago Press, 2019), 15.

16. Frazier, *People of Chaco*, 257.

17. Jill Neitzel, "The Organization, Function, and Population of Pueblo Bonito," in *Pueblo Bonito: Center of the Chacoan World*, ed. Jill E. Neitzel (Washington, D.C.: Smithsonian Books, 2003), 149.

18. Wikimedia Commons, "Pueblo Bonito—Chaco Culture National Historical Park, New Mexico, US," https://commons.wikimedia.org/wiki/File:PUEBLO_BON ITO_-_Chaco_Culture_National_Historical_Park_,_New_Mexico,_US_-_panora mio_%2816%29.jpg#:~:text=https%3A//web.archive.org/web/20161031145553/http% 3A//www.panoramio.com/photo/122549093.

19. See Lekson, *Chaco Meridian*; and Ruth M. Van Dyke, *The Chaco Experience: Landscape and Ideology at the Center Place* (Santa Fe: School for Advanced Research Press, 2007).

20. W. James Judge, "Chaco's Golden Century," in *In Search of Chaco: New Approaches to an Archaeological Enigma*, ed. David Grant Noble (Santa Fe: School of American Research Press, 2004), 1.

21. R. Gwinn Vivian and Bruce Hilpert, *The Chaco Handbook* (Salt Lake City: University of Utah Press, 2012), 10.

22. Edward A. Jolie and Laurie D. Webster, "A Perishable Perspective on Chacoan Social Identities," in *Chaco Revisited*, ed. Carrie C. Heitman and Stephen Plog (Tucson: University of Arizona Press, 2015), 96–133; W. H. Wills, "Cultural Identity and the Archaeological Construction of Historical Narratives: An Example from Chaco Canyon," *Journal of Archaeological Method and Theory* 16 (2009): 283–319, https://doi.org/10.1080/00231940.2017.1343109.

23. Frazier, *People of Chaco*, 181–182, 235–36. Judge, "Chaco's Golden Century," 2–6.

24. Tainter, *Collapse of Complex Societies*, 187.

25. Tainter, *Collapse of Complex Societies*, 37.

26. Tainter, *Collapse of Complex Societies*, 53.

27. For a helpful, multidisciplinary overview of more than four hundred articles and books on these questions, see Danilo Brozović, "Societal Collapse: A Literature Review," *Futures: The Journal of Policy, Planning and Futures Studies* 145 (2023), https://doi.org/10.1016/j.futures.2022.103075.

28. Donna Glowacki, *Living and Leaving: A Social History of Regional Depopulation in Thirteenth-Century Mesa Verde* (Tucson: University of Arizona Press, 2015), 3 (italics in original). See Shmuel Eisenstadt, "Beyond Collapse," in *The Collapse of Ancient States and Civilizations*, ed. Norman Yoffee and George L. Cowgill (Tucson: University of Arizona Press, 1988).

29. Ronald K. Faulseit, "Collapse, Resilience, and Transformation in Complex Societies," in *Beyond Collapse: Archaeological Perspectives on Resilience, Revitalization, and Transformation in Complex Systems*, ed. Ronald K. Faulseit (Carbondale: Southern Illinois University Press, 2015), 5. See also Joseph Tainter, "Why Collapse Is So Difficult to Understand," in *Beyond Collapse*.

Although most archaeologists use the term *collapse* in the ways just described, the term has been criticized. In *Questioning Collapse*, for instance, Patricia Ann McAnany and Norman Yoffee question the validity of the concept of collapse, asserting that "when closely examined, the overriding human story is one of survival and regeneration" and "rarely did societies collapse in an absolute and apocalyptic sense." To make this argument, they cite the claim made by Shmuel Eisenstadt that "societal collapse seldom occurs if collapse is taken to mean 'the complete end of those political systems and their accompanying civilizational framework'" and assert that when Joseph Tainter went looking for "archaeological evidence of societal 'overshoot' and collapse," he "arrived at a conclusion similar to Eisenstadt's: there wasn't any." Three things should be said here. First, as Tainter himself points out in his review of *Questioning Collapse*, neither McAnany and Yoffee nor the other authors in their collection define *collapse*. Without a clear definition of the term, it seems difficult to judge in any reasonable way whether it might apply to any given set of evidence. As Tainter puts it, "There seems little sense in questioning collapse if one has not defined it." Second, insofar as they do define the concept of collapse, relying on Eisenstadt, it's a concept tendentiously constructed so as to be useless. It's not clear what it would mean for a society to "absolutely" collapse unless it was completely eradicated by another society. What's more, their point relies on misreading Eisenstadt, who contends that human social organization needs to be viewed holistically and that collapse should be seen not as an absolute, but rather as part of "the larger problem of how social boundaries are restructured and reconstructed." That is, Eisenstadt's helpful point is that collapse is one way that human social organization changes, rather than the end of human social organization. This is no way invalidates the concept of collapse, but rather, as Eisenstadt notes, it helps us see what is worthwhile in continuing to study it. Third, McAnany and Yoffee misappropriate Tainter by suggesting that he somehow repudiates his own earlier work on societal collapse. It's true that in the article McAnany and Yoffee cite, Tainter questions the evidence for "overshoot" in the archaeological record as a primary driver of societal collapse, but he does so without in any way rejecting his own definition of societal collapse, while also leaving open the question of whether or not agricultural "intensification can continue indefinitely." The question of overshoot and collapse is, as he says, "a question in the wild." Patricia Ann McAnany and Norman Yoffee, "Why We Question Collapse," in *Questioning Collapse: Human Resilience, Ecological Vulnerability, and the Aftermath of Empire*, ed. Patricia Ann McAnany and Norman Yoffee (Cambridge: Cambridge University Press, 2010); Joseph Tainter, review of *Questioning Collapse: Human Resilience, Ecological Vulnerability, and the Aftermath of Empire* by Patricia A. McAnany and Norman Yoffee, *Human Ecology* 38, no. 5 (2010): 709–10; Shmuel Eisenstadt, "Beyond Collapse," 236; Joseph Tainter, "Archaeology of Overshoot and Collapse," *Annual Review of Anthropology* 35, no. 1 (2006): 59–74, https://doi.org/10.1146/annurev.anthro.35.081.

30. Tainter, *Collapse of Complex Societies*, 4. See also Colin Renfrew, *Approaches to Social Archaeology* (Cambridge: Harvard University Press, 1984); and Faulseit, "Collapse, Resilience, and Transformation."

31. See Vaclav Smil, *Energy and Civilization: A History* (Cambridge: MIT Press, 2017); Timothy Mitchell, *Carbon Democracy* (Verso, 2013); Vaclav Smil, *Energy in*

Nature and Society: General Energetics of Complex Systems (Cambridge: MIT Press, 2008); Helmut Haberl, "The Global Socioeconomic Energetic Metabolism as a Sustainability Problem," *Energy* 31, no. 1 (2006): 87–99; and Marina Fischer-Kowalski and Helmut Haberl, "Sustainable Development: Socio-Economic Metabolism and Colonization of Nature," *International Social Science Journal*, English edition 50, no. 158 (1998): 573–87, https://doi.org/10.1111/1468-2451.00169. This point is also central to sophisticated accounts of ecological Marxism. See for instance Kohei Saito, *Karl Marx's Ecosocialism: Capital, Nature, and the Unfinished Critique of Political Economy* (New York: NYU Press, 2017); Paul Burkett, *Marx and Nature: A Red and Green Perspective* (Chicago: Haymarket Books, 2014); and John Bellamy Foster, *Marx's Ecology: Materialism and Nature*.(New York: Monthly Review Press, 2000).

32. Tainter, *Collapse of Complex Societies*, 91.
33. Tainter, *Collapse of Complex Societies*, 92.
34. Tainter, *Collapse of Complex Societies*, 119–120.
35. Guy Middleton, *Understanding Collapse: Ancient History and Modern Myths* (Cambridge: Cambridge University Press, 2017), 30.
36. Tainter, *Collapse of Complex Societies*, 121.
37. Tainter, *Collapse of Complex Societies*, 186. It's worth noting that Chacoan collapse remains a topic of archaeological debate. While the evidence for the abandonment of Chaco Canyon is clear, current consensus holds that the Chacoan political system survived in some form for another two hundred years or so at Aztec Ruins, a colony, imitator, or dissident outpost built approximately fifty miles north on the Rio Animas. Aztec collapsed rather violently in its turn, which was either the end of the Chacoan political system or, according to the Chaco Meridian hypothesis advanced by Stephen Lekson, the beginning of an almost four-hundred-mile trek south to a new home in Paquimé, in what is now the state of Chihuahua, Mexico. Whatever really happened, it seems clear from Lekson's own evidence that even if Chacoan elites migrated to Paquimé, the political system that emerged there was distinct from the earlier Chacoan hegemony, even if in some sense its successor. See Lekson, *Chaco Meridian*; Glowacki, *Living and Leaving*; Paul F. Reed, *Chaco's Northern Prodigies: Salmon, Aztec, and the Ascendancy of the Middle San Juan Region after AD 1100* (Salt Lake City: University of Utah Press, 2008).
38. Tainter, *Collapse of Complex Societies*, 184.
39. Tainter, *Collapse of Complex Societies*, 186.
40. Tainter, *Collapse of Complex Societies*, 187.
41. Robert Redfield, "CDC Washington Testimony February 27, 2020," https://www.cdc.gov/washington/testimony/2020/t20200127.htm; Anthony S. Fauci et al., "Covid-19—Navigating the Uncharted," *New England Journal of Medicine* 382, no. 13 (2020): 1268–1269, doi: 10.1056/NEJMe2002387. Nicole Acevedo and Minyvonne Burke, "Washington State Man Becomes First U.S. Death from Coronavirus," NBC News, February 29, 2020, https://www.nbcnews.com/news/us-news/1st-coronavirus-death-u-s-officials-say-n1145931.

42. Leah Asmelash, "The Surgeon General Wants Americans to Stop Buying Face Masks," CNN, March 2, 2020, https://www.cnn.com/2020/02/29/health/face-masks-coronavirus-surgeon-general-trnd/index.html.

43. Bruce Y. Lee, "One Year After Coronavirus Pandemic Declared: How Many Deaths from COVID-19?," *Forbes*, March 13, 2021, https://www.forbes.com/sites/brucelee/2021/03/13/one-year-after-coronavirus-pandemic-declared-how-many-deaths-from-covid-19/?sh=7f30e3586900.

44. Mitchell, *Carbon Democracy*, 15.

45. On wicked problems, see C. West Churchman, "Guest Editorial: Wicked Problems," *Management Science* 14, no. 4 (1967): B141–B142; and Charles Lindblom, "The Science of Muddling Through," *Public Administration Review* 19, no. 2 (1959): 79–88. On super-wicked problems, see Kelly Levin et al., "Overcoming the Tragedy of Super Wicked Problems: Constraining Our Future Selves to Ameliorate Global Climate Change," *Policy Sciences* 45 (2012): 123–152; and Kelly Levin et al., "Playing It Forward: Path Dependency, Progressive Incrementalism, and the 'Super Wicked' Problem of Global Climate Change," paper presented at International Studies Association Convention, Chicago, IL, February 28–March 3, 2007. Bedour Alagraa, "The Interminable Catastrophe," *Offshoot*, March 1, 2021, https://offshootjournal.org/the-interminable-catastrophe/.

Introduction

1. James E. Hansen, Pushker Kharecha, and Makiko Sato, "Comments on Global Warming Acceleration, Sulfur Emissions, Observations," May 16, 2024, http://www.columbia.edu/~jeh1/mailings/2024/MayEmail.2024.05.16.pdf; James E. Hansen et al. "Global Warming in the Pipeline," *Oxford Open Climate Change* 3, no. 1 (2023): kgad008; Nico Wunderling et al., "Climate Tipping Point Interactions and Cascades: A Review," *Earth System Dynamics* 15, no. 1 (2024): 41–74; H. Damon Matthews and Seth Wynes, "Current Global Efforts Are Insufficient to Limit Warming to 1.5°C," *Science* 376, no. 6600 (2022): 1404–1409; Robert S. Pindyck, *Climate Future: Averting and Adapting to Climate Change* (Oxford: Oxford University Press, 2022); June-Yi Lee et al., "Chapter 4: Future Global Climate: Scenario-Based Projections and Near-Term Information," in *Climate Change 2021: The Physical Science Basis. Working Group I Contribution to the Sixth Assessment Report of the Intergovernmental Panel on Climate Change*, ed. Valérie Masson-Delmotte et al. (Cambridge: Cambridge University Press, 2021) (see especially 4.3); Chen Zhou et al., "Greater Committed Warming After Accounting for The Pattern Effect," *Nature Climate Change* 11, no. 2 (2021): 132–136, https://doi.org/10.1038/s41558-020-00955-x; Peiran R. Liu and Adrian E. Raftery, "Country-Based Rate of Emissions Reductions Should Increase by 80% Beyond Nationally Determined Contributions to Meet the 2°C Target," *Communications Earth & Environment* 2, no. 1 (2021), 29; Victor Brovkin et al., "Past Abrupt Changes, Tipping Points and Cascading Impacts in the Earth System," *Nature Geoscience* 14, no. 8 (2021): 550–558; S. C. Sherwood et al., "An Assessment of Earth's

Climate Sensitivity Using Multiple Lines of Evidence," *Reviews of Geophysics* 58, no. 4 (2020), e2019RG000678; Timothy M. Lenton et al., "Climate Tipping Points—Too Risky to Bet Against," *Nature* 575, no. 7784 (2019): 592–595, doi: 10.1038/d41586-019-03595-0; Adrian E. Raftery et al., "Less Than 2 °C Warming by 2100 Unlikely," *Nature Climate Change* 7, no. 9 (2017): 637–641, https://doi.org/10.1038/nclimate3352; Richard A. Betts et al., "When Could Global Warming Reach 4°C?," *Philosophical Transactions of the Royal Society A: Mathematical, Physical and Engineering Sciences* 369, no. 1934 (2011): 67–84; Mark New et al., "Four Degrees and Beyond: The Potential for a Global Temperature Increase of Four Degrees and Its Implications," *Philosophical Transactions of the Royal Society A: Mathematical, Physical and Engineering Sciences* 369, no. 1934 (2011): 6–19; Kevin Anderson and Alice Bows, "Beyond 'Dangerous' Climate Change: Emission Scenarios for a New World," *Philosophical Transactions of the Royal Society A: Mathematical, Physical and Engineering Sciences* 369, no. 1934 (2011): 20–44; and V. Ramanathan and Y. Feng. "On Avoiding Dangerous Anthropogenic Interference with the Climate System: Formidable Challenges Ahead," *Proceedings of the National Academy of Sciences* 105, no. 38 (2008): 14245–14250.

2. Diana Reckien et al., "Navigating the Continuum Between Adaptation and Maladaptation," *Nature Climate Change* 13, no. 9 (2023): 907–918; Sarah Kehler and S. Jeff Birchall, "Climate Change Adaptation: How Short-Term Political Priorities Trample Public Well-Being," *Environmental Science & Policy* 146 (2023): 144–150; World Bank, *Turn Down the Heat: Why A 4°C Warmer World Must Be Avoided* (Washington, D.C.: World Bank Group, 2012), http://documents.worldbank.org/curated/en/865571468149107611/Turn-down-the-heat-why-a-4-C-warmer-world-must-be-avoided; Mark Stafford Smith et al., "Rethinking Adaptation for a 4°C World," *Philosophical Transactions of the Royal Society A: Mathematical, Physical and Engineering Sciences* 369, no. 1934 (2011): 196–216; W. Neil Adger and Jon Barnett, "Four Reasons for Concern About Adaptation to Climate Change," *Environment and Planning A* 41, no. 12 (2009): 2800–2805; Stéphane Hallegatte, "Strategies to Adapt to an Uncertain Climate Change," *Global Environmental Change* 19, no. 2 (2009): 240–247.

3. Christopher H. Trisos et al., "The Projected Timing of Abrupt Ecological Disruption from Climate Change," *Nature* 580, no. 7804 (2020): 496–501. See also National Research Council, Committee on Abrupt Climate Change, *Abrupt Climate Change: Inevitable Surprises* (Washington, D.C.: National Academy Press, 2002).

4. Coal use reached new highs in 2022 and 2023. International Energy Agency, *Coal 2023: Analysis and Forecast to 2026*, https://www.iea.org/reports/coal-2023. Andrew L. Fanning et al., "The Social Shortfall and Ecological Overshoot of Nations," *Nature Sustainability* 5, no. 1 (2022): 26–36.

5. Thorfinn Stainforth and Bartosz Brzezinski, "More Than Half of All CO2 Emissions Since 1751 Emitted in the Last 30 Years," Institute for European Environmental Policy, April 29, 2020, https://ieep.eu/news/more-than-half-of-all-co2-emissions-since-1751-emitted-in-the-last-30-years.

6. John Mecklin, "2023 Doomsday Clock Statement," *Bulletin of the Atomic Sci-*

entists, https://thebulletin.org/doomsday-clock/current-time/. On the instability of the global financial system, see Adam Tooze, *Crashed: How a Decade of Financial Crises Changed the World* (New York: Penguin Books, 2018); and Adam Tooze, *Shutdown: How Covid Shook the World's Economy* (New York: Penguin Books, 2021).

7. Daniel Steel et al., "Climate Change and the Threat to Civilization," *Proceedings of the National Academy of Sciences* 119, no. 42 (2022): 1–e2210525119, https://doi.org/10.1073/pnas.2210525119.

8. "Consensus is growing that elevated CO_2 concentration is the immediate kill mechanism of most and perhaps all mass extinctions." W. Jackson Davis, "Mass Extinctions and Their Relationship with Atmospheric Carbon Dioxide Concentration: Implications for Earth's Future," *Earth's Future* 11, no. 6 (2023): e2022EF003336. See also James W. B. Rae et al., "Atmospheric CO_2 over the Past 66 Million Years from Marine Archives," *Annual Review of Earth and Planetary Sciences* 49 (2021): 609–641; Yuyang Wu et al., "Six-Fold Increase of Atmospheric pCO_2 During the Permian–Triassic Mass Extinction," *Nature Communications* 12, no. 1 (2021): 2137; Manfredo Capriolo et al., "Deep CO_2 in the End-Triassic Central Atlantic Magmatic Province," *Nature Communications* 11, no. 1 (2020): 1670; Hana Jurikova et al., "Permian–Triassic Mass Extinction Pulses Driven by Major Marine Carbon Cycle Perturbations," *Nature Geoscience* 13, no. 11 (2020): 745–750; and B. Van de Schootbrugge et al., "Carbon Cycle Perturbation and Stabilization in the Wake of the Triassic-Jurassic Boundary Mass-Extinction Event," *Geochemistry, Geophysics, Geosystems* 9, no. 4 (2008).

9. The literature on climate in history is voluminous. For a cursory introduction, see Geoffrey Parker, *Global Crisis: War, Climate Change and Catastrophe in the Seventeenth Century* (New Haven: Yale University Press, 2013); Robert Costanza et al., eds., *Sustainability or Collapse?: An Integrated History and Future of People on Earth* (Cambridge: MIT Press, 2007); Roderick J McIntosh et al., *The Way the Wind Blows: Climate History and Human Action* (New York: Columbia University Press, 2000); Emmanuel Le Roy Ladurie, *Times of Feast, Times of Famine: A History of Climate Since the Year 1000* (New York: Farrar, Straus and Giroux, 1988); and Gustaf Utterström, "Climatic Fluctuations and Population Problems in Early Modern History," *Scandinavian Economic History Review* 3, no. 1 (1955): 3–47.

10. Vaclav Smil, *How the World Really Works: The Science Behind How We Got Here and Where We're Going* (New York: Viking, 2022); Simon P. Michaux, *The Mining of Minerals and the Limits to Growth* (Espoo, FI: Geological Survey of Finland, 2021); Louis Delannoy et al., "Peak Oil and the Low-Carbon Energy Transition: A Net-Energy Perspective," *Applied Energy* 304 (2021): 117843; Megan K. Seibert and William E. Rees, "Through the Eye of a Needle: An Eco-heterodox Perspective on the Renewable Energy Transition." *Energies* 14, no. 15 (2021): 4508; Simon P. Michaux, "Assessment of the Extra Capacity Required of Alternative Energy Electrical Power Systems to Completely Replace Fossil Fuels," GTK Open Work File internal report, serial no. 42/2021 (2021), https://tupa.gtk.fi/raportti/arkisto/42_2021.pdf; Jason

Hickel and Giorgos Kallis, "Is Green Growth Possible?" *New Political Economy* 25, no. 4 (2020): 469–86, https://doi.org/10.1080/13563467.2019.1598964; Helmut Haberl et al., "A Systematic Review of the Evidence on Decoupling of GDP, Resource Use and GHG Emissions, Part II: Synthesizing the Insights," *Environmental Research Letters* 15, no. 6 (2020): 65003, https://doi.org/10.1088/1748-9326/ab842a; Dominik Wiedenhofer et al., "A Systematic Review of the Evidence on Decoupling of GDP, Resource Use and GHG Emissions, Part I: Bibliometric and Conceptual Mapping," *Environmental Research Letters* 15, no. 6 (2020): 63002, https://doi.org/10.1088/1748-9326/ab8429.

11. Vaclav Smil, "What We Need to Know About the Pace of Decarbonization," *Substantia* 3, no. 2, suppl. 1 (2019): 13–28, doi: 10.13128/Substantia-XXX.

12. I am not the first to make this claim. See Marcello Di Paola and Sven Nyholm, "Climate Change and Anti-Meaning," *Ethical Theory and Moral Practice* 26, no. 5 (2023): 709–724; Paul N. Edwards, "Is Climate Change Ungovernable?," *Proceedings of the Stanford Existential Risks Conference* 2023, 133–146, https://doi.org/10.25740/yc096zw4572; Susanne C. Moser, "The Work After 'It's Too Late' (to Prevent Dangerous Climate Change)," *Wiley Interdisciplinary Reviews: Climate Change* 11, no. 1 (2020): e606; Robert J. Brulle and Kari Marie Norgaard, "Avoiding Cultural Trauma: Climate Change and Social Inertia," *Environmental Politics* 28, no. 5 (2019), doi: 10.1080/09644016.2018.1562138; Amitav Ghosh, *The Great Derangement: Climate Change and the Unthinkable* (Chicago: University of Chicago Press, 2016); Timothy Clark, *Ecocriticism on the Edge: The Anthropocene as a Threshold Concept* (London: Bloomsbury, 2015); Dale Jamieson, *Reason in a Dark Time: Why the Struggle Against Climate Change Failed—and What It Means for Our Future* (Oxford: Oxford University Press, 2014); Gernot Wagner and Richard J. Zeckhauser, "Climate Policy: Hard Problem, Soft Thinking," *Climatic Change* 110, no. 3–4 (2012): 507–521; Clive Hamilton, *Requiem for a Species: Why We Resist the Truth About Climate Change* (Washington, D.C.: Earthscan, 2010).

13. Economist Robert Pyndick has gone so far as to argue that flaws in climate models make them close to useless and that "IAM-based analyses of climate policy create a perception of knowledge and precision that is illusory and can fool policymakers into thinking that the forecasts the models generate have some kind of scientific legitimacy." Robert S. Pindyck, "The Use and Misuse of Models for Climate Policy," *Review of Environmental Economics and Policy* 11, no. 1 (2017): 100–114, doi: 10.1093/reep/rew012. See also V. Balaji et al., "Are General Circulation Models Obsolete?," *Proceedings of the National Academy of Sciences* 119, no. 47 (2022): e2202075119; Casey J. Wall et al., "Assessing Effective Radiative Forcing from Aerosol–Cloud Interactions over the Global Ocean," *Proceedings of the National Academy of Sciences* 119, no. 46 (2022): 1–e2210481119, https://doi.org/10.1073/pnas.2210481119; Jorge Sebastián Moraga et al., "Uncertainty in High-Resolution Hydrological Projections: Partitioning the Influence of Climate Models and Natural Climate Variability," *Hydrological Processes* 36, no. 10 (2022): e14695, https://doi.org/10

.1002/hyp.14695; Zeke Hausfather et al., "Climate Simulations: Recognize the 'Hot Model' Problem," *Nature* 605, no. 7908 (2022): 26–29; Yue Dong et al., "Two-Way Teleconnections between the Southern Ocean and the Tropical Pacific via a Dynamic Feedback," *Journal of Climate* 35, no. 19 (2022): 6267–6282, https://doi.org/10.1175/JCLI-D-22-0080.1; Yue Dong et al., "Biased Estimates of Equilibrium Climate Sensitivity and Transient Climate Response Derived from Historical CMIP6 Simulations," *Geophysical Research Letters* 48 (2021): e2021GL095778, https://doi.org/10.1029/2021GL095778; Charles F. Manski, Alan H. Sanstad, and Stephen J. DeCanio, "Addressing Partial Identification in Climate Modeling and Policy Analysis," *Proceedings of the National Academy of Sciences* 118, no. 15 (2021): e2022886118; Mark D. Zelinka et al., "Causes of Higher Climate Sensitivity in CMIP6 Models," *Geophysical Research Letters* 47, no. 1 (2020): e2019GL085782; Clara Deser, "Certain Uncertainty: The Role of Internal Climate Variability in Projections of Regional Climate Change and Risk Management," *Earth's Future* 8, no. 12 (2020): e2020EF001854, https://doi: 10.1029/2020EF001854; Zeke Hausfather et al., "Evaluating the Performance of Past Climate Model Projections," *Geophysical Research Letters* 47, no. 1 (2020): e2019GL085378; Nicholas Stern, "Economics: Current Climate Models Are Grossly Misleading," *Nature* 530 (2016): 407–409, doi: 10.1038/530407a; Simone Fatichi et al., "Uncertainty Partition Challenges the Predictability of Vital Details of Climate Change," *Earth's Future* 4, no. 5 (2016): 240–251, doi: 10.1002/2015EF000336; Mark Maslin and Patrick Austin, "Climate Models at Their Limit?," *Nature* 486 (2012): 183–184, https://doi.org/10.1038/486183a; Tapio Schneider, "Uncertainty in Climate-Sensitivity Estimates," *Nature* 446, no. 7131 (2007): E1; Jouni Raäisaänen, "How Reliable Are Climate Models?" *Tellus A: Dynamic Meteorology and Oceanography* 59, no. 1 (2007): 2–29, doi: 10.1111/j.1600-0870.2006.00211.x.

14. Deyshawn Moser et al., "Facing Global Environmental Change: The Role of Culturally Embedded Cognitive Biases," *Environmental Development* 44 (2022): 100735.

15. On the concept of planetary boundaries for human flourishing, see Johan Rockström et al., "Safe and Just Earth System Boundaries," *Nature* (2023): 1–10; Frank Biermann and Rakhyun E. Kim, "The Boundaries of the Planetary Boundary Framework: A Critical Appraisal of Approaches to Define a 'Safe Operating Space' for Humanity," *Annual Review of Environment and Resources* 45, no. 1 (2020): 497–521; Daniel W. O'Neill et al., "A Good Life for All Within Planetary Boundaries," *Nature Sustainability* 1, no. 2 (2018): 88–95; Johan Rockström et al., "A Safe Operating Space for Humanity." *Nature* 461, no. 7263 (2009): 472–475; Johan Rockström et al., "Planetary Boundaries: Exploring the Safe Operating Space for Humanity," *Ecology and Society* 14, no. 2 (2009).

16. Bruce C. Glavovic et al., "The Tragedy of Climate Change Science," *Climate and Development* 14, no. 9 (2022): 829–833, doi: 10.1080/17565529.2021.2008855; Luke Kemp et al., "Climate Endgame: Exploring Catastrophic Climate Change Scenarios," *Proceedings of the National Academy of Sciences* 119, no. 34 (2022): 1–e2108146119,

https://doi.org/10.1073/pnas.2108146119; Christopher Lyon, et al., "Climate Change Research and Action Must Look Beyond 2100," *Global Change Biology* 28, no. 2 (2022): 349–361; Naomi Oreskes, "IPCC, You've Made Your Point: Humans Are a Primary Cause of Climate Change," *Scientific American*, November 1, 2021, https://www.scientificamerican.com/article/ipcc-youve-made-your-point-humans-are-a-primary-cause-of-climate-change/.

17. The first four IPCC reports showed an approximately 50-50 split between mentions of temperature outcomes below 2°C and those above. The 2013 AR5 showed a jump of about 10% in favor of the lower temperatures, and the 2021 AR6 showed an astonishing 20% shift, from approximately 60-40 in favor of lower temperatures to approximately 80-20. Florian Jehn et al., "Focus of the IPCC Assessment Reports Has Shifted to Lower Temperatures," *Earth's Future* 10, no. 5 (2022): e2022EF002876, https://doi.org/10.1029/2022EF002876.

18. As Paul Edwards writes, "No matter how dire much of it may appear, political analysis of climate issues generally tends toward optimism." Edwards, "Is Climate Change Ungovernable?"

19. Eduardo Viveiros De Castro, "Cosmological Deixis and Amerindian Perspectivism," *Journal of the Royal Anthropological Institute* (1998): 469–488, 469. For more on perspectivism, see Michela Massimi and Casey D. McCoy, *Understanding Perspectivism: Scientific Challenges and Methodological Prospects* (New York: Taylor & Francis, 2020); Evandro Agazzi, "Scientific Realism Within Perspectivism and Perspectivism Within Scientific Realism," *Axiomathes* 26 (2016): 349–365; Ronald N. Giere, *Scientific Perspectivism* (Chicago: University of Chicago Press, 2006); and Friedrich Nietzsche, *The Gay Science; with a Prelude in Rhymes and an Appendix of Songs*, trans. Walter Kaufmann (New York: Random House, 1974).

20. Stephen M. Gardiner, "Ethics and Global Climate Change," *Ethics* 114, no. 3 (2004): 555–600, https://doi.org/10.1086/382247, 556.

21. Jamieson, *Reason in a Dark Time*, xii.

22. Plato, *Republic*, 6.496D, trans. Paul Shorey.

23. "2020 Atlantic Hurricane Season Takes Infamous Top Spot for Busiest on Record," National Oceanic and Atmospheric Administration, November 10, 2020, https://www.noaa.gov/news/2020-atlantic-hurricane-season-takes-infamous-top-spot-for-busiest-on-record; Blacki Migliozzi et al., "Record Wildfires on the West Coast Are Capping a Disastrous Decade," *The New York Times*, September 24, 2020, https://www.nytimes.com/interactive/2020/09/24/climate/fires-worst-year-california-oregon-washington.html; Kate Ramsayer, "2020 Arctic Sea Ice Minimum at Second Lowest on Record," NASA Global Climate Change, September 21, 2020, https://climate.nasa.gov/news/3023/2020-arctic-sea-ice-minimum-at-second-lowest-on-record; Liz Kimbrough, "As 2020 Amazon Fire Season Winds Down, Brazil Carbon Emissions Rise," *Mongabay*, November 16, 2020, https://news.mongabay.com/2020/11/as-2020-amazon-fire-season-winds-down-brazil-carbon-emissions-rise/; Jonathan Watts, "Arctic Methane Deposits 'Starting to Release,' Scientists Say," *The

Guardian, October 27, 2020, https://www.theguardian.com/science/2020/oct/27/sleeping-giant-arctic-methane-deposits-starting-to-release-scientists-find; "July 2020 Was Record Hot for N. Hemisphere, 2nd Hottest for Planet," National Oceanic and Atmospheric Administration, August 13, 2020, https://www.noaa.gov/news/july-2020-was-record-hot-for-n-hemisphere-2nd-hottest-for-planet; Andrea Thompson, "NASA Says 2020 Tied for Hottest Year on Record," *Scientific American*, January 14, 2021, https://www.scientificamerican.com/article/2020-will-rival-2016-for-hottest-year-on-record/; Damien Cave, "Great Barrier Reef's Bleaching and Dying," *The New York Times*, April 6, 2020, https://www.nytimes.com/2020/04/06/world/australia/great-barrier-reefs-bleaching-dying.html; Moira Warburton, "Canada's Last Fully Intact Arctic Ice Shelf Collapses," *Reuters*, August 6, 2020, https://www.reuters.com/article/us-climate-change-canada/canadas-last-fully-intact-arctic-ice-shelf-collapses-idUSKCN2523JH; Alexandra Witze, "Why Arctic Fires are Bad News for Climate Change," *Nature* 585 (September 17, 2020): 336–337, https://doi.org/10.1038/d41586-020-02568-y.

24. Euan G. Nisbet et al., "Very Strong Atmospheric Methane Growth in the 4 Years 2014–2017: Implications for the Paris Agreement," *Global Biogeochemical Cycles* 33 (2019): 318–342, https://doi.org/10.1029/2018GB006009.

25. Tooze, *Shutdown*, 293–294.

Chapter 1

1. Frank Kermode, *The Sense of an Ending: Studies in the Theory of Fiction* (Oxford: Oxford University Press, 2000), 5.

2. Richard Webster, "Frank Kermode's 'The Sense of an Ending,'" *The Critical Quarterly* 16, no. 4 (1974): 311. See also Hayden White, "Historical Fictions: Kermode's Idea of History," *The Critical Quarterly* 54, no. 1 (2012): 43–59; Joseph Frank, "His Sense of an Ending: In Memory of Frank Kermode," *Common Knowledge* 17, no. 3 (2011): 427–432; Margaret Doody, "Finales, Apocalypses, Trailings-Off," *Raritan* 15, no. 3 (1996): 24–46; and Roy Pascal, "Narrative Fictions and Reality: A Comment on Frank Kermode's 'The Sense of an Ending,'" *Novel: A Forum on Fiction* 11, no. 1 (1977): 40–50.

3. Kermode was, however, one of the transitional figures who helped bring French theory into the study of English literature. *Aevum* is a theological term Kermode adapted from scholastic philosophy denoting angelic time.

4. Terry Eagleton, *Literary Theory: An Introduction* (Minneapolis: University of Minnesota, 2008), 78.

5. Kermode, *The Sense of an Ending*, 99.

6. Jiawei Da et al., "Low CO_2 Levels of the Entire Pleistocene Epoch," *Nature Communications* 10 (2019): 4342, https://doi.org/10.1038/s41467-019-12357-5.

7. Hannah Ritchie, Pablo Rosado, and Max Roser, "CO_2 and Greenhouse Gas Emissions," OurWorldInData.Org, 2023, https://OurWorldInData.org/co2-and-greenhouse-gas-emissions. "Climate at a Glance: Global Time Series," NOAA Na-

tional Centers for Environmental Information, accessed July 25, 2023, https://www.ncei.noaa.gov/access/monitoring/climate-at-a-glance/global/time-series.

8. See Michael Lawrence et al., "Global Polycrisis: The Causal Mechanisms of Crisis Entanglement," *Global Sustainability* 7 (2024): e6, https://doi.org/10.1017/sus.2024.1; and Steven Schneider, "Climate Change and the World Predicament: A Case Study for Interdisciplinary Research," *Climatic Change* 1, no. 1 (1977): 21–43, https://doi.org/10.1007/BF00162775. On the term *impasse*, see Alexander F. Gazmararian and Dustin Tingley, *Uncertain Futures: How to Unlock the Climate Impasse* (Cambridge University Press, 2023); Marco Grasso and J. Timmons Roberts, "A Compromise to Break the Climate Impasse," *Nature Climate Change* 4, no. 7 (2014): 543–549; Chad Shomura, "Vital Impasse: Animacy Hierarchies, Irredeemability, and a Life Otherwise," *American Quarterly* 73, no. 4 (2021): 835–855; Mark Simpson and Imre Szeman, "Impasse Time," *South Atlantic Quarterly* 120, no. 1 (2021): 77–89; Mark Simpson, "Plastic in the Time of Impasse," in *Plastics, Environment, Culture and the Politics of Waste*, ed. Tatiana Prorokova-Konrad (Edinburgh: Edinburgh University Press, 2023), 338–360; and Casey A. Williams, "Climate Impasse, Fossil Hegemony, and the Modern Crisis of Imagination," PhD diss., Duke University, 2022, Proquest.

9. Stephen Gardiner calls this "moral corruption." Stephen Gardiner, *A Perfect Moral Storm: The Ethical Tragedy of Climate Change* (Oxford: Oxford University Press, 2011).

10. Kermode, *The Sense of an Ending*, 7.

11. Charles Taylor's "social imaginary" and Eviatar Zerubavel's "social mindscape" express similar ideas, as does Paul Bains's "semiotic reality," though it's important to note that the Heideggerian and Uexküllian sense of "world" I'm sketching here includes not merely social reality but reality as it is experienced as a total phenomenological whole. Charles Taylor, *Modern Social Imaginaries* (Durham: Duke University Press, 2004); Eviatar Zerubavel, *Social Mindscapes* (Cambridge: Harvard University Press, 1999); Paul Bains, "Umwelten," *Semiotica* 134, no. 134 (2001): 137–167, https://doi.org/10.1515/semi.2001.020. See also Jakob von Uexküll, *A Foray into the Worlds of Animals and Humans, with A Theory of Meaning*, trans. Joseph D. O'Neil (Minneapolis: University of Minnesota Press, 2010).

12. Mikhail M. Bakhtin, "Forms of Time and of the Chronotope in the Novel: Notes toward a Historical Poetics," in *The Dialogic Imagination: Four Essays* (Austin: University of Texas Press, 1981), 84–85.

13. Edward O. Wilson, *The Social Conquest of Earth* (New York: Liveright, 2012), 213, 217.

14. Kermode, *The Sense of an Ending*, 17.

15. Kermode, *The Sense of an Ending*, 8–9.

16. Kermode, *The Sense of an Ending*, 98.

17. Kermode, *The Sense of an Ending*, 29.

18. Alison McQueen, *Political Realism in Apocalyptic Times* (Cambridge: Cambridge University Press, 2017), 24, 26.

19. Kermode, *The Sense of an Ending*, 37.

20. Vaihinger was important for Wallace Stevens, which may be what drew Kermode to him.

21. Hans Vaihinger, *The Philosophy of "As If"* (New York: Routledge, 2021), 7.

22. Patricia Lockwood, *No One Is Talking about This* (New York: Riverhead Books, 2021), 143.

23. Kermode, *The Sense of an Ending*, 38.

24. Wallace Stevens, "Of Modern Poetry," *Collected Poetry and Prose* (New York: Library of America, 1997), 218–219.

25. For a rich literary portrayal of just this, see Thomas Mann's tetralogy *Joseph and His Brothers* (1933–1943), especially the first two volumes.

26. Friedrich Nietzsche, *The Will to Power*, trans. Walter Kaufmann and R. J. Hollingdale (Random House, 1967), § 481.

27. This challenge is addressed with care by thinkers in the American pragmatist tradition, particularly Charles Peirce, William James, and Richard Rorty.

28. Kermode, *The Sense of an Ending*, 164.

29. On humanism as self-domestication, see Peter Sloterdijk, "Rules for the Human Zoo: A Response to the Letter on Humanism," *Environment and Planning D: Society and Space* 27, no. 1 (2009): 12–28.

30. Kermode, *The Sense of an Ending*, 3.

31. Kermode, *The Sense of an Ending*, 124.

32. Zygmunt Bauman, *Legislators and Interpreters: On Modernity, Post-modernity, and Intellectuals* (Ithaca: Cornell University Press, 1987), 17. See also Barbara Ehrenreich and John Ehrenreich's well-known and useful analysis of the Professional-Managerial Class (PMC), which they identify as "salaried mental workers who do not own the means of production and whose major function in the social division of labor may be described broadly as the reproduction of capitalist culture and capitalist class relations." The PMC might be seen as the Fordist expansion of the intellectual class identified by Zygmunt Bauman, with the difference that while earlier intellectuals largely mediated bourgeois self-understanding (Bauman's "legislators"), the PMC (Bauman's "interpreters") adds the additional roles of managing and producing working class culture and mediating between the bourgeoisie and the proletariat. The fate of the PMC is ambiguous and beyond the scope of this note, but it is worth noting how far the analysis goes toward explaining contemporary class divisions in the United States so often misidentified as racial or cultural. Barbara Ehrenreich and John Ehrenreich, "The Professional-Managerial Class," *Radical America* 11, no. 2 (1977): 7–31; and "The New Left: A Case Study in Professional-Managerial Class Radicalism," *Radical America* 11, no. 3 (1977): 7–22.

33. Karl Marx, "Preface to *A Contribution to the Critique of Political Economy*," in *Marx: Later Political Writings*, ed. Terrell Carver (Cambridge: Cambridge University Press, 1996), 158–162, 160.

34. Walter Benjamin, *Illuminations*, trans. Harry Zohn (New York: Schocken, 1978), 253.

35. Gilles Deleuze and Félix Guattari, *A Thousand Plateaus: Capitalism and Schizophrenia*, vol. 2 (Minneapolis: University of Minnesota Press, 1987), 91.

36. "Language is not first of all a communication system. Language is first of all a way of modeling the world according to possibilities envisioned as alternative to what is given in sensation or experienced in perception." John Deely, "Umwelt," *Semiotica* 134, no. 134 (2001): 125–135, https://doi.org/10.1515/semi.2001.019.

37. W. V. Quine, "Main Trends in Recent Philosophy: Two Dogmas of Empiricism," *The Philosophical Review* 60, no. 1 (1951): 20–43, https://doi.org/10.2307/2181906, 39.

38. Friedrich Nietzsche, *Daybreak: Thoughts on the Prejudices of Morality*, trans. R. J. Hollingdale (Cambridge: Cambridge University Press, 1997), §564.

39. Alfred North Whitehead, *Process and Reality*, corr. ed. (New York: Free Press, 1978), 167.

40. Whitehead, *Process and Reality*, 15.

41. Bauman, *Legislators and Interpreters*, 49.

42. Bauman, *Legislators and Interpreters*, 93. See also Norbert Elias, *The Civilizing Process: The History of Manners*, trans. Edmund Jephcott (New York: Urizen Books, 1978), 3–4.

43. Bauman, *Legislators and Interpreters*, 147.

44. Sylvia Wynter, "Unsettling the Coloniality of Being/Power/Truth/Freedom: Towards the Human, After Man, Its Overrepresentation—An Argument," *CR: The New Centennial Review* 3, no. 3 (2003): 257–337, https://doi.org/10.1353/ncr.2004.0015, 271.

45. Wynter, "Unsettling the Coloniality of Being/Power/Truth/Freedom," 269.

46. On the debate, see, in addition to those following, Lars Kirkhusmo Pharo, "The Council of Valladolid (1550–1551): A European Disputation About the Human Dignity of Indigenous Peoples of the Americas," in *The Cambridge Handbook of Human Dignity: Interdisciplinary Perspectives*, ed. Marcus Düwell et al. (Cambridge: Cambridge University Press, 2014), 95–100; and Ángel Losada, "The Controversy Between Sepúlveda and Las Casas in the Junta of Valladolid," in *Bartolomé de Las Casas in History: Toward an Understanding of the Man and His Work*, ed. Juan Friede and Benjamin Keen (DeKalb: Northern Illinois University Press, 1971).

47. Lewis Hanke, *All Mankind Is One: A Study of the Disputation Between Bartolomé de Las Casas and Juan Ginés de Sepúlveda in 1550 on the Intellectual and Religious Capacity of the American Indians* (DeKalb: Northern Illinois University Press, 1974), 9.

48. "The judges at Valladolid . . . fell into argument with one another and reached no collective decision." Hanke, *All Mankind Is One*, 113. See also Francisco Castilla Urbano, "The Debate of Valladolid (1550–1551): Background, Discussions, and Results of the Debate Between Juan Ginés de Sepúlveda and Bartolomé de las Casas," in *A Companion to Early Modern Spanish Imperial Political and Social Thought* (Leiden: Brill, 2020), 245–247.

49. David M. Lantigua, *Infidels and Empires in a New World Order: Early Modern Spanish Contributions to International Legal Thought* (Cambridge: Cambridge University Press, 2020), 141.

50. Bauman, *Legislators and Interpreters*, 93.

51. Lantigua, *Infidels and Empires*, 147.

52. Lantigua, *Infidels and Empires*, 176.

53. The *Apologética* was composed sometime between 1527 and 1561, though the issue of whether it was composed before or after Valladolid is an important scholarly question. John L. Phelan, "The Apologetic History of Fray Bartolomé de las Casas," *Hispanic American Historical Review* 49, no. 1 (February 1969): 94–99, https://doi.org/10.1215/00182168-49.1.94. Anthony Pagden, *The Fall of Natural Man: The American Indian and the Origins of Comparative Ethnology* (Cambridge: Cambridge University Press, 1982).

54. Lantigua, *Infidels and Empires*, 162 et passim.

55. It was this basis in theological rather than humanitarian reasoning that allowed Las Casas to argue for transporting African slaves to the Americas (although he later recanted this position), since the heathen slaves in question had already been enslaved by other heathen Africans. For Las Casas, a heathen Mohammedan or African who had rejected Christ and was thus an enemy of the church was legally distinct from a pagan Indian who lived in ignorance of Christ. See Lawrence Clayton, "Bartolomé de las Casas and the African Slave Trade," *History Compass* 7, no. 6 (2009): 1526–1541, https://doi.org/10.1111/j.1478-0542.2009.00639.x.

56. "We are somewhat at a loss to understand the real object of this meeting," writes historian Henry R. Wagner. "By now the conquest of the Indies was a *fait accompli*; no one could have sanctioned abandoning those lands, even had the conquests been declared unjust. The only real question of importance was how the Indians were to be governed. Why then was the junta called?" Perhaps, as Wagner suggests, the debate may have been intended to salve the emperor's troubled conscience. Nevertheless, there was clearly more at stake, as Wynter and others argue: namely, reweaving the narrative of European Christendom into a narrative of rational European civilization. Henry R. Wagner and Helen Rand Parish, *The Life and Writings of Bartolome de Las Casas* (Albuquerque: University of New Mexico Press, 1967), 177–178.

57. Wynter, "Unsettling the Coloniality of Being/Power/Truth/Freedom," 273.

58. Sylvia Wynter, "The Ceremony Must Be Found: After Humanism," *Boundary 2* 12/13, no. 3 (1984): 56, https://doi.org/10.2307/302808.

59. Wynter, "Unsettling the Coloniality of Being/Power/Truth/Freedom," 328–329. See also Aimé Césaire, "Poetry and Knowledge," *Sulfur* 5, no. 5 (1982): 24–25, which Wynter cites here.

60. See Zimitri Erasmus, "Sylvia Wynter's Theory of the Human: Counter-, Not Post-Humanist," *Theory, Culture & Society* 37, no. 6 (2020): 47–65, https://doi.org/10.1177/0263276420936333; and Bedour Alagraa, "Homo Narrans and the Science of the Word: Toward a Caribbean Radical Imagination," *Critical Ethnic Studies* 4, no. 2 (2018): 166–167, https://doi.org/10.5749/jcritethnstud.4.2.0164.

61. "Neither science nor the arts can be complete without combining their sepa-

rate strengths. Science needs the intuition and metaphorical power of the arts, and the arts needs the fresh blood of science." Edward O. Wilson, *Consilience: The Unity of Knowledge* (New York: Knopf, 1998), 211.

62. Shoshana Zuboff, *The Age of Surveillance Capitalism: The Fight for a Human Future at the New Frontier of Power* (PublicAffairs, 2019). See also Jacob Seigel, "Invasion of the Fact-Checkers," *Tablet*, March 21, 2022, https://www.tabletmag.com/sections/news/articles/invasion-fact-checkers; and Jacob Seigel, "Learn this Term: Whole of Society," *Tablet*, July 25, 2024, https://www.tabletmag.com/sections/news/articles/whole-society-american-politics.

63. Frantz Fanon, *The Wretched of the Earth*, trans. Richard Philcox (New York: Grove, 2005), 145.

64. Karl Marx, *The Eighteenth Brumaire of Louis Bonaparte* (New York: International Publishing, 1994), 15.

65. Michel Foucault, *The Order of Things: An Archaeology of the Human Sciences* (New York: Pantheon Books, 1971), 422.

66. For a careful analysis of this impulse, see Dipesh Chakrabarty, "Postcolonial Studies and the Challenge of Climate Change," *New Literary History* 43, no. 1 (2012): 1–18, https://doi.org/10.1353/nlh.2012.0007.

67. Claire Colebrook, "What Is the Anthropo-political?," in *Twilight of the Anthropocene Idols*, ed. J. Hillis Miller et al. (London: Open Humanities Press, 2016), 84.

68. See John D. Niles, *Homo Narrans: The Poetics and Anthropology of Oral Literature* (Philadelphia: University of Pennsylvania Press, 1999); and Kurt Ranke, "Problems of Categories in Folk Prose," trans. Carl Lindahl, *Folklore Forum* 14, no. 1 (1981).

69. Daniel Kahneman, *Thinking, Fast and Slow* (New York: Farrar, Straus and Giroux, 2011), 114.

70. Kahneman, *Thinking, Fast and Slow*, 74–78 et passim; Fernando Blanco, "Positive and Negative Implications of the Causal Illusion," *Consciousness and Cognition* 50 (2017): 56–68, https://doi.org/10.1016/j.concog.2016.08.012; see also David Hume, *An Enquiry Concerning Human Understanding* (Oxford: Oxford University Press, 2007).

71. Amos Tversky and Daniel Kahneman, "Judgment Under Uncertainty: Heuristics and Biases," *Science* 185, no. 4157 (1974): 1124–1131, https://doi.org/10.1126/science.185.4157.1124.

72. See Martie G. Haselton et al., "The Evolution of Cognitive Bias," in *The Handbook of Evolutionary Psychology*, ed. David M. Buss (Hoboken: John Wiley, 2005), 724–746.

73. Jorge Luis Borges, "Funes, His Memory," in *Collected Fiction*, trans. Andrew Hurley (New York: Viking, 1998), 131. According to neuroscientist Rodrigo Quian Quiroga, "Borges described very precisely the problems of distorted memory capacities well before neuroscience caught up." Rodrigo Quian Quiroga, "In Retrospect: Funes the Memorious," *Nature* 464, no. 4 (February 2010): 611.

74. Borges, "Funes, His Memory," 135.

75. Funes is, of course, still constrained by the biological affordances of human cognition. I leave it to Ted Chiang or Jeff VanderMeer to imagine a Funes who can see in infrared, hear subsonic frequencies, and taste other dimensions.

76. Wynter, "Unsettling the Coloniality of Being/Power/Truth/Freedom," 269.

77. Samuel Delany, *Times Square Red, Times Square Blue* (New York: New York University Press, 1999), 162.

78. Whereas *sapientia* means knowing in the broad sense of prudence, judgment, understanding, and wisdom, *scientia* is closer to *cognitio* and has to do with the intellectual knowledge of objects. Ludwig von Doederlein, *Döderlein's Handbook of Latin Synonymes*, trans. Henry Hamilton Arnold (W. F. Draper, 1875), Project Gutenberg, 40–41, 141.

79. Wynter, "Unsettling the Coloniality of Being/Power/Truth/Freedom," 275.

80. Francis Bacon, *The New Organon, Book I: Aphorisms*, ed. Lisa Jardine and Michael Silverthorne (Cambridge: Cambridge University Press, 2000), 33.

81. Deborah Kelemen et al., "Professional Physical Scientists Display Tenacious Teleological Tendencies: Purpose-Based Reasoning as a Cognitive Default," *Journal of Experimental Psychology: General* 142, no. 4 (2013): 1074–1083, https://doi.org/10.1037/a0030399; Thomas Kuhn, *The Structure of Scientific Revolutions*, 50th anniv. ed. (Chicago: University of Chicago Press, 2012); Shu Li et al., "Every Science/Nature Potter Praises His Own Pot—Can We Believe What He Says Based on His Mother Tongue?," *Journal of Cross-Cultural Psychology* 42, no. 1 (2011): 125–130, https://doi.org/10.1177/0022022110383425; Amos Tversky and Daniel Kahneman, "Belief in the Law of Small Numbers," *Psychological Bulletin* 76, no. 2 (1971): 105–110, https://doi.org/10.1037/h0031322.

82. Karen E. Fields and Barbara J. Fields, *Racecraft: The Soul of Inequality in American Life* (New York: Verso, 2012), 199.

83. Naomi Oreskes, *Why Trust Science?* (Princeton: Princeton University Press, 2019), 151.

84. Carolyn Merchant, *The Death of Nature: Women, Ecology, and the Scientific Revolution* (New York: HarperCollins, 1980), 193.

85. Karl Popper, *Unended Quest: An Intellectual Autobiography*, rev. ed. (La Salle, IL: Open Court, 1976), 79.

86. Oreskes, *Why Trust Science?*, 32–39. Quine, "Two Dogmas of Empiricism." Pierre Maurice Marie Duhem, *The Aim and Structure of Physical Theory* (Princeton: Princeton University Press, 1954), 180–218. See also Kuhn, *Structure of Scientific Revolutions*, 145–146.

87. Duhem, *Aim and Structure of Physical Theory*, 183.

88. Hans Jonas, *The Imperative of Responsibility* (Chicago: University of Chicago Press, 1985), 164. "Human universals—of which hundreds have been identified—consist of those features of culture, society, language, behavior, and mind that, so far as the record has been examined, are found among all peoples known to ethnog-

raphy and history." Donald E. Brown, "Human Universals, Human Nature and Human Culture," *Daedalus* 133, no. 4 (2004): 47–54, https://doi.org/10.1162/00115260 42365645, 47. See also Donald E. Brown, *Human Universals* (Philadelphia: Temple University Press, 1991).

89. Matthew Arnold, "Literature and Science," *Four Essays on Life and Letters* (New York: Appleton-Century-Crofts, 1947).

90. Oreskes, *Why Trust Science?*, 55.

91. Oreskes, *Why Trust Science?*, 246.

92. Oreskes, *Why Trust Science?*, 249.

93. See Matthew W. Slaboch, *A Road to Nowhere: The Idea of Progress and Its Critics* (Philadelphia: University of Pennsylvania Press, 2017); Christopher Lasch, *The True and Only Heaven: Progress and Its Critics* (New York: W. W. Norton, 1991); Robert A. Nisbet, *History of the Idea of Progress* (New York: Basic Books, 1980); Sidney Pollard, *The Idea of Progress: History and Society* (New York: Basic Books, 1969); Karl Löwith, *Meaning in History: Theological Implications of the Philosophy of History* (Chicago: University of Chicago Press, 1949); J. B. Bury, *The Idea of Progress: An Inquiry into Its Origin and Growth* (New York: Macmillan, 1920).

94. For a thoughtful analysis of idea of the "judgment of history," see Joan Wallach Scott, *On the Judgment of History* (New York: Columbia University Press, 2020).

95. Georg Henrik von Wright, "The Myth of Progress," in *The Tree of Knowledge and Other Essays* (New York: E. J. Brill, 1993), 202–228, 211.

96. Ludwig Wittgenstein, *Culture and Value*, ed. G. H. von Wright, trans. Peter Winch (Chicago: University of Chicago Press, 1980), 7c.

97. Joseph A. Tainter, "Resources and Cultural Complexity: Implications for Sustainability," *Critical Reviews in Plant Sciences* 30 (2011): 24–34.

98. Eric Voegelin, *The New Science of Politics, an Introduction* (Chicago: University of Chicago Press, 1952), 117–132 et passim. See also Kathleen Davis, "The Sense of an Epoch," in *The Legitimacy of the Middle Ages*, ed. Andrew Cole and D. Vance Smith (Durham: Duke University Press, 2010), 43.

99. Löwith, *Meaning in History*, 19.

100. Jean-Antoine-Nicolas de Caritat Condorcet, *Outlines of an Historical View of the Progress of the Human Mind: Being a Posthumous Work of the Late M. de Condorcet* (London: Printed for J. Johnson, 1795), https://hdl.handle.net/2027/njp .32101004143127, 4.

101. Lasch, *The True and Only Heaven*, 47.

102. Adam Smith, *An Inquiry into the Nature and Causes of the Wealth of Nations*, abr. with commentary by Laurence Dickey, Book 4, Chapter 2 (Indianapolis: Hackett, 1993), 130.

103. Kermode, *The Sense of an Ending*, 8.

104. The term *bright-siding* comes from Barbara Ehrenreich, *Bright-Sided: How the Relentless Promotion of Positive Thinking Has Undermined America* (New York: Metropolitan Books, 2009). A few examples of climate bright-siding from 2020 to

2022: Chris Turner, *How to Be a Climate Optimist* (Toronto: Random House Canada, 2022); German Lopez, "Climate Optimism: We Have Reason for Hope on Climate Change," *New York Times*, April 3, 2022, https://www.nytimes.com/2022/04/03/briefing/climate-optimism-ukraine-week-ahead.html; Saul Griffith, "We Can Beat Climate Change If We Do One Thing Fast," *Time*, November 4, 2021, https://time.com/6113719/optimistic-fighting-climate-change/; Rebecca Solnit, "Dare We Hope? Here's My Cautious Case for Climate Optimism," *The Guardian*, May 1, 2021, https://www.theguardian.com/commentisfree/2021/may/01/climate-change-environment-hope-future-optimism-success; Bill Gates's *How to Avoid a Climate Disaster: The Solutions We Have and the Breakthroughs We Need* (New York: Vintage, 2021); Paul Hawken's *Regeneration: Ending the Climate Crisis in One Generation* (New York: Penguin, 2021); Katharine Hayhoe's *Saving Us: A Climate Scientist's Case for Hope and Healing in a Divided World* (New York: Atria, 2021); Christiana Figueres and Tom Rivett-Carnac's *The Future We Choose: Surviving the Climate Crisis* (New York: Alfred A. Knopf, 2020); Ayana Elizabeth Johnson and Katharine Keeble Wilkinson's *All We Can Save: Truth, Courage, and Solutions for the Climate Crisis* (New York: One World, 2020); and David Miller's *Solved: How the World's Great Cities Are Fixing the Climate Crisis* (Toronto: University of Toronto Press, 2020).

105. David Spratt, "Always Look on Bright Side of Life," Climate Code Red, April 2012, http://www.climatecodered.org/2012/04/always-look-on-bright-side-of-life.html.

106. Jem Bendell, "The Biggest Mistakes in Climate Communications, Part 2: Climate Brightsiding," Brave New Europe, September 15, 2022, https://braveneweurope.com/jem-bendell-the-biggest-mistakes-in-climate-communications-part-2-climate-brightsiding.

107. Pierre Friedlingstein et al., "Global Carbon Budget 2022," *Earth System Science Data* 14 (2022): 4811–4900, https://doi.org/10.5194/essd-14-4811-2022; Vally Koubi, "Climate Change and Conflict," *Annual Review of Political Science* 22 (2019): 343–360; Solomon M. Hsiang et al., "Civil Conflicts Are Associated with the Global Climate," *Nature* 476, no. 7361 (2011): 438–441; Thomas F. Homer-Dixon et al., "Environmental Change and Violent Conflict," *Scientific American* 268, no. 2 (1993): 38–45.

108. Bjorn Lomborg, *False Alarm: How Climate Change Panic Costs Us Trillions, Hurts the Poor, and Fails to Fix the Planet* (New York: Basic Books, 2020), 9. Lomborg does not identify the "experts" referred to here.

109. A small army of critics and reviewers have taken Pinker to task for his numerous scholarly sins, from cherry-picking data and glib generalization to crudely misrepresenting the complexity of the Enlightenment. See David Bell, "The PowerPoint Philosophe," *The Nation*, March 7, 2018, https://www.thenation.com/article/archive/waiting-for-steven-pinkers-enlightenment/; Ian Goldin, "The Limitations of Steven Pinker's Optimism," *Nature*, February 16, 2018, https://www.nature.com/articles/d41586-018-02148-1; John Gray, "Unenlightened Thinking: Steven Pinker's Embarrass-

ing New Book Is a Feeble Sermon for Rattled Liberals," *The New Statesman*, February 22, 2018, https://www.newstatesman.com/culture/2018/02/enlightenment-now-the-case-for-reason-science-humanism-and-progress-review-steven-pinker; Gary Gutting, "Never Better: Steven Pinker's Narrow Enlightenment," *Commonweal*, May 04, 2018, 20–23; Aaron Hanlon, "Steven Pinker's New Book on the Enlightenment Is a Huge Hit. Too Bad It Gets the Enlightenment Wrong," *Vox*, May 17, 2018, https://www.vox.com/the-big-idea/2018/5/17/17362548/pinker-enlightenment-now-two-cultures-rationality-war-debate; Jeremy Lent, "Steven Pinker's Ideas Are Fatally Flawed: These Eight Graphs Show Why," *Open Democracy*, May 21, 2018, https://www.opendemocracy.net/en/transformation/steven-pinker-s-ideas-are-fatally-flawed-these-eight-graphs-show-why/. See also Philip Dwyer and Mark Micale, *The Darker Angels of Our Nature: Refuting the Pinker Theory of History and Violence* (London: Bloomsbury Publishing, 2021).

110. Steven Pinker, interviewed by Gareth Cook, "The Secret Behind One of the Greatest Success Stories in All of History," *Scientific American*, February 15, 2018, https://www.scientificamerican.com/article/the-secret-behind-one-of-the-greatest-success-stories-in-all-of-history/. See also Steven Pinker, *Enlightenment Now* (New York: Penguin Books, 2018).

111. See Dale Jamieson, "Is There Progress in Morality?" *Utilitas* 14, no. 3 (November 2002), 318–338; von Wright, "The Myth of Progress."

112. Dag Prawitz, "Progress in Philosophy," in *The Idea of Progress*, ed. Arnold Burgen, Peter McLaughlin, and Jürgen Mittelstraß (Berlin: Walter de Gruyter, 1997), 139–154, 144.

113. Hannah Ritchie et al. "Energy Production and Consumption," OurWorldinData.Org, 2020, https://ourworldindata.org/energy-production-consumption. Data based on Energy Institute, *Statistical Review of World Energy* (2023), https://www.energyinst.org/statistical-review; Vaclav Smil, *Energy Transitions: Global and National Perspectives* (Westport, CT: Praeger, 2017).

114. It is ironic that the very carbon-intensive industrial civilization that has fulfilled the promise of cosmopolitan, rights-based, identarian progressivism has also brought about the conditions leading to its destruction, a point made by Claire Colebrook, "We Have Always Been Post-Anthropocene: The Anthropocene Counterfactual," in *Anthropocene Feminism*, ed. Richard Grusin (Minneapolis: University of Minnesota Press, 2017), 1–20.

115. J. R. McNeill and Peter Engelke, *The Great Acceleration* (Cambridge: Belknap Press, 2016), 4–5. Emphasis added.

116. "Since 1980, the prevalence of obesity has doubled in more than 70 countries and has continuously increased in most other countries." Pascal Bovet, Arnaud Chiolero, and Jude Gedeon, "Health Effects of Overweight and Obesity in 195 Countries," *The New England Journal of Medicine* 377, no. 15 (2017): 1495–1496.

117. See Mario Giampietro, "On the Circular Bioeconomy and Decoupling: Implications for Sustainable Growth," *Ecological Economics* 162 (2019): 143–156; Joseph A.

Tainter, "Social Complexity and Sustainability," *Ecological Complexity* 3 (2006): 91–103.

118. Tooze, *Shutdown*, 291–292. See also David M. Cutler and Lawrence H. Summers, "The COVID-19 Pandemic and the $16 Trillion Virus," *JAMA* 324, no. 15 (2020): 1495–1496, doi: 10.1001/jama.2020.19759.

119. See Ugo Bardi, Sara Falsini, and Ilaria Perissi, "Toward a General Theory of Societal Collapse: A Biophysical Examination of Tainter's Model of the Diminishing Returns of Complexity," *BioPhysical Economics and Resource Quality* 4, no. 1 (2019).

120. Joseph Tainter, "Collapse and Sustainability: Rome, the Maya, and the Modern World," *Archaeological Papers of the American Anthropological Association* 24 (2014): 201.

121. Peter Turchin et al., "Quantitative Historical Analysis Uncovers a Single Dimension of Complexity That Structures Global Variation in Human Social Organization," *Proceedings of the National Academy of Sciences* 115, no. 2 (2018): E144, https://doi.org/10.1073/pnas.1708800115.

122. Miguel A. Centeno et al., "Globalization and Fragility: A Systems Approach to Collapse," in *How Worlds Collapse: What History, Systems, and Complexity Can Teach Us About Our Modern World and Fragile Future*, ed. Miguel A. Centeno et al., (New York: Routledge Press, 2023), 7.

123. The term "great simplification" may be attributed to podcast host and activist Nate Hagens (https://www.thegreatsimplification.com/).

124. Thomas Homer-Dixon et al., "Synchronous Failure," *Ecology and Society* 20, no. 3 (2015): 6, https://doi.org/10.5751/ES-07681-200306.

125. Tainter, *Collapse of Complex Societies*, 214.

Chapter 2

1. David Wallace-Wells, "The Uninhabitable Earth," *New York*, July 9, 2017, https://nymag.com/intelligencer/2017/07/climate-change-earth-too-hot-for-humans.html.

2. "New York Magazine Traffic Analytics," *SimilarWeb*, September 2023, https://www.similarweb.com/website/nymag.com/#overview.

3. Historian of science Zachary Loeb has offered a series of insightful thoughts on what people call "doomism" at his blog *Librarian Shipwreck*. See especially Zachary Loeb, "Be Afraid! But Not That Afraid?—On Climate Doom," *Librarian Shipwreck* (blog), August 16, 2019, https://librarianshipwreck.wordpress.com/2019/08/16/be-afraid-but-not-that-afraid-on-climate-doom. See also Kyle Paoletta, "The Incredible Disappearing Doomsday," *Harper's*, April 2023, https://harpers.org/archive/2023/04/the-incredible-disappearing-doomsday-climate-catastrophists-new-york-times-climate-change-coverage.

4. These things tend to happen in waves. For a 2022 wave of anti-"doomism," see Seth Borenstein, "No Obituary for Earth: Scientists Fight Climate Doom Talk," *Associated Press*, April 4, 2022, https://apnews.com/article/fighting-climate-doom-d4

7f2ea47bc428656b7be1f48771b75d; Cara Buckley, "'OK Doomer' and the Climate Advocates Who Say It's Not Too Late," *New York Times*, March 22, 2022, https://www.nytimes.com/2022/03/22/climate/climate-change-ok-doomer.html; SueEllen Campbell, "Recent Readings on Climate 'Doomerism' and Science," *Yale Climate Connections*, April 22, 2022, https://yaleclimateconnections.org/2022/04/recent-readings-on-climate-doomerism-and-science/; SueEllen Campbell, "On Climate 'Doomism': Heart and Mind Reasons to Resist It," *Yale Climate Connections*, May 6, 2022, https://yaleclimateconnections.org/2022/05/on-climate-doomism-heart-mind-reasons-to-resist-it/; Annie Goodykoontz, "The Spread of 'Climate Doom' on TikTok Is Hurting the Climate Justice Movement—and Gen Z," *USA Today*, September 22, 2022, https://www.usatoday.com/story/opinion/voices/2022/09/22/climate-doom-tiktok-gen-z-eco-anxiety/8004862001/?gnt-cfr=1; and Philippa Nuttall, "Don't Listen to the Climate Doomists," *The New Statesman*, August 2022, https://www.newstatesman.com/environment/climate/2022/08/climate-doomism-dont-listen-toxic-narrative. See also Luke Kemp et al., "Climate Endgame: Exploring Catastrophic Climate Change Scenarios," *Proceedings of the National Academy of Sciences* 119, no. 34 (2022): 1–e2108146119, https://doi.org/10.1073/pnas.2108146119.

5. Michael Mann, "Not a big fan of the doomist framing of new @NYMag article ('The Uninhabitable Earth')," Twitter, July 10, 2017, https://twitter.com/MichaelEMann/status/884415889625554944. Early criticisms of Wallace-Wells included Andrew Dessler (@AndrewDessler), "Like the author, I'm very concerned about climate change. But there's some nonsense in this article," Twitter, July 10, 2017, https://twitter.com/AndrewDessler/status/884446573379416064; Zack Labe (@ZLabe), "Climate change is a real/present issue but I think we can reach a broader audience by talking about impacts/solutions rather than hyperboles," Twitter, July 9, 2017, https://twitter.com/zlabe/status/884255678302269441; Alex Steffen (@AlexSteffen), "Despair is not action; 'we're doomed' is not a critique; 'unless we change' is not a solution, without a vision of what change looks like," Twitter, July 10, 2017, https://twitter.com/AlexSteffen/status/884440490996752385; Ramez Naam (@ramez), "6. Finally, through combo of exaggeration and hopelessness, it turns away those in the middle we need to persuade. It makes action *harder*," Twitter, July 10, 2017, https://twitter.com/ramez/status/884486583734358017; Jon Foley (@GlobalEcoGuy), "This is a deeply irresponsible article, cherry-picking doomsday scenarios," Twitter, July 10, 2017, https://twitter.com/GlobalEcoGuy/status/884453087678578689 (Foley's tweet is no longer accessible, but it is quoted in Eric Holthaus, "Stop Scaring People About Climate Change. It Doesn't Work," *Grist*, July 10, 2017, https://grist.org/climate-energy/stop-scaring-people-about-climate-change-it-doesnt-work/). See also Chris Mooney, "Scientists Challenge Magazine Story about 'Uninhabitable Earth,'" *Washington Post*, July 12, 2017, https://www.washingtonpost.com/news/energy-environment/wp/2017/07/12/scientists-challenge-magazine-story-about-uninhabitable-earth/.

6. Michael Mann, "Since this New York Magazine article ('The Uninhabitable

Earth') is getting so much play this morning, I figured I should comment on it," Facebook, July 10, 2017, https://www.facebook.com/MichaelMannScientist/posts/since-this-new-york-magazine-article-the-uninhabitable-earth-is-getting-so-much-/1470539096335621/.

7. Jenny Offill, *The Weather* (New York: Knopf Doubleday, 2020), 67.

8. Michael E. Mann, Susan Joy Hassol, and Tom Toles, "Doomsday Scenarios Are as Harmful as Climate Change Denial," *The Washington Post*, July 12, 2017, https://www.washingtonpost.com/opinions/doomsday-scenarios-are-as-harmful-as-climate-change-denial/2017/07/12/880ed002-6714-11e7-a1d7-9a32c91c6f40_story.html. Smith and Leiserowitz survey some relevant literature on fear appeals and negative messaging but too heavily weight cases where such appeals "can" be counterproductive, present speculation as if it were evidence, and rely on flawed studies such as O'Neill and Nicholson-Cole's "Fear Won't Do It" (more on which in a couple paragraphs). In any case, the science on fear appeals and negative messaging, particularly when it comes to climate change, is complex, contentious, and evolving, thus reference to a single outdated study is simply inadequate. Nicholas Smith and Anthony Leiserowitz, "The Role of Emotion in Global Warming Policy Support and Opposition," *Risk Analysis* 34, no. 5 (2014): 937–948, https://doi.org/10.1111%2Frisa.12140.

9. Andrew Freedman, "No, New York Mag: Climate Change Won't Make the Earth Uninhabitable by 2100," *Mashable*, July 10, 2017, https://mashable.com/2017/07/10/new-york-mag-climate-story-inaccurate-doomsday-scenario/.

10. Robinson Meyer, "Are We as Doomed as That New York Magazine Article Says?," *The Atlantic*, July 10, 2017, https://www.theatlantic.com/science/archive/2017/07/is-the-earth-really-that-doomed/533112/.

11. Emmanuel Vincent, "Scientists Explain What New York Magazine Article on 'The Uninhabitable Earth' Gets Wrong," *Climate Feedback*, July 12, 2017, https://climatefeedback.org/evaluation/scientists-explain-what-new-york-magazine-article-on-the-uninhabitable-earth-gets-wrong-david-wallace-wells/.

12. Emily Atkin, "The Power and Peril of 'Climate Disaster Porn,'" *The New Republic*, July 10, 2017, https://newrepublic.com/article/143788/power-peril-climate-disaster-porn.

13. Saffron O'Neill and Sophie Nicholson-Cole, "'Fear Won't Do It': Promoting Positive Engagement with Climate Change Through Visual and Iconic Representations," *Science Communication* 30, no. 3 (March 2009): 355–379, doi: 10.1177/1075547008329201.

14. O'Neill and Nicholson-Cole, "'Fear Won't Do It,'" 365.

15. John R. Hibbing, Kevin B. Smith, and John R. Alford, "Differences in Negativity Bias Underlie Variations in Political Ideology," *Behavioral and Brain Sciences* 37, no. 3 (2014): 297–307, doi: 10.1017/S0140525X13001192.

16. O'Neill and Nicholson-Cole, "'Fear Won't Do It,'" 375.

17. Kathryn Stevenson and Nils Peterson, "Motivating Action through Fostering

Climate Change Hope and Concern and Avoiding Despair among Adolescents," *Sustainability* 8, no. 1 (2015): 6, https://doi.org/10.3390/su8010006.

18. Holthaus, "Stop Scaring People About Climate Change."

19. Paul Bain et al., "Co-Benefits of Addressing Climate Change Can Motivate Action Around the World," *Nature Climate Change* 6, no. 2 (2016): 154–157, https://doi.org/10.1038/nclimate2814.

20. Dean Ornish, "Love, Not Fear, Will Help Us Fix Climate Change," *Time*, September 30, 2014, https://time.com/3450002/love-not-fear-will-help-us-fix-climate-change/.

21. Paul C. Stern, "Fear and Hope in Climate Messages," *Nature Climate Change* 2, no. 8 (2012): 572–573, https://doi.org/10.1038/nclimate1610.

22. Stern, "Fear and Hope in Climate Messages," 572.

23. Paul G. Bain et al., "Promoting Pro-Environmental Action in Climate Change Deniers," *Nature Climate Change* 2, no. 8 (2012): 600–603, https://doi-org.proxy.library.nd.edu/10.1038/nclimate1532.

24. Will Steffen and Wolfgang Knorr look at another case of tendentious "fact-checking" in examining *Climate Feedback*'s response to Jonathan Franzen's *New Yorker* article "What If We Stopped Pretending?" Will Steffen and Wolfgang Knorr, "Fact Checking the Climate Crisis: Franzen vs. Facebook on False News," *Initiative for Leadership and Sustainability*, February 19, 2020, https://iflas.blogspot.com/2020/02/fact-checking-climate-crisis-franzen-vs.html. See Jonathan Franzen, "What If We Stopped Pretending?," *The New Yorker*, September 8, 2019, https://www.newyorker.com/culture/cultural-comment/what-if-we-stopped-pretending; and Jacob Siegel, "Invasion of the Fact-Checkers," *Tablet*, March 21, 2022, https://www.tabletmag.com/sections/news/articles/invasion-fact-checkers. ·

25. See Robert J. Brulle and Kari Marie Norgaard, "Avoiding Cultural Trauma: Climate Change and Social Inertia," *Environmental Politics* 28, no. 5 (2019), doi: 10.1080/09644016.2018.1562138; Gerdien De Vries, "Public Communication as A Tool to Implement Environmental Policies," *Social Issues and Policy Review* 14, no. 1 (2020): 244–272, doi: 10.1111/sipr.12061; George Marshall, *Don't Even Think About It* (New York: Bloomsbury, 2015); Sander Van der Linden, Edward Maibach, and Anthony Leiserowitz, "Improving Public Engagement with Climate Change: Five 'Best Practice' Insights from Psychological Science," *Perspectives on Psychological Science* 10, no. 6 (2015): 758–763, doi: 10.1177/1745691615598516; Per Espen Stoknes, *What We Think About When We Try Not to Think About Global Warming* (White River Junction, VT: Chelsea Green, 2015); Kari Norgaard, *Living in Denial* (Cambridge: MIT Press, 2011); Robert Gifford. "The Dragons of Inaction: Psychological Barriers That Limit Climate Change Mitigation and Adaptation," *American Psychologist* 66, no. 4 (2011): 290–302, doi: 10.1037/a0023566; Brigitte Nerlich, Nelya Koteyko, and Brian Brown, "Theory and Language of Climate Change Communication," *Wiley Interdisciplinary Reviews: Climate Change* 1, no. 1 (2010): 97–110, https://doi.org/10.1002/wcc.2.

26. Robin Bayes et al., "When and How Different Motives Can Drive Motivated

Political Reasoning," *Political Psychology* 41, no. 5 (2020): 1031–1052, https://doi.org/10.1111/pops.12663.

27. For a provocative reflection on some of the problems inherent in science communication, see Earle E. Spamer, "Know Thyself: Responsible Science and the Lectotype of Homo Sapiens Linnaeus, 1758," *Proceedings of the Academy of Natural Sciences of Philadelphia* 149 (1999): 109–114, https://www.jstor.org/stable/4065043.

28. Stephen Schneider, *Global Warming: Are We Entering the Greenhouse Century?* (San Francisco: Sierra Club Books, 1989), xi.

29. Chris Russill, "Stephen Schneider and the 'Double Ethical Bind' of Climate Change Communication," *Bulletin of Science, Technology & Society* 30, no. 1 (2010): 60–69, doi: 10.1177/0270467609355055.

30. Nerlich, Koteyko, and Brown, "Theory and Language of Climate Change Communication."

31. Stephen Vaisey and Omar Lizardo, "Cultural Fragmentation or Acquired Dispositions? A New Approach to Accounting for Patterns of Cultural Change," *Socius* 2 (2016): 2378023116669726. https://doi.org/10.1177/2378023116669726.

32. It should be noted that the topic remains contentious. See Sedona Chinn and P. Sol Hart, "Climate Change Consensus Messages Cause Reactance," *Environmental Communication* 17, no. 1 (2023): 51–59, doi: 10.1080/17524032.2021.1910530; Sander Van der Linden, Anthony Leiserowitz, and Edward Maibach, "The Gateway Belief Model: A Large-Scale Replication," *Journal of Environmental Psychology* 62 (2019): 49–58, https://doi.org/10.1016/j.jenvp.2019.01.009; Graham Dixon, Jay Hmielowski, and Yanni Ma, "More Evidence of Psychological Reactance to Consensus Messaging: A Response to van der Linden, Maibach, and Leiserowitz (2019)," *Environmental Communication* 17, no. 1 (2019): 9–15, https://doi.org/10.1080/17524032.2019.1671472; Graham Dixon and Austin Hubner, "Neutralizing the Effect of Political Worldviews by Communicating Scientific Agreement: A Thought-Listing Study," *Science Communication* 40, no. 3 (2018): 393–415, https://doi.org/10.1177/1075547018769907; Toby Bolsen and James Druckman, "Do Partisanship and Politicization Undermine the Impact of a Scientific Consensus Message About Climate Change?," *Group Processes & Intergroup Relations* 21, no. 3 (2018): 389–402, https://doi-org.proxy.library.nd.edu/10.1177/1368430217737855; John Cook and Stephan Lewandowsky, "Rational Irrationality: Modeling Climate Change Belief Polarization Using Bayesian Networks," *Topics in Cognitive Science* 8, no. 1 (2016): 160–179, https://doi.org/10.1111/tops.12186.

33. Yanni Ma, Graham Dixon, and Jay D. Hmielowski, "Psychological Reactance from Reading Basic Facts on Climate Change: The Role of Prior Views and Political Identification," *Environmental Communication* 13, no. 1 (2019): 71–86, doi: 10.1080/17524032.2018.1548369.

34. Matthew J. Hornsey and Kelly S. Fielding, "A Cautionary Note About Messages of Hope: Focusing on Progress in Reducing Carbon Emissions Weakens Mitigation Motivation," *Global Environmental Change* 39 (2016): 26–34, doi: 10.1016/j.gloenvcha.2016.04.003.

35. Bethany Albertson and Joshua William Busby, "Hearts or Minds? Identifying Persuasive Messages on Climate Change," *Research & Politics* 2, no. 1 (2015): 1–9, doi: 10.1177/2053168015577712. In addition, a 2017 master's thesis at the University of Copenhagen found that while both fear and hope messages motivated pro-environmental behavior, fear appeals, or what the author called "impact framing," is "a more successful way than solutions to influence people's mitigating behaviour." Henry James Evans, "How Do Different Framings of Climate Change Affect Pro-environmental Behaviour?" (master's thesis, University of Copenhagen, 2017), https://www.ind.ku.dk/publikationer/studenterserien/studenterserien-alle/studenterserie_91_Henry_James_Evans.pdf.

36. Joshua Ettinger et al., "Climate of Hope or Doom and Gloom? Testing the Climate Change Hope vs. Fear Communications Debate Through Online Videos," *Climatic Change* 164, no. 19 (2021), https://doi.org/10.1007/s10584-021-02975-8.

37. Jacob B. Rode et al., "Influencing Climate Change Attitudes in the United States: A Systematic Review and Meta-Analysis," *Journal of Environmental Psychology* 76 (2021): 101623, https://doi.org/10.1016/j.jenvp.2021.101623.

38. Madalina Vlasceanu et al. "Addressing Climate Change with Behavioral Science: A Global Intervention Tournament in 63 Countries," *Science Advances* 10, no. 6 (2024): eadj5778.

39. Jan G. Voelkel et al. "A Large Experimental Test of the Effectiveness of Highly Cited Climate Change Messages [registered Report Stage 1 Protocol]," figshare, May 13, 2024, https://doi.org/10.6084/m9.figshare.25807429.v1.

40. Linh Vu et al., "Ignorance by Choice: A Meta-Analytic Review of the Underlying Motives of Willful Ignorance and Its Consequences," *Psychological Bulletin* 149, no. 9–10 (2023): 611; Tali Sharot and Cass R. Sunstein, "How People Decide What They Want to Know," *Nature Human Behaviour* 4, no. 1 (2020): 14–19, https://doi.org/10.1038/s41562-019-0793-1; Ziva Kunda, "The Case for Motivated Reasoning," *Psychological Bulletin* 108, no. 3 (November 1990): 480–498, https://doi.org/10.1037/0033-2909.108.3.480; Charles G. Lord et al., "Biased Assimilation and Attitude Polarization: The Effects of Prior Theories on Subsequently Considered Evidence," *Journal of Personality and Social Psychology* 37, no. 11 (1979): 2098–2109, doi: 10.1037/0022-3514.37.11.2098.

41. Debra Javeline and Gregory Shufeldt, "Scientific Opinion in Policymaking: The Case of Climate Change Adaptation," *Policy Sciences* 47, no. 2 (2014): 121–139, https://doi.org/10.1007/s11077-013-9187-9.

42. Frances C. Moore et al., "Rapidly Declining Remarkability of Temperature Anomalies May Obscure Public Perception of Climate Change," *Proceedings of the National Academy of Sciences* 116, no. 11 (2019): 4905–4910; Masashi Soga and Kevin J. Gaston, "Shifting Baseline Syndrome: Causes, Consequences, and Implications," *Frontiers in Ecology and the Environment* 16, no. 4 (2018): 222–230; Robert K. Kaufmann et al., "Spatial Heterogeneity of Climate Change as an Experiential Basis for Skepticism," *Proceedings of the National Academy of Sciences* 114, no. 1 (2017):

67–71; Van der Linden, Maibach, and Leiserowitz, "Improving Public Engagement with Climate Change"; Sarah K. Papworth et al., "Evidence for Shifting Baseline Syndrome in Conservation," *Conservation Letters* 2, no. 2 (2009): 93–100; Daniel Pauly, "Anecdotes and the Shifting Baseline Syndrome of Fisheries," *Trends in Ecology & Evolution* 10, no. 10 (1995): 430.

43. Daniel Kahneman et al., "Anomalies: The Endowment Effect, Loss Aversion, and Status Quo Bias," *The Journal of Economic Perspectives* 5, no. 1 (1991): 193–206, https://doi.org/10.1257/jep.5.1.193.

44. Brian Kennedy and Meg Hefferon, "What Americans Know About Science," Pew Research Center, March 28, 2019, https://www.pewresearch.org/science/2019/03/28/what-americans-know-about-science/.

45. Dan Kahan, "They've Already Gotten the Memo! What The Public (Rs & Ds) Think 'Climate Scientists Believe,'" *The Cultural Cognition Project at Yale Law School* (2014), https://www.culturalcognition.net/blog/2014/6/23/theyve-already-gotten-the-memo-what-the-public-rs-ds-think-c.html; Bo MacInnis and Jon A. Krosnick, *Climate Insights 2020: Partisan Divide* (Resources for the Future, 2020), https://www.rff.org/publications/reports/climateinsights2020-partisan-divide/.

46. Brian Kennedy and Alec Tyson, "Americans' Trust in Scientists, Positive Views of Science Continue to Decline," Pew Research Center, November 14, 2023, https://www.pewresearch.org/science/2023/11/14/americans-trust-in-scientists-positive-views-of-science-continue-to-decline/.

47. Thomas G. Safford et al., "Questioning Scientific Practice: Linking Beliefs About Scientists, Science Agencies, and Climate Change," *Environmental Sociology* 6, no. 2 (2019): 1–13, doi: 10.1080/23251042.2019.1696008.

48. Jessica Gall Myrick and Suzannah Evans Comfort, "The Pope May Not Be Enough: How Emotions, Populist Beliefs, and Perceptions of an Elite Messenger Interact to Influence Responses to Climate Change Messaging," *Mass Communication & Society* 23, no. 1 (2020): 1–21, doi: 10.1080/15205436.2019.1639758.

49. Robert A. Huber, "The Role of Populist Attitudes in Explaining Climate Change Skepticism and Support for Environmental Protection," *Environmental Politics* 29, no. 6 (2020): 959–982. Matthew Lockwood, "Right-Wing Populism and the Climate Change Agenda: Exploring the Linkages," *Environmental Politics* 27, no. 4 (2018): 712–732.

50. Risa Palm, Toby Bolsen, and Justin T. Kingsland, "'Don't Tell Me What to Do': Resistance to Climate Change Messages Suggesting Behavior Changes," *Weather, Climate, and Society* 12, no. 4 (2020): 827–835, https://doi-org.proxy.library.nd.edu/10.1175/WCAS-D-19-0141. Erik C. Nisbet et al., "The Partisan Brain: How Dissonant Science Messages Lead Conservatives and Liberals to (Dis)Trust Science," *The Annals of the American Academy of Political and Social Science* 658, no. 1 (2015): 36–66, doi: 10.1177/0002716214555474. See also Tali Sharot, *The Influential Mind* (New York: Henry Holt, 2017): 29–30; and Camelia Kuhnen, "Asymmetric Learning from Financial Information," *The Journal of Finance* 70, no. 5 (2015): 2029–2062; https://doi.org/10.1111/jofi.12223.

51. For a discussion of why, see Jacob B. Rode and Peter H. Ditto, "Can the Partisan Divide in Climate Change Attitudes Be Bridged? A Review of Experimental Interventions," in *The Psychology of Political Polarization*, ed. Jan-Willem van Prooijen (New York: Routledge, 2021), 148–168. See also Anthony Leiserowitz et al., *Climate Change in the American Mind: December 2020* (New Haven: Yale Program on Climate Change Communication, 2021); and Robert J. Antonio and Robert J. Brulle, "The Unbearable Lightness of Politics: Climate Change Denial and Political Polarization," *Sociological Quarterly* 52, no. 2 (2011): 195–202, https://doi.org/10.1111/j.1533-8525.2011.01199.x.

52. Aaron McCright and Riley Dunlap's research on the "anti-reflexivity thesis" offers a nuanced understanding of *which* kinds of science conservatives and liberals trust, especially distinguishing between science focused on environmental and public health impacts versus science focused on economic production. For instance, their research shows that while "conservatives report much less trust in impact scientists," they report "greater trust in production scientists than their liberal counterparts." Aaron M. McCright, Katharine Dentzman, and Thomas Dietz, "The Influence of Political Ideology on Trust in Science," *Environmental Research Letters* 8, no. 4 (2013): 044029, doi: 10.1088/1748-9326/8/4/044029. On conservative skepticism and climate science in general, see Robert J. Brulle, "Critical Reflections on the March for Science," *Sociological Forum* 33, no. 1 (2018): 255–258, doi: 10.1111/socf.12398; Matthew Motta, "The Polarizing Effect of the March for Science on Attitudes Toward Scientists," *PS: Political Science & Politics* 51, no. 4 (2018): 782–788, doi: 10.1017/S1049096518000938; Lawrence C. Hamilton et al. "Trust in Scientists on Climate Change and Vaccines," *SAGE Open* 5, no. 3 (2015), doi: 10.1177/2158244015602752; Anthony A. Leiserowitz et al., "Climategate, Public Opinion, and the Loss of Trust," *The American Behavioral Scientist* 57, no. 6 (2013): 818–837, doi: 10.1177/0002764212458272; Aaron M. McCright and Riley Dunlap, "The Politicization of Climate Change and Polarization in the American Public's Views of Global Warming, 2001–2010," *Sociological Quarterly* 52, no. 2 (2011): 155–194, doi: 10.1111/j.1533-8525.2011.01198.x. On recent skepticism toward science with regard to COVID-19, see Gordon Pennycook et al., "Beliefs about COVID-19 in Canada, the United Kingdom, and the United States: A Novel Test of Political Polarization and Motivated Reasoning," *Personality and Social Psychology Bulletin* 48, no. 5 (2022): 750–765; Matthew T. Ballew et al., *American Public Responses to COVID-19, April 2020* (New Haven: Yale Program on Climate Change Communication, 2020); Emma Grey Ellis, "The Coronavirus Outbreak Is a Petri Dish for Conspiracy Theories," *Wired*, February 4, 2020, https://www.wired.com/story/coronavirus-conspiracy-theories; Jay J. van Bavel et al., "Using Social and Behavioural Science to Support COVID-19 Pandemic Response," *Nature Human Behaviour* 4, no. 5 (2020): 460–471, doi: 10.1038/s41562-020-0884-z.

53. This goes back to foundational research by Kent D. Van Liere and Riley E. Dunlap, "The Social Bases of Environmental Concern: A Review of Hypotheses, Explanations and Empirical Evidence," *Public Opinion Quarterly* 44, no. 2 (1980):

181–197, doi: 10.1086/268583. Also see Dan M. Kahan, "Climate-Science Communication and the Measurement Problem," *Political Psychology* 36 (2015): 1–43.

54. Kahan, "They've Already Gotten the Memo!"

55. Matthew J. Hornsey et al., "Meta-Analyses of the Determinants and Outcomes of Belief in Climate Change," *Nature Climate Change* 6, no. 6 (2016): 622–626, doi: 10.1038/nclimate2943.

56. Dan M. Kahan and Jonathan C. Corbin, "A Note on the Perverse Effects of Actively Open-Minded Thinking on Climate-Change Polarization," *Research & Politics* 3, no. 4 (October 2016): 2053168016676705, doi: 10.1177/2053168016676705; Dan M. Kahan et al., "Motivated Numeracy and Enlightened Self-Government," *Behavioural Public Policy* 1, no. 1 (2017): 54–86, doi: 10.1017/bpp.2016.2.

57. Hamilton et al., "Trust in Scientists"; Gordon Gauchat, "Politicization of Science Sphere: A Study of Public Trust in the United States, 1974 to 2010," *American Sociological Review* 77, no. 2 (2012), https://doi.org/10.1177/0003122412438225; Kahan et al., "Motivated Numeracy;" and Matthew S. Nurse and Will J. Grant, "I'll See It When I Believe It: Motivated Numeracy in Perceptions of Climate Change Risk," *Environmental Communication* 14, no. 2 (2020): 184–201, https://doi.org/10.1080/17524032.2019.1618364.

58. Dan M. Kahan, quoted in Elizabeth Kolbert, "Rethinking How We Think About Climate Change," *Audubon Magazine*, September–October 2014, https://www.audubon.org/magazine/september-october-2014/rethinking-how-we-think-about-climate.

59. Matthew J. Hornsey, "The Role of Worldviews in Shaping How People Appraise Climate Change," *Current Opinion in Behavioral Sciences* 42 (2021): 36–41, https://doi.org/10.1016/j.cobeha.2021.02.021.

60. Lawrence C. Hamilton et al., "Flood Realities, Perceptions and the Depth of Divisions on Climate." *Sociology* 50, no. 5 (2016): 913–933, doi:10.1177/0038038516648547; Lawrence Hamilton et al., "Wildfire, Climate, and Perceptions in Northeast Oregon," *Regional Environmental Change* 16, no. 6 (2016): 1819–1832, doi:10.1007/s10113-015-0914-y.

61. P. Sol Hart and Erik C. Nisbet, "Boomerang Effects in Science Communication: How Motivated Reasoning and Identity Cues Amplify Opinion Polarization About Climate Mitigation Policies," *Communication Research* 39, no. 6 (2012): 701–723, doi: 10.1177/0093650211416646.

62. Leiserowitz et al., *Climate Change in the American Mind: December 2020*.

63. Nisbet et al., "The Partisan Brain."

64. Justin Farrell, "Corporate Funding and Ideological Polarization About Climate Change," *Proceedings of the National Academy of Sciences* 113, no. 1 (2016): 92–97, https://doi.org/10.1073/pnas.1509433112.

65. Geoff Dembicki, *The Petroleum Papers: Inside the Far-Right Conspiracy to Cover Up Climate Change* (Vancouver, B.C.: Greystone Books, 2022); Geoffrey Supran and Naomi Oreskes, "Rhetoric and Frame Analysis of ExxonMobil's Climate Change

Communications," *One Earth* 4, no. 5 (2021): 696–719, https://doi.org/10.1016/j.one ear.2021.04.014; Robert J. Brulle, "The Climate Lobby: A Sectoral Analysis of Lobbying Spending on Climate Change in the USA, 2000 to 2016," *Climatic Change* 149, no. 3–4 (2018): 289–303, https://doi.org/10.1007/s10584-018-2241-z; Geoffrey Supran and Naomi Oreskes, "Assessing ExxonMobil's Climate Change Communications (1977–2014)," *Environmental Research Letters* 12, no. 8 (2017): 084019, doi: 10.1088/1748-9326/aa815f/; Robert J. Brulle, "Institutionalizing Delay: Foundation Funding and the Creation of U.S. Climate Change Counter-Movement Organizations," *Climatic Change* 122, no. 4 (2014): 681–694, https://doi.org/10.1007/s10584-013-1018-7; Naomi Oreskes and Erik Conway, *Merchants of Doubt: How a Handful of Scientists Obscured the Truth on Issues from Tobacco Smoke to Global Warming* (New York: Bloomsbury Press, 2010).

66. Supran and Oreskes, "Rhetoric and Frame Analysis of ExxonMobil's Climate Change Communications," 712. See also Julie Doyle, "Where Has All the Oil Gone? BP Branding and the Discursive Elimination of Climate Change Risk," in *Culture, Environment and Ecopolitics*, ed. Nick Heffernan and David A. Wragg (Newcastle upon Tyne: Cambridge Scholars Publishing, 2011), 200–225.

67. Wändi Bruine de Bruin et al., "Public Understanding of Climate Change Terminology," *Climatic Change* 167, no. 3–4 (2021): 37, doi: 10.1007/s10584-021-03183-0.

68. In addition to Wiedenhofer, Haberl, Hickel, and others cited elsewhere, see Selçuk Gürçam, "The Neoliberal Initiative of the Aviation Industry to Fight the Climate Crisis: Greenwashing," *International Journal of Environment and Geoinformatics* 9, no. 3 (2022): 178–186, https://doi.org/10.30897/ijegeo.1083921; Matthew Megura and Ryan Gunderson, "Better Poison Is the Cure? Critically Examining Fossil Fuel Companies, Climate Change Framing, And Corporate Sustainability Reports," *Energy Research & Social Science* 85 (2022): 102388, https://doi.org/10.1016/j.erss.2021.102388; Rex Weyler, "The Great Carbon Capture Scam," Greenpeace, June 1, 2022, https://www.greenpeace.org/international/story/54079/great-carbon-capture-scam/; James Temple, "Carbon Removal Hype Is Becoming a Dangerous Distraction," *MIT Technology Review*, July 8, 2021, https://www.technologyreview.com/2021/07/08/1027908/carbon-removal-hype-is-a-dangerous-distraction-climate-change/.

69. Katharine Hayhoe (@Khayhoe), "We need to push back just as strongly on doomerism as we do on denial because they both accomplish exactly the same thing: inaction," Twitter, July 30, 2022, https://twitter.com/KHayhoe/status/1553573600547979266.

70. Mann, Hassol, and Toles, "Doomsday Scenarios Are as Harmful as Climate Change Denial." See also Mann, *The New Climate War* (New York: Public Affairs, 2021), especially ch. 8. Mann's chapter on "doomism," titled "The Truth Is Bad Enough," is full of factual inaccuracies, ad hominem attacks, and unfounded insinuation. He lumps in journalists and writers who emphasize the dangers of climate change with genuine apocalypticists and conflates pessimistic and monitory assess-

ments with arguments in favor of inaction, a regrettably common fallacy among "anti-doomist" critics.

71. Joseph Romm, "Apocalypse Not: Oscars, Media, and the Myths of Climate Message," *Energy Central*, February 25, 2013, https://energycentral.com/c/ec/apocalypse-not-oscars-media-and-myths-climate-message.

72. Florian Jehn et al., "Betting on the Best Case: Higher End Warming is Underrepresented in Research," *Environmental Research Letters* 16, no. 8 (2021): 084036. Also see Kemp et al., "Climate Endgame"; and Florian Jehn et al., "Focus of the IPCC Assessment Reports Has Shifted to Lower Temperatures," *Earth's Future* 10, no. 5 (2022): e2022EF002876, https://doi.org/10.1029/2022EF002876.

73. Ronald W. Rogers, "Cognitive and Physiological Processes in Fear Appeals and Attitude Change: A Revised Theory of Protection Motivation," in *Social Psychophysiology: A Sourcebook*, ed. John T. Cacioppo and Richard E. Petty (New York: Guilford Press, 1983), 153–176; Ronald W. Rogers, "A Protection Motivation Theory of Fear Appeals and Attitude Change," *The Journal of Psychology* 91, no. 1 (1975): 93–114, https://doi.org/10.1080/00223980.1975.9915803.

74. Roy Baumeister et al., "Bad Is Stronger Than Good," *Review of General Psychology* 5, no. 4 (2001): 323–370, https://doi.org/10.1037//1089-2680.5.4.323.

75. Kim Witte and Mike Allen, "A Meta-Analysis of Fear Appeals: Implications for Effective Public Health Campaigns," *Health Education & Behavior* 27, no. 5 (2000): 591–615, https://doi.org/10.1177/109019810002700506.

76. This meta-analysis of 127 articles concluded that "fear appeals consistently work." Melanie B. Tannenbaum et al., "Appealing to Fear: A Meta-Analysis of Fear Appeal Effectiveness and Theories," *Psychological Bulletin* 141, no. 6 (2015): 1178–1204, https://doi.org/10.1037/a0039729. See also Craig A. Harper et al., "Functional Fear Predicts Public Health Compliance in the COVID-19 Pandemic," *International Journal of Mental Health and Addiction* 19 (2021): 1875–1888, doi: 10.1007/s11469-020-00281-5; Gjalt-Jorn Ygram Peters, Robert A. C. Ruiter, and Gerjo Kok, "Threatening Communication: A Critical Re-Analysis and a Revised Meta-Analytic Test of Fear Appeal Theory," *Health Psychology Review* 7, Suppl. 1 (2013): S8–S31, https://doi.org/10.1080/17437199.2012.703527; Natascha de Hoog et al., "The Impact of Vulnerability to and Severity of a Health Risk on Processing and Acceptance of Fear-Arousing Communications," *Review of General Psychology* 11, no. 3 (2007): 258–85, https://doi.org/10.1037/1089-2680.11.3.258; Paul S. Martin, "Inside the Black Box of Negative Campaign Effects: Three Reasons Why Negative Campaigns Mobilize," *Political Psychology* 25, no. 4 (2004): 545–562, https://doi.org/10.1111/j.1467-9221.2004.00386.x; Sarah Milne, Paschal Sheeran, and Sheina Orbell. "Prediction and Intervention in Health-Related Behavior: A Meta-Analytic Review of Protection Motivation Theory," *Journal of Applied Social Psychology* 30, no. 1 (2000): 106–143. https://doi.org/10.1111/j.1559-1816.2000.tb02308.x; Donna L. Floyd, Steven Prentice-Dunn, and Ronald W. Rogers, "A Meta-Analysis of Research on Protection Motivation Theory," *Journal of Applied Social Psychology* 30, no. 2 (2000): 407–429, https://doi.org/10.1111/j.1559-1816.2000.tb02323.x.

77. Edward W. Maibach et al., "Communication and Marketing as Climate Change–Intervention Assets," *American Journal of Preventive Medicine* 35, no. 5 (2008) 488–500, https://doi.org/10.1016/j.amepre.2008.08.016.

78. Joseph P. Reser and Graham L. Bradley, "Fear Appeals in Climate Change Communication," in *The Oxford Encyclopedia of Climate Change Communication*, vol. 2, ed. Matthew Nisbet (Oxford: Oxford University Press, 2018), 614. See also "The Wages of Fear?," where political scientist Alison McQueen presents a concise, empirically grounded philosophical argument that climate change fear appeals are resistant to the kinds of criticism usually levied against them, appeals to "hope" have serious drawbacks, and "we should take our bearings from Aristotle in an effort to cultivate fear more responsibly" (170). McQueen covers some of the same ground I do and offers a robust Aristotelian argument for the civic value of responsible fear appeals. Alison McQueen, "The Wages of Fear?: Toward Fearing Well About Climate Change," *Philosophy and Climate Change*, ed.. Mark Budolfson, Tristam McPherson, and David Plunkett (Oxford: Oxford University Press, 2021), 152–177.

79. Witte and Allen, "A Meta-Analysis of Fear Appeals," 595.

80. Roy Baumeister and John Tierney, *The Power of Bad: How the Negativity Effect Rules Us and How We Can Rule It* (New York: Penguin, 2019). See also Adam Shriver, "The Asymmetrical Contributions of Pleasure and Pain to Subjective Well-Being," *Review of Philosophy and Psychology* 5 (2014): 135–153, https://doi.org/10.1007/s13164-013-0171-2; Benjamin E. Hilbig, "Sad, Thus True: Negativity Bias in Judgments of Truth," *Journal of Experimental Social Psychology* 45, no. 4 (2009): 983–986, https://doi.org/10.1016/j.jesp.2009.04.012; Kevin J. Flannelly et al., "Beliefs, Mental Health, and Evolutionary Threat Assessment Systems in the Brain," *The Journal of Nervous and Mental Disease* 195, no. 12 (2007): 996–1003, https://doi.org/10.1097/NMD.0b013e3 1815c19b1; Paul Rozin and Edward B. Royzman, "Negativity Bias, Negativity Dominance, and Contagion," *Personality and Social Psychology Review* 5, no. 4 (2001): 296–320, https//doi.org/10.1207/S15327957PSPR0504_2; Daniel Kahneman and Amos Tversky, "Choices, Values, and Frames," *American Psychologist* 39, no. 4 (1984), 341–350. https://doi.org/10.1037/0003-066X.39.4.341.

81. Reser and Bradley, "Fear Appeals in Climate Change Communication," 615. For more on negative messaging and fear appeals, see Tobias Brosch, "Affect and Emotions as Drivers of Climate Change Perception and Action: A Review," *Current Opinion in Behavioral Sciences* 42 (2021): 15–21, https://doi.org/10.1016/j.cobeha.2021 .02.001; Anne M. Van Valkengoed and Linda Steg, "Meta-Analyses of Factors Motivating Climate Change Adaptation Behaviour," *Nature Climate Change* 9, no. 2 (2019): 158–163, https://doi.org/10.1038/s41558-018-0371-y; Chrysantus Awagu and Debra Z. Basil, "Fear Appeals: The Influence of Threat Orientations," *Journal of Social Marketing* 6, no. 4 (2016): 361–376, https://doi.org/10.1108/JSOCM-12-2014-0089; Lauren Feldman and P. Sol Hart, "Using Political Efficacy Messages to Increase Climate Activism: The Mediating Role of Emotions," *Science Communication* 38, no. 1 (2016): 99–127, https://doi.org/10.1177/1075547015617941; Stuart Neil Soroka, *Negativity in Democratic Politics : Causes and Consequences* (Cambridge: Cambridge University

Press, 2014); Kevin Arceneaux, "Cognitive Biases and the Strength of Political Arguments," *American Journal of Political Science* 56, no. 2 (2012): 271–285, https://doi.org/10.1111/j.1540-5907.2011.00573.x; Kim Leslie Fridkin and Patrick J. Kenney, "Do Negative Messages Work?: The Impact of Negativity on Citizens' Evaluations of Candidates," *American Politics Research* 32, no. 5 (September 2004): 570–605, https://doi.org/10.1177/1532673X03260834; Gerard Hastings et al., "Fear Appeals in Social Marketing: Strategic and Ethical Reasons for Concern," *Psychology and Marketing* 21, no. 11 (2004): 961–986, https://doi.org/10.1002/mar.20043; Michael D. Cobb and James H. Kuklinski, "Changing Minds: Political Arguments and Political Persuasion," *American Journal of Political Science* 41, no. 1 (1997): 88–121, https://doi.org/10.2307/2111710.

82. Thomas Lowe et al., "Does Tomorrow Ever Come? Disaster Narrative and Public Perceptions of Climate Change," *Public Understanding of Science* 15, no. 4 (2006): 435–457, https//:doi.org/10.1177/0963662506063796.

83. Anthony Lane in the *New Yorker* called the film "a shambles of dud writing and dramatic inconsequence," and *Salon.com*'s Stephanie Zacharek called it "a big, dumb movie, another Hollywood entertainment that . . . leaves us feeling thick and stupid." Anthony Lane, "Cold Comfort," *New Yorker*, May 30, 2004, https://www.newyorker.com/magazine/2004/06/07/cold-comfort-4; Stephanie Zacharek, "The Day After Tomorrow," *Salon*, May 28, 2004, https://www.salon.com/2004/05/28/day_after_tomorrow/.

84. The gap between intention and action is a consistent challenge for such research. It is much easier to measure "intention" in a survey than it is to collect long-term data about climate change "actions," especially since whatever actions one takes in response to climate change are likely to be complexly overdetermined by multiple motivations.

85. Lorraine Whitmarsh et al., "Climate Anxiety: What Predicts It and How Is It Related to Climate Action?," *Journal of Environmental Psychology* 83 (2022): 101866, https://doi.org/10.1016/j.jenvp.2022.101866.

86. Charles A. Ogunbode et al., "Climate Anxiety, Wellbeing and Pro-Environmental Action: Correlates of Negative Emotional Responses to Climate Change in 32 Countries," *Journal of Environmental Psychology* 84 (2022): 101887, https://doi.org/10.1016/j.jenvp.2022.101887.

87. Wen Xue et al., "Combining Threat and Efficacy Messaging to Increase Public Engagement with Climate Change in Beijing, China," *Climatic Change* 137, no. 1–2 (2016): 43–55, https://doi.org/10.1007/s10584-016-1678-1.

88. Matthew Ballew et al., "Is Distress About Climate Change Associated with Climate Action?," Yale Program on Climate Change Communication, August 3, 2023, https://climatecommunication.yale.edu/publications/distress-about-climate-change-and-climate-action/. Naoko Kaida and Kosuke Kaida, "Facilitating Pro-Environmental Behavior: The Role of Pessimism and Anthropocentric Environmental Values," *Social Indicators Research* 126, no. 3 (2016): 1243–1260, https://doi.org/10

.1007/s11205-015-0943-4; Naoko Kaida and Kosuke Kaida, "Pro-Environmental Behavior Correlates with Present and Future Subjective Well-Being," *Environment, Development and Sustainability* 18 (2016): 111–127, https://doi.org/10.1007/s10668-015-9629-y; Feldman and Hart, "Using Political Efficacy Messages to Increase Climate Activism."

89. Lauren Feldman and P. Sol Hart, "Is There Any Hope? How Climate Change News Imagery and Text Influence Audience Emotions and Support for Climate Mitigation Policies," *Risk Analysis* 38, no. 3 (2018): 585–602, https://doi.org/10.1111/risa.12868.

90. Matthew J. Hornsey and Kelly S. Fielding, "Understanding (and Reducing) Inaction on Climate Change," *Social Issues and Policy Review* 14, no. 1 (2020): 3–35, doi: 10.1111/sipr.12058; Matthew J. Hornsey et al., "Evidence for Motivated Control: Understanding the Paradoxical Link Between Threat and Efficacy Beliefs About Climate Change," *Journal of Environmental Psychology* 42 (2015): 57–65, doi: 10.1016/j.jenvp.2015.02.003; Hornsey and Fielding, "A Cautionary Note."

91. Brandi S. Morris et al., "Optimistic vs. Pessimistic Endings in Climate Change Appeals," *Humanities and Social Sciences Communications* 7, no. 1 (2020), https://doi-org.proxy.library.nd.edu/10.1057/s41599-020-00574-z.

92. Kim Witte, "Putting the Fear Back into Fear Appeals: The Extended Parallel Process Model," *Communication Monographs* 59 no. 4 (1992): 329–349, doi: 10.1080/03637759209376276

93. Witte and Allen, "A Meta-Analysis of Fear Appeals." Erin K. Maloney et al., "Fear Appeals and Persuasion: A Review and Update of the Extended Parallel Process Model," *Social and Personality Psychology Compass* 5, no. 4 (2011): 206–219, https://doi.org/10.1111/j.1751-9004.2011.00341.x. Suzanne Moser makes a similar argument in Suzanne Moser, "More Bad News: The Risk of Neglecting Emotional Responses to Climate Change Information," in *Creating a Climate for Change*, ed. Suzanne Moser (Cambridge: Cambridge University Press, 2007), 64–80. See also Xin Ma et al., "Promoting Behaviors to Mitigate the Effects of Climate Change: Using the Extended Parallel Process Model at the Personal and Collective Level in China," *Environmental Communication* 17, no. 4 (2023): 353–369. doi:10.1080/17524032.2023.2181134.

94. Nicholas A. Valentino et al., "Efficacy, Emotions and the Habit of Participation," *Political Behavior* 31, no. 3 (2009): 307–330, https://doi.org/10.1007/s11109-008-9076-7. A meta-analysis of studies on HIV prevention from Earl and Albarracín offers a contrasting example, in that threat messaging did not increase condom use and in fact led to decreased condom use over time, while HIV counseling and testing led to increased condom use. Albarracín and Earl don't offer an explanation for why fear appeals failed in this case. Allison Earl and Dolores Albarracín, "Nature, Decay, and Spiraling of the Effects of Fear-Inducing Arguments and HIV Counseling and Testing," *Health Psychology* 26, no. 4 (2007): 496–506, https://doi.org/10.1037/0278-6133.26.4.496.

95. Maloney et al., "Fear Appeals and Persuasion."

96. John R. Hibbing et al., "Differences in Negativity Bias Underlie Variations in Political Ideology," *Behavioral and Brain Sciences* 37, no. 3 (2014): 297–307, https://doi.org/10.1017/S0140525X13001192; Feldman and Hart, "Is There Any Hope?" For an interesting revision of this theory, see Shona M. Tritt et al., "Ideological Reactivity: Political Conservatism and Brain Responsivity to Emotional and Neutral Stimuli," *Emotion* 16, no. 8 (2016): 1172, https://doi.org/10.1037/emo0000150.

97. Fridkin and Kenney, "Do Negative Messages Work?"

98. During the COVID-19 pandemic, for instance, concern about COVID pushed climate change out of focus. Oleg Smirnov and Pei-Hsun Hsieh, "COVID-19, Climate Change, and the Finite Pool of Worry in 2019 to 2021 Twitter Discussions," *Proceedings of the National Academy of Sciences* 119, no. 43 (2022): e2210988119. https://doi.org/10.1073/pnas.2210988119.

99. Cass R. Sunstein, "The Availability Heuristic, Intuitive Cost-Benefit Analysis, and Climate Change," *Climatic Change* 77 (2006): 195–210; https://doi-org.proxy.library.nd.edu/10.1007/s10584-006-9073-y; M. Granger Morgan et al., *Risk Communication: A Mental Models Approach* (Cambridge: Cambridge University Press, 2002).

100. Christopher B. Mann et al., "Do Negatively Framed Messages Motivate Political Participation? Evidence from Four Field Experiments," *American Politics Research* 48, no. 1 (January 2020): 3–21, https://doi.org/10.1177/1532673X19840732; Jennifer Corns, "Rethinking the Negativity Bias," *Review of Philosophy and Psychology* 9, no. 3 (2018): 607–625; Jonah Berger, Alan T. Sorensen, and Scott J. Rasmussen, "Positive Effects of Negative Publicity: When Negative Reviews Increase Sales," *Marketing Science* 29, no. 5 (2010): 815–827, https://doi.org/10.1287/mksc.1090.0557; Hastings et al., "Fear Appeals in Social Marketing." A provocatively titled 2018 attempt to re-evaluate fear appeals in the domain of cigarette packaging by Gerjo Kok et al., "Ignoring Theory and Misinterpreting Evidence: The False Belief in Fear Appeals," was strongly refuted in the very journal issue in which it appeared. See Gerjo Kok et al., "Ignoring Theory and Misinterpreting Evidence: The False Belief in Fear Appeals," *Health Psychology Review* 12, no. 2 (2018): 111–125; and response articles in that same issue.

101. Marjolaine Martel-Morin and Erick Lachapelle, "Code Red for Humanity or Time for Broad Collective Action? Exploring the Role of Positive and Negative Messaging in (De)motivating Climate Action," *Frontiers in Communication* 7 (2022): 252, https://doi.org/10.3389/fcomm.2022.968335.

102. Sonia Hélène Merkel et al., "Climate Change Communication: Examining the Social and Cognitive Barriers to Productive Environmental Communication," *Social Science Quarterly* 101, no. 5 (2020): 2085–2100, https://doi.org/10.1111/ssqu.12843.

103. Matthew Feinberg and Robb Willer, "Apocalypse Soon? Dire Messages Reduce Belief in Global Warming by Contradicting Just-World Beliefs," *Psychological Science* 22, no. 1 (2011): 34–38, https://doi.org/10.1177/0956797610391911.

104. Jan G. Voelkel et al., "The Effects of Dire and Solvable Messages on Belief in Climate Change: A Replication Study," PsyArXiv.org (2021), https://doi.org/10.31234/osf.io/prk9f.

105. Kat Kerlin, "Eric Post: Arctic Awe and Anxiety on Climate Coping in the Classroom and Nature's Generosity," *UC Davis News*, August 8, 2022, https://www.ucdavis.edu/climate/news/eric-post-arctic-awe-and-anxiety; Jeff Tollefson, "Top Climate Scientists Are Sceptical That Nations Will Rein In Global Warming," *Nature* 599, no. 7883 (2021): 22–24, https://doi.org/10.1038/d41586-021-02990-w; Daniel Gilford et al., "The Emotional Toll of Climate Change on Science Professionals," *Eos* 100 (2019): 6, https://doi.org/10.1029/2019EO137460; Susan Clayton, "Mental Health Risk and Resilience Among Climate Scientists," *Nature Climate Change* 8, no. 4 (2018): 260–261, https://doi.org/10.1038/s41558-018-0123-z; Daniel Oberhaus, "Climate Change Is Giving Us 'Pre-Traumatic Stress,'" *VICE* (2017), https://www.vice.com/en/article/vvzzam/climate-change-is-giving-us-pre-traumatic-stress; Emma Lawrance et al., "The Impact of Climate Change on Mental Health and Emotional Wellbeing: A Narrative Review of Current Evidence and Its Implications," *International Review of Psychiatry* 34, no. 5 (2022): 443–498, doi:10.1080/09540261.2022.2128725.

106. Sharon Dunwoody, "Science Journalism: Prospects in the Digital Age," *Routledge Handbook of Public Communication of Science and Technology*, ed. Massimiano Bucchi and Brian Trench (London: Routledge, 2021), 14–32; Josh Anderson and Anthony Dudo, "A View from the Trenches: Interviews with Journalists About Reporting Science News," *Science Communication* 45, no. 1 (2023): 39–64, https://doi.org/10.1177/10755470221149156,

107. Zeynep Tufekci, "An Object Lesson from Covid on How to Destroy Public Trust," *New York Times*, June 8, 2024, https://www.nytimes.com/2024/06/08/opinion/covid-fauci-hearings-health.html.

108. Myanna Lahsen and Esther Turnhout, "How Norms, Needs, and Power in Science Obstruct Transformations Towards Sustainability," *Environmental Research Letters* 16, no. 2 (2021): 025008, https://doi.org/10.1088/1748-9326/abdcf0; Michael Oppenheimer et al., *Discerning Experts: The Practices of Scientific Assessment for Environmental Policy* (Chicago: University of Chicago Press, 2019); Debra Javeline, et al., "Expert Opinion on Climate Change and Threats to Biodiversity," *Bioscience* 63, no. 8 (2013): 666–673, https://doi.org/10.1525/bio.2013.63.8.9; Jehn et al., "Betting on the Best Case."

109. Meyer, "Are We as Doomed as That New York Magazine Article Says?"

110. Reiner Grundmann, "'Climategate' and the Scientific Ethos," *Science, Technology, & Human Values* 38, no. 1 (2013): 67–93, https://doi.org/10.1177/0162243911432318, 86.

111. Andrew Revkin, "Hacked E-Mail Is New Fodder for Climate Dispute," *The New York Times*, November 20, 2009, https://www.nytimes.com/2009/11/21/science/earth/21climate.html.

112. Leiserowitz et al., "Climategate, Public Opinion, and the Loss of Trust." See

also Edward Maibach et al., "The Legacy of Climategate: Undermining or Revitalizing Climate Science and Policy?," *WIREs Climate Change* 3 (2012): 289–295, doi: 10.1002/wcc.168.

113. For more on government and corporate efforts to censor climate science, see Christopher Reddy, *Science Communication in a Crisis: An Insider's Guide* (New York: Routledge, 2023), ch. 8; James Hansen, "Uncensored Science is Crucial for Global Conservation," in *Conservation Science and Advocacy for a Planet in Peril: Speaking Truth to Power*, ed. Dominick DellaSala (Cambridge: Elsevier, 2021), xxv–lvi. Mark Stander Bowen, *Censoring Science: Inside the Political Attack on Dr. James Hansen and the Truth of Global Warming* (New York: Penguin, 2008).

114. Holdren quoted in James Kanter and Andrew C. Revkin, "World Scientists Near Consensus on Warming," *New York Times*, January 24, 2007, https://www.nytimes.com/2007/01/30/world/30climate.html.

115. Roger Pielke et al., "Lifting the Taboo on Adaptation," *Nature* 445, no. 7128 (2007): 597–598, https://doi.org/10.1038/445597a.

116. Albert Gore, *Earth in the Balance: Ecology and the Human Spirit* (New York: Houghton Mifflin, 1992), 240.

117. William F. Lamb et al., "Discourses of Climate Delay," *Global Sustainability* 3 (2020): e17, https://doi.org/10.1017/sus.2020.13.

118. Debra Javeline and Sophia N. Chau, "The Unexplored Politics of Climate Change Adaptation," in *Handbook of US Environmental Policy*, ed. David Konisky (Cheltenham: Edward Elgar, 2020), 357–372.

119. Joseph Aldy and Richard Zeckhauser, "Three Prongs for Prudent Climate Policy," *Southern Economic Journal* 87, no. 1 (2020): 3–29, https://doi.org/10.3386/w26991; Robyn S. Wilson et al., "From Incremental to Transformative Adaptation in Individual Responses to Climate-Exacerbated Hazards," *Nature Climate Change* 10, no. 3 (2020): 200–208, https://doi.org/10.1038/s41558-020-0691-6; Tollefson, "Top Climate Scientists Are Sceptical."

120. Charles S. Carver and Michael F. Scheier, "Optimism, Pessimism, and Self-Regulation," in *Optimism and Pessimism: Implications for Theory, Research, and Practice*, ed. Edward C. Chang (Washington, D.C.: APA, 2001), 31–52.

121. Daniel Kahneman, "A Perspective on Judgment and Choice: Mapping Bounded Rationality," *American Psychologist* 58, no. 9 (2003): 697.

122. See Rebecca Solnit, *A Paradise Built in Hell: The Extraordinary Communities That Arise in Disaster* (New York: Penguin, 2010); and Carl Schmitt, *Political Theology: Four Chapters on the Concept of Sovereignty* (Chicago: University of Chicago Press, 2005).

123. Examples are numerous. Limiting our consideration to the United States in the twentieth century, think of the confluence of revolutionary politics and Cold War crises that defined the 1960s, the violent domestic racial politics of World War II, the intense political fights over Roosevelt's New Deal during the Great Depression, and the conflict between labor and big business during World War I.

124. Meghan A. Duffy, "Why We Should Preach to the Climate Change Choir: The Importance of Science Communication That Engages People Who Already Accept Climate Change," *The American Naturalist* 198, no. 3 (2021): 433–436, https://doi.org/10.1086/715153. Geoff Dembicki, "Preaching to the Choir," The Breakthrough Institute, August 23, 2013, https://thebreakthrough.org/issues/energy/preaching-to-the-choir.

125. Dylan Bugden, "Does Climate Protest Work? Partisanship, Protest, and Sentiment Pools," *Socius* 6 (2020), https://doi.org/2378023120925949.

126. Rebecca Solnit, "Preaching to the Choir," *Harper's Magazine*, November 2017.

127. Graham Dixon, Jay Hmielowski, and Yanni Ma, "Improving Climate Change Acceptance Among U.S. Conservatives Through Value-Based Message Targeting," *Science Communication* 39, no. 4 (2017): 520–534, https://doi.org/10.1177/1075547017715473. See also Mathew Goldberg et al., "Shifting Republican Views on Climate Change Through Targeted Advertising," *Nature Climate Change* 11 (2021): 573–577, https://doi.org/10.1038/s41558-021-01070-1.

128. Matthew Feinberg and Robb Willer, "The Moral Roots of Environmental Attitudes," *Psychological Science* 24, no. 1 (2013): 56–62, https://doi.org/10.1177/0956797612449177.

129. Christopher Wolsko, Hector Ariceaga, and Jesse Seiden, "Red, White, and Blue Enough to Be Green: Effects of Moral Framing on Climate Change Attitudes and Conservation Behaviors," *Journal of Experimental Social Psychology* 65 (2016): 7–19.

130. Rob Nixon, *Slow Violence* (Cambridge: Harvard University Press, 2011).

131. Matthew J. Hornsey, "The Role of Worldviews in Shaping How People Appraise Climate Change," *Current Opinion in Behavioral Sciences* 42 (2021): 36–41; Irina Feygina et al., "System Justification, the Denial of Global Warming, and the Possibility of 'System-Sanctioned Change,'" *Personality & Social Psychology Bulletin* 36, no. 3 (2010): 326–338, https://doi.org/10.1177/0146167209351435; Aaron M. McCright et al., "Examining the Effectiveness of Climate Change Frames in the Face of a Climate Change Denial Counter-Frame," *Topics in Cognitive Science* 8, no. 1 (2016): 76–97.

132. Thomas Bernauer and Liam F. McGrath, "Simple Reframing Unlikely to Boost Public Support for Climate Policy," *Nature Climate Change* 6, no. 7 (2016): 680–683.

133. Hilary Byerly et al., "A Story Induces Greater Environmental Contributions Than Scientific Information Among Liberals but Not Conservatives," *One Earth* 4, no. 4 (2021): 545–552, https://doi.org/10.1016/j.oneear.2021.03.004.

134. Lukas P. Fesenfeld et al., "The Role and Limits of Strategic Framing for Promoting Sustainable Consumption and Policy," *Global Environmental Change* 68 (2021): 102266, https://www.sciencedirect.com/science/article/pii/S0959378021000455, doi: 10.1016/j.gloenvcha.2021.102266. It should be noted that none of the frames offered by Fesenfield are recognizably "conservative."

135. Laura M. Arpan et al., "Politics, Values, and Morals: Assessing Consumer Responses to the Framing of Residential Renewable Energy in the United States," *Energy Research & Social Science* 46 (2018): 321–331; Samantha K. Stanley et al., "The Effects of a Temporal Framing Manipulation on Environmentalism: A Replication and Extension," *PLOS ONE* 16, no. 2 (2021): e0246058, doi: 10.1371/journal.pone.0246058.

136. David Roberts, "Is It Worth Trying to 'Reframe' Climate Change? Probably Not," *Vox*, updated February 27, 2017, https://www.vox.com/2016/3/15/11232024/reframe-climate-change.

137. Brulle and Norgaard, "Avoiding Cultural Trauma."

138. Elinor Ostrom et al., "Revisiting the Commons: Local Lessons, Global Challenges," *Science* 284 (1999): 278–282, doi: 10.1126/science.284.5412.278.

139. Garrett Hardin, "The Tragedy of the Commons," *Science* 162, no. 3859 (1968): 1243–1248, http://www.jstor.org/stable/1724745, 1248.

140. Kenneth Boulding, "The Economics of the Coming Spaceship Earth," in *Environmental Quality in a Growing Economy: Essays from the Sixth RFF Forum*, ed. Henry Jarrett (Baltimore: Johns Hopkins University Press, 1966): 3–14; R. Buckminster Fuller, *Operating Manual for Spaceship Earth* (New York: Simon and Schuster, 1969).

141. Garrett Hardin, "Living on a Lifeboat," *Bioscience* 24, no. 10 (1974): 561–568, https://doi.org/10.2307/1296629, 561.

142. Pentti Linkola, *Can Life Prevail? A Revolutionary Approach to the Environmental Crisis* (London: Arktos, 2011), 131. Evangelos D. Protopapadakis identifies Linkola as an ecofascist, though Protopapadakis's definition of ecofascism is quite broad. See Evangelos D. Protopapadakis, "Environmental Ethics and Linkola's Ecofascism: An Ethics Beyond Humanism," *Frontiers of Philosophy in China* 9, no. 4 (2014): 586–601, https://doi.org/10.3868/s030-003-014-0048-3.

143. Guillaume Faye, *Convergence of Catastrophes*, trans. E. Christian Kopff (London: Arktos, 2012), 27.

144. Parag Khanna, "What Comes After the Coming Climate Anarchy?," *Time*, August 15, 2022, https://time.com/6206111/climate-change-anarchy-what-comes-next/.

145. Tarrant's manifesto, "The Great Replacement," has been suppressed and is not easily available. At the time of writing, it could be found here: https://dl1.cuni.cz/mod/resource/view.php?id=522947. It may be difficult to still find online. See also Rakib Ehsan and Paul Stott, *Far-Right Terrorist Manifestos: A Critical Analysis* (London: Henry Jackson Society, 2020), https://henryjacksonsociety.org/wp-content/uploads/2020/02/HJS-Terrorist-Manifesto-Report-WEB.pdf.

146. Patrick Crusius, "The Inconvenient Truth." At the time of writing, Crusius's manifesto could be found here: https://randallpacker.com/wp-content/uploads/2019/08/The-Inconvenient-Truth.pdf.

147. Maxine Joselow, "Suspect in Buffalo Rampage Cited 'Ecofascism' to Justify

Actions," *Washington Post*, May 17, 2022, https://www.washingtonpost.com/politics/2022/05/17/suspect-buffalo-rampage-cited-ecofascism-justify-actions/. Gendron's manifesto, "You Wait for a Signal While Your People Wait for You," has also been suppressed and is not easily available. I was able to find what appeared to be a copy here: https://www.hoplofobia.info/wp-content/uploads/2022/05/PG-Manifesto.pdf.

148. Alexander Ross and Emmi Bevensee, "Confronting the Rise of Eco-Fascism Means Grappling with Complex Systems," *CARR Research Insight* 3 (London: Centre for Analysis of the Radical Right, 2020). See also Sam Moore and Alex Roberts's very helpful *The Rise of Ecofascism: Climate Change and the Far Right* (Cambridge: Polity, 2022), which as of this writing is the best contemporary account of racialist right-wing ecological politics.

149. Jake Hanrahan, "Inside the Unabomber's Odd and Furious Online Revival," *Wired*, August 1, 2018, https://www.wired.co.uk/article/unabomber-netflix-tv-series-ted-kaczynski; Kate Aronoff, "Green Tea Party," *Harper's Magazine*, May 30, 2017; Sarah Manavis, "Eco-Fascism: The Ideology Marrying Environmentalism and White Supremacy Thriving Online," *The New Statesman*, September 21, 2018.

150. Imogen Richards et al., "Eco-Fascism Online: Conceptualizing Far-Right Actors' Response to Climate Change on Stormfront," *Studies in Conflict and Terrorism* (2022): 1–27, https://doi.org/10.1080/1057610X.2022.2156036.

151. Brian Hughes et al., "Ecofascism: An Examination of the Far-Right/Ecology Nexus in the Online Space," *Terrorism and Political Violence* 34, no. 5 (2022): 997–1023, https://doi.org/10.1080/09546553.2022.2069932, 998. Hughes et al. argue that there a clear distinction between ecofascism, which they define as "first and foremost an imaginary and cultural expression of mystical, anti-humanist Romanticism," and more conventional "far right ecologism," which term they adopt from Balša Lubarda, "Beyond Ecofascism? Far-Right Ecologism (FRE) as a Framework for Future Inquiries," *Environmental Values* 29, no. 6 (2020): 713–732, https://doi.org/10.3197/096327120X15752810323922.

152. Hughes et al., "Ecofascism," 998.

153. Marc Sageman, *Turning to Political Violence: The Emergence of Terrorism* (Philadelphia: University of Pennsylvania Press, 2017), 17.

154. Sageman, *Turning to Political Violence*, 17.

155. Sean Fleming, "The Unabomber and the Origins of Anti-tech Radicalism," *Journal of Political Ideologies* 27, no. 2 (2022): 207–225, doi: 10.1080/13569317.2021.1921940. See also Ole Martin Moen, "The Unabomber's Ethics," *Bioethics* 33, no. 2 (2019): 223–229.

156. The literature on the environmental and ecological aspects of Nazism, as well as the racist and imperialist roots of ecological thinking, is substantial and deserves to be better known. See, for starters, Timothy Snyder, *Black Earth: The Holocaust as History and Warning* (New York: Tim Duggan Books, 2015); Janet Biehl and Peter Staudenmaier, *Ecofascism Revisited: Lessons from the German Experience* (New York: New Compass Press, 2011); Frank Uekötter, *The Green and the Brown: A*

History of Conservation in Nazi Germany (Cambridge: Cambridge University Press, 2006); and Thomas M. Lekan, *Imagining the Nation in Nature: Landscape Preservation and German Identity, 1885–1945* (Cambridge: Harvard University Press, 2004). These are not merely historical questions. For instance, the work of one of the leading early thinkers combining ecological thinking and eugenicist racism, Ludwig Klages, is today kept in print by the reactionary publisher Arktos Press.

157. See Moore and Roberts, *Rise of Ecofascism*, ch. 3; as well as Jedediah Purdy, *After Nature: A Politics for the Anthropocene* (Cambridge: Harvard University Press, 2015); Graham Macklin, "The Extreme Right, Climate Change and Terrorism," *Terrorism and Political Violence* 34, no. 5 (2022): 979–996, doi: 10.1080/09546553.2022.2069928.

158. Some of the best accounts of these issues can be found in Keith Makato Woodhouse, *The Ecocentrists: A History of Radical Environmentalism* (New York: Columbia University Press, 2018). See also Bron Taylor, "Religion, Violence and Radical Environmentalism: From Earth First! to the Unabomber to the Earth Liberation Front," *Terrorism and Political Violence* 10, no. 4 (1998): 1–42, doi: 10.1080/09546559808427480; George Bradford, "How Deep is Deep Ecology?" *Fifth Estate* 22, no. 3 (Fall 1987), https://jstor.org/stable/community.28036651; George Sessions, *Deep Ecology for the Twenty-First Century* (Boston: Shambhala, 1994). A key dispute between Murray Bookchin and Dave Foreman can be traced in Bill Devall, "A Spanner in the Woods: An Interview with Dave Foreman," *Simply Living* 2, no. 12 (1986): 3–4; Murray Bookchin, "Social Ecology versus Deep Ecology: A Challenge for the Ecology Movement," *Green Perspectives: Newsletter of the Green Program Project*, nos. 4–5 (summer 1987), http://dwardmac.pitzer.edu/Anarchist_Archives/bookchin/socecovdeepeco.html; and Steve Chase, ed., *Defending the Earth: A Dialogue Between Murray Bookchin and Dave Foreman* (Boston: South End Press, 1991).

159. Biehl and Staudenmaier, *Ecofascism Revisited*, 41.

160. Blair Taylor, "Alt-Right Ecology: Ecofascism and Far-Right Environmentalism in the United States," in *The Far Right and the Environment*, ed. Bernhard Forchtner (New York: Routledge, 2019), 275–292.

161. Moore and Roberts, *Rise of Ecofascism*, 48. See also Bernhard Forchtner, "Climate Change and the Far Right," *WIREs Climate Change* 10, no. 5 (2019), https://doi.org/10.1002/wcc.604. Joakim Kulin et al., "Nationalist Ideology, Rightwing Populism, and Public Views About Climate Change in Europe," *Environmental Politics* 30, no. 7 (2021): 1111–1134.

162. Balša Lubarda, "Beyond Ecofascism?"

163. See my own "Climate Change Is Not World War," *New York Times*, September 18, 2019, https://www.nytimes.com/2019/09/18/opinion/climate-change-mobilization.html.

164. Mona Ali, "Militarized Adaptation," *Phenomenal World*, January 25, 2023, https://www.phenomenalworld.org/analysis/militarized-adaptation/; Pierre Charbonnier, "La naissance de l'écologie de guerre," *Le Grand Continent*, March 18, 2022,

https://legrandcontinent.eu/fr/2022/03/18/la-naissance-de-lecologie-de-guerre/; Thea Riofrancos, "The Security-Sustainability Nexus: Lithium Onshoring in the Global North," *Global Environmental Politics* 23, no. 1 (2023): 20–41, https://doi.org/10.1162/glep_a_00668. See also Geoff Mann and Joel Wainwright, *Climate Leviathan* (New York: Verso, 2018).

165. C. Boyden Gray, "American Energy, Chinese Ambition, and Climate Realism," *American Affairs*, November 20, 2021, https://americanaffairsjournal.org/2021/11/american-energy-chinese-ambition-and-climate-realism/; Sophia Kalantzakos, "The Race for Critical Minerals in an Era of Geopolitical Realignments," *The International Spectator* 55, no. 3 (2020): 1–16. https://doi.org/10.1080/03932729.2020.1786926.

166. Ali, "Militarized Adaptation."

167. Snyder, *Black Earth*, 327.

168. Ramez Naam, Twitter post, July 10, 2017, https://twitter.com/ramez/status/884486583734358017.

169. Pew Research Center, "Political Independents: Who They Are, What They Think," March 2019, https://www.pewresearch.org/politics/2019/03/14/political-independents-who-they-are-what-they-think/.

170. Gallup, "Party Affiliation," Gallup, accessed June 23, 2024, https://news.gallup.com/poll/15370/party-affiliation.aspx.

171. Shoshana Zuboff, *The Age of Surveillance Capitalism: The Fight for a Human Future at the New Frontier of Power* (PublicAffairs, 2019); Benjamin I. Page et al., *Billionaires and Stealth Politics* (Chicago: University of Chicago Press, 2018); Nancy MacLean, *Democracy in Chains: The Deep History of the Radical Right's Stealth Plan for America* (New York: Viking, 2017); Jane Mayer, *Dark Money: The Hidden History of the Billionaires Behind the Rise of the Radical Right* (New York: Doubleday, 2016); Martin Gilens and Benjamin I. Page, "Testing Theories of American Politics: Elites, Interest Groups, and Average Citizens," *Perspectives on Politics* 12, no. 3 (2014): 564–581, doi:10.1017/S1537592714001595; Martin Gilens, *Affluence and Influence: Economic Inequality and Political Power in America* (Princeton: Princeton University Press, 2012).

172. Daniel Aldana Cohen, "New York Mag's Climate Disaster Porn Gets It Painfully Wrong," *Jacobin*, July 10, 2017, https://jacobinmag.com/2017/07/climate-change-new-york-magazine-response. Cohen's critique of Wallace-Wells's "The Uninhabitable Earth" doesn't really engage with his arguments or evidence and essentially asks "What about capitalism?" It's a poor critique but a worthwhile point.

173. Sam Zacher, "Polarization of the Rich: The New Democratic Allegiance of Affluent Americans and the Politics of Redistribution," *Perspectives on Politics* 22, no. 2 (2024): 338–356, https://doi.org/10.1017/S1537592722003310.

174. Daniel Aldana Cohen, "New York City as 'Fortress of Solitude' After Hurricane Sandy: A Relational Sociology of Extreme Weather's Relationship to Climate Politics," *Environmental Politics* 30, no. 5 (2021): 687–707, https://doi.org/10.1080/09644016.2020.1816380.

175. Cohen, "New York City as 'Fortress of Solitude,'" 15.
176. Cohen, "New York City as 'Fortress of Solitude,'" 5.
177. Naomi Klein, *The Shock Doctrine: The Rise of Disaster Capitalism* (New York: Henry Holt, 2007).
178. Cohen, "New York City as 'Fortress of Solitude,'" 4.
179. Alyssa Battistoni and Geoff Mann, "Climate Bidenomics," *New Left Review* 143 (Sept.–Oct. 2023), 55–77.
180. Gilles Deleuze and Félix Guattari, *Anti-Oedipus: Capitalism and Schizophrenia*, vol. 1 (Minneapolis: University of Minnesota Press, 1983), 151.
181. Christian Parenti, *Tropic of Chaos* (New York: PublicAffairs, 2011), 11.
182. Andreas Malm, *How to Blow Up a Pipeline* (New York: Verso, 2020), 67.
183. Adam Tooze, "Ecological Leninism," *London Review of Books* 43, no. 22 (November 18, 2021), https://www.lrb.co.uk/the-paper/v43/n22/adam-tooze/ecological-leninism.
184. Chenoweth and Stepan look at more than three hundred political campaigns across the twentieth century. Of these, only about half of the nonviolent campaigns succeeded, compared to about a quarter of the violent campaigns. Thus, one can conclude that sometimes nonviolent civil resistance works and sometimes it doesn't. Likewise, according to their data, violent resistance sometimes works, too, though less often. At one point Chenoweth suggested a "3.5% rule" for nonviolent civil resistance, arguing that "that no government has withstood a challenge of 3.5% of their population mobilized against it during a peak event." But she has since qualified the claim, observing that it is a historical description, not a prediction or norm, and at least two exceptions, one from within her dataset and one from new research, suggest that the 3.5% rule is at best a rule of thumb. There are numerous serious challenges in adopting Chenoweth's rule for organizing on climate change, however, as Malm points out, as does journalist and activist Nafeez Ahmed in his article "The Flawed Social Science Behind Extinction Rebellion's Change Strategy." Most importantly, as Ahmed writes, the "cases studied by Chenoweth involved 'resistance to repressive regimes or occupations, or in support of secession'—in other words, they involved resistance to regimes that actively invoked domestic violence against opposition forces, which therefore drew on an already existing groundswell of discontent. **Not only did very few of these cases involve overthrow of a democracy, but none of them also involved successful nonviolent efforts to overthrow or change a Western liberal democracy**" (emphasis mine). Mobilizing 3.5% of the population against a repressive regime or occupation is a very different prospect than mobilizing 3.5% of the population in favor of a massive program of social transformation that opponents might see as repressive. Erica Chenoweth and Maria J. Stephan, *Why Civil Resistance Works: The Strategic Logic of Nonviolent Conflict* (New York: Columbia University Press, 2011). Erica Chenoweth, "Questions, Answers, and Some Cautionary Updates Regarding the 3.5% Rule," Carr Center for Human Rights Policy at the John F. Kennedy School of Government at Harvard University, April 2020, https://

www.hks.harvard.edu/sites/default/files/2024-05/Erica%20Chenoweth_2020-005.pdf. Nafeez Ahmed, "The Flawed Social Science Behind Extinction Rebellion's Change Strategy," *Resilience*, October 31, 2019, https://www.resilience.org/stories/2019-10-31/the-flawed-social-science-behind-extinction-rebellions-change-strategy/.

185. Malm, *How to Blow Up a Pipeline*, 61.

186. Malm, *How to Blow Up a Pipeline*, 63.

187. Malm, *How to Blow Up a Pipeline*, 94. On subsistence emissions and luxury emissions, see Henry Shue, "Subsistence Emissions and Luxury Emissions," *Law & Policy* 15, no. 1 (1993): 39–60.

188. Andreas Malm, "The End of the Road: Andreas Malm on Ende Gelände," *Salvage Zone*, May 16, 2016, https://salvage.zone/the-end-of-the-road-andreas-malm-on-the-ende-gelande-protests/.

189. Malm, *How to Blow Up a Pipeline*, 161.

190. J. Glenn Gray, *The Warriors: Reflections on Men in Battle* (Lincoln: University of Nebraska Press, 1970), 161; René Girard, *Battling to the End: Conversations with Benoit Chantre* (Ann Arbor: Michigan State University Press, 2009). See also Carl von Clausewitz, *On War* (Princeton: Princeton University Press, 1989), 77.

191. This definition is consistent with the US criminal code, which defines violence as "(a) an offense that has as an element the use, attempted use, or threatened use of physical force against the person or property of another, or (b) any other offense that is a felony and that, by its nature, involves a substantial risk that physical force against the person or property of another may be used in the course of committing the offense" (18 U.S.C. § 16). For a careful discussion of the legal definition of violence and the ambiguous conceptualization of destruction of property in such discourses, see Connor Sunderman, "Violence Against Property: The Breaking Point of Federal Crime of Violence Classifications," *Columbia Law Review* 122, no. 3 (2022): 755–792. See also Hillel R. Smith, "The Federal 'Crime of Violence' Definition: Overview and Judicial Developments," *Congressional Research Service Report* R45220 (June 8, 2018): 1–17.

192. See, to begin with, the four-volume *Cambridge World History of Violence*, ed. Philip Dwyer and Joy Damousi (Cambridge: Cambridge University Press, 2020); Siniša Malešević, *The Rise of Organised Brutality: A Historical Sociology of Violence* (Cambridge: Cambridge University Press, 2017); Miguel A., Centeno and Elaine Enriquez, *War and Society* (Malden, MA: Polity, 2016); Bruce B. Lawrence and Aisha Karim, eds. *On Violence: A Reader* (Durham: Duke University Press, 2007); and William Vollman's seven-volume, 3,352-page essay on violence, *Rising Up and Rising Down: Some Thoughts on Violence, Freedom and Urgent Means* (San Francisco: McSweeney's, 2003).

193. See, for instance, Ernst Jünger, *Storm of Steel* (New York: Penguin, 2003).

194. Malm, *How to Blow Up a Pipeline*, 110.

195. Sageman, *Turning to Political Violence*, 41.

196. Gray, *The Warriors*, 27.

197. My reflections here closely follow those first developed in my essay "Memories of My Green Machine: Posthumanism at War," *Theory & Event* 13, no. 1 (2010).

198. Gray, *The Warriors*, 53.

199. Gray, *The Warriors*, 28–29. See also Ernst Jünger, *On Pain* (Candor, NY: Telos Press Publishing, 2008).

200. Malm, *How to Blow Up a Pipeline*, 119. See Marshall Curry, dir., *If a Tree Falls: A Story of the Earth Liberation Front* (Brooklyn: Marshall Curry Productions, 2011); Jeremy Varon, *Bringing the War Home: The Weather Underground, the Red Army Faction, and Revolutionary Violence in the Sixties and Seventies* (Berkeley: University of California Press, 2004), ch. 4.

201. Sageman, *Turning to Political Violence*, 42.

202. Bue Rübner Hansen, "The Kaleidoscope of Catastrophe—On the Clarities and Blind Spots of Andreas Malm," *Viewpoint Magazine*, April 14, 2021, https://viewpointmag.com/2021/04/14/the-kaleidoscope-of-catastrophe-on-the-clarities-and-blind-spots-of-andreas-malm/.

203. Sageman, *Turning to Political Violence*, 11–12.

204. Paul Joosse, "Elves, Environmentalism, and 'Eco-Terror': Leaderless Resistance and Media Coverage of the Earth Liberation Front," *Crime, Media, Culture* 8, no. 1 (2012): 75–93, https://doi.org/10.1177/1741659011433366.

205. Malm, *How to Blow Up a Pipeline*, 152–153. For a more nuanced history of radical environmentalism in the US, see Keith Woodhouse's *The Ecocentrists*. Also, see Bue Rübner Hansen's critique of Malm's historically inaccurate dismissal of 1990s ecotage in "The Kaleidoscope of Catastrophe."

206. See Jeffrey M. Jones, "Confidence in U.S. Institutions Down; Average at New Low," Gallup, July 5, 2022, https://news.gallup.com/poll/394283/confidence-institutions-down-average-new-low.aspx; Chris Jackson et al., "Very Few Americans Believe Political Violence Is Acceptable," Ipsos, August 22, 2022, https://www.ipsos.com/en-us/very-few-americans-believe-political-violence-acceptable; Kaleigh Rogers and Zoha Qamar, "What Americans Think About Political Violence," Five ThirtyEight, November 4, 2022, https://fivethirtyeight.com/features/what-americans-think-about-political-violence/; Sean J. Westwood et al., "Current Research Overstates American Support for Political Violence," *Proceedings of the National Academy of Sciences* 119, no. 12 (2022): e2116870119, https://doi.org/10.1073/pnas.2116870119.

207. Malm, *How to Blow Up a Pipeline*, 118.

208. Hansen, "The Kaleidoscope of Catastrophe."

209. Kim Stanley Robinson, *Ministry for the Future* (New York: Orbit, 2020). I don't have the time or patience to pick apart the cartoon terrorism in Robinson's novel, but it is of a piece with his generally wooden characters and creaky plots.

210. Max Ajl, "Andreas Malm's Corona, Climate, Chronic Emergency," *Brooklyn Rail*, November 2020, https://brooklynrail.org/2020/11/field-notes/Corona-Climate-Chronic-Emergency.

211. Hansen, "The Kaleidoscope of Catastrophe."

212. Katarzyna Jasko et al., "A Comparison of Political Violence by Left-Wing, Right-Wing, and Islamist Extremists in the United States and the World," *Proceedings of the National Academy of Sciences* 119, no. 30 (2022): 1–e2122593119, https://doi.org/10.1073/pnas.2122593119.

213. Sageman, *Turning to Political Violence*, 29. Radicalization is complex and contextual, and the literature on political violence deep. In addition to Sageman, see Lorenzo Bosi et al., *Dynamics of Political Violence: A Process-Oriented Perspective on Radicalization and the Escalation of Political Conflict* (Burlington: Ashgate Publishing, 2014); Erica Chenoweth et al., eds. *The Oxford Handbook of Terrorism*, (Oxford: Oxford University Press, 2019).

214. Sageman, *Turning to Political Violence*, 29.

215. Malm, *How to Blow Up a Pipeline*, 141–142.

216. Malm, *How to Blow Up a Pipeline*, 142.

217. Michael E. Zimmerman, *Contesting Earth's Future: Radical Ecology and Postmodernity* (Berkeley: University of California Press, 1994), 322.

Chapter 3

1. Peter Singer, "Famine, Affluence, and Morality," *Philosophy & Public Affairs* 1, no. 3 (1972): 229–243, 231. Godwin similarly writes, in 1793, "Does any person in distress apply to me for relief? It is my duty to grant it, and I commit a breach of duty in refusing. If the principle be not of universal application, it is because, in conferring a benefit upon an individual, I may in some instances inflict an injury of superior magnitude upon myself or society. Now the same justice, that binds me to any individual of my fellow men, binds me to the whole." William Godwin, *An Enquiry Concerning Political Justice*, book 2, ii (Oxford: Oxford University Press, 2013), 56

2. Singer, "Famine, Affluence, and Morality," 232.

3. Garrett Hardin, "Living on a Lifeboat," *Bioscience* 24, no. 10 (1974): 561–568, https://doi.org/10.2307/1296629.

4. Kwame Anthony Appiah points out that the problem with Singer's argument "isn't that it says we have incredible obligations to foreigners; the problem is that it claims we have incredible obligations." Kwame Anthony Appiah, *Cosmopolitanism: Ethics in a World of Strangers* (New York: W. W. Norton, 2006), 160. See also John Kekes, "On the Supposed Obligation to Relieve Famine," *Philosophy* 77, no. 4 (2002): 503–517; Anton Markoč, "Draining the Pond: Why Singer's Defense of the Duty to Aid the World's Poor Is Self-Defeating," *Philosophical Studies* 177, no. 7 (2020): 1953–1970.

5. National Center for Charitable Statistics, "The Nonprofit Sector in Brief 2019," Urban Institute, June 18, 2020, https://nccs.urban.org/nccs/resources/sector-brief-2019/.

6. While Hardin recognizes the role of differential labor costs in driving immigration policy, he doesn't pursue this line of thought.

7. Avram Hiller, "A 'Famine, Affluence, and Morality' for Climate Change?," *Public Affairs Quarterly* 28, no. 1 (2014): 19–39.

8. Boulding, "The Economics of the Coming Spaceship Earth," in *Environmental Quality in a Growing Economy: Essays from the Sixth RFF Forum*, ed. Henry Jarrett (Baltimore: Johns Hopkins University Press, 1966), 9.

9. Hardin, "Living on a Lifeboat," 561.

10. Jeff VanderMeer, *Annihilation* (New York: Farrar, Straus, and Giroux, 2014).

11. For ethical consideration of individual inefficacy, see James Garvey, "Climate Change and Causal Inefficacy: Why Go Green When It Makes No Difference?," *Royal Institute of Philosophy Supplements* 69 (2011): 157–174; Shelly Kagan, "Do I Make a Difference?," *Philosophy & Public Affairs* (2011): 105–141; and Theresa Scavenius, "Climate Change and Moral Excuse: The Difficulty of Assigning Responsibility to Individuals," *Journal of Agricultural and Environmental Ethics* 31 (2018): 1–15.

12. Aside from whatever deeper philosophical problems utilitarianism faces (which are numerous), generic utilitarianism of the "most happiness for the most people" faces unanswerable difficulties when confronting planetary ecological crisis generally and climate change more specifically. First, the negative impact of human overpopulation on nonrenewable resources, planetary habitability, and psychological well-being forecloses simple maximalism and demands of utilitarianism an "ideal global population for maximum happiness," an unknowable and incalculable number, which even if it were known would lead to policies most people with commitments to basic dignity and the value of human life would consider ethically dubious at best, such as compulsory birth control, mandated abortions, and forced euthanasia. Second, there is no real psychological, philosophical, or economic basis on which to compare the happiness of future generations against the happiness of the present. The ways social scientists measure aggregate happiness in contemporary society are highly questionable; the idea that we can plausibly compare such spurious data against happiness in other historical periods in any robust or quantifiable way is wholly unsupportable; and projecting such malarky into the future doesn't even rise to the level of meaningful intellectual activity. Third, any utilitarian response to climate change would need to begin by determining the social cost of carbon, but there is no consensus on how to calculate such a number and it's not even clear that it's possible. Finally, as philosopher John Broome points out, there are much more cost-effective and reliable ways to contribute to aggregate global well-being than by fighting climate change—for instance, by supporting malaria prevention. "Consequently," Broome writes, "it seems utilitarianism does not require you to reduce your emissions except to the extent that you can do so almost costlessly" (Broome, "Utilitarianism and Climate Change," in Richard Yetter Chappell, Darius Meissner, and William MacAskill, eds., *An Introduction to Utilitarianism*, 2003, https://www.utilitarianism.net/guest-essays/utilitarianism-and-climate-change). Confronted with trajectories of declining well-being ranging from human extinction to ecological disruption and unable to say anything meaningful about

how our choices today will impact the happiness of future generations, utilitarianism offers little in terms of ethical guidance. Indeed, the challenges climate change poses for utilitarianism are so severe that Dale Jamieson has puckishly argued that the only consistent and logical utilitarian position is to take up virtue ethics (Dale Jamieson, "When Utilitarians Should Be Virtue Theorists," *Utilitas* 19, no. 2 (2007): 160–183, https://doi.org/10.1017/S0953820807002452).

What's more, as environmental philosophers such as J. Baird Callicott, John Rodman, Eric Katz, and Holmes Rolston III have argued, utilitarianism's typical anthropocentricism discounts the value of nonhuman life, particularly insofar as it's impossible to judge the happiness of an ecosystem, a rock, a cloud, or a herd of elk. Yet replacing the value of "happiness" with some more abstract sense of wellbeing such as "flourishing" or "general health" doesn't help much, since once you leave the realm of human values, there's no basis on which to make comparative ethical judgments. In terms of sheer biomass, a planet teeming with maggots might have greater aggregate "well-being" than one harboring highly complex and diverse fauna, but who's to say quantity matters more than quality, or vice versa? Extrapolating out to cosmic scales, there's ultimately no reason to favor any particular transient concatenation of matter and energy over any other, since it's all just flux.

On some of the problems climate change poses for utilitarianism, see Maddalena Ferranna, "Discounting under Risk: Utilitarianism vs. Prioritarianism," in *Philosophy and Climate Change*, ed. Mark Budolfson, Tristam McPherson, and David Plunkett (Oxford: Oxford University Press, 2021); Tim Mulgan, "Utilitarianism and Our Obligations to Future People," in *The Cambridge Companion to Utilitarianism*, ed. Ben Eggleston and Dale E. Miller (Cambridge: Cambridge University Press, 2014); Holmes Rolston III, "Respect for Life: Counting What Singer Finds of No Account," in *Singer and His Critics*, ed. Dale Jamieson, (Malden, MA: Blackwell Publishers, 1999); Eric Katz, *Nature as Subject: Human Obligation and Natural Community* (Lanham, MD: Rowman & Littlefield, 1997); John Broome, *Counting the Cost of Global Warming: A Report to the Economic and Social Research Council on Research* (Cambridge: White Horse, 1992); J. Baird Callicott, "Animal Liberation: A Triangular Affair," *Environmental Ethics* 2, no. 4 (1980): 311–338; John Rodman, "The Liberation of Nature?," *Inquiry* 20, no. 1–4 (1977): 83–131. On deeper problems with utilitarianism, see, among others, Christopher Woodard, *Taking Utilitarianism Seriously* (Oxford: Oxford University Press, 2019); Roger Crisp, "Taking Stock of Utilitarianism," *Utilitas* 26, no. 3 (2014): 231–249; J. J. C. Smart and Bernard Williams, *Utilitarianism: For and Against* (Cambridge: Cambridge University Press, 1973).

13. Dale Jamieson, *Reason in a Dark Time: Why the Struggle Against Climate Change Failed—and What It Means for Our Future* (Oxford: Oxford University Press, 2014); Stephen Gardiner, *A Perfect Moral Storm: The Ethical Tragedy of Climate Change* (Oxford: Oxford University Press, 2011); Broome, *Counting the Cost of Global Warming*; Damian Bridge, "The Ethics of Climate Change: A Systematic Literature Review," *Account Finance* 62 (2022): 2651–2665, https://doi-org.proxy.library.nd.edu

/10.1111/acfi.12877; Stephen Mark Gardiner, ed., *Climate Ethics: Essential Readings* (Oxford: Oxford University Press, 2010); Dale Jamieson, "Ethics, Public Policy, and Global Warming," *Science, Technology, & Human Values* 17, no. 2 (1992): 139–153, https://doi.org/10.1177/016224399201700201.

14. Dipesh Chakrabarty, "Postcolonial Studies and the Challenge of Climate Change," *New Literary History* 43, no. 1 (2012): 1–18, https://doi.org/10.1353/nlh.2012.0007.

15. Richard Heede, "Tracing Anthropogenic Carbon Dioxide and Methane Emissions to Fossil Fuel and Cement Producers, 1854–2010," *Climatic Change* 122, no. 1 (2014): 229–241; Paul Griffin and C. R. Heede, "The Carbon Majors Database," *CDP Carbon Majors Report 2017* 14 (2017).

16. Claire Colebrook, "What Is the Anthropo-political?," in *Twilight of the Anthropocene Idols*, ed. J. Hillis Miller et al., 81–125 (London: Open Humanities Press, 2016), https://doi.org/10.26530/OAPEN_588463, 115.

17. Hannah Arendt, *The Human Condition* (Chicago: University of Chicago Press, 1958).

18. Hans Jonas, *The Imperative of Responsibility* (Chicago: University of Chicago Press, 1985), 1–24.

19. See Jaron Lanier, *Dawn of the New Everything: Encounters with Reality and Virtual Reality* (New York: Henry Holt, 2017); Glenn Greenwald, *No Place to Hide: Edward Snowden, the NSA, and the U.S. Surveillance State* (Henry Holt, 2014); Lewis Brinkmann et al. "Machine Culture," *Nature Human Behaviour* 7, no. 11 (2023): 1855–1868; Joseph Firth et al., "The 'Online Brain': How the Internet May Be Changing Our Cognition," *World Psychiatry* 18, no. 2 (2019): 119–129; Sandra González-Bailón and Yphtach Lelkes, "Do Social Media Undermine Social Cohesion? A Critical Review," *Social Issues and Policy Review* 17, no. 1 (2023): 155–180; and Shoshana Zuboff, *The Age of Surveillance Capitalism: The Fight for a Human Future at the New Frontier of Power* (PublicAffairs, 2019).

20. Paul Gilroy offers a sophisticated analysis of this impasse in his 2014 Tanner Lectures. Paul Gilroy, "The Black Atlantic and the Re-enchantment of Humanism," The Tanner Lectures on Human Values, lecture presented at Yale University, February 21, 2014, https://tannerlectures.utah.edu/_resources/documents/a-to-z/g/Gilroy%20manuscript%20PDF.pdf.

21. Jonathan Watts, "Brazil's New Foreign Minister Believes Climate Change Is a Marxist Plot," *The Guardian* 11, no. 8 (2018); Kathryn Yusoff, *A Billion Black Anthropocenes or None* (Minneapolis: University of Minnesota Press, 2018).

22. Dipesh Chakrabarty, *The Climate of History in a Planetary Age* (Chicago: University of Chicago Press, 2021).

23. See Joseph Henrich, *The WEIRDest People in the World?* (New York: Penguin, 2020); and Joseph Henrich et al.. "The Weirdest People in the World?," *Behavioral and Brain Sciences* 33, no. 2–3 (2010): 61–83.

24. Howard Winant, *Racial Conditions: Politics, Theory, Comparisons* (Minneapolis: University of Minnesota Press, 1994), xiii.

25. T. S. Eliot, "Burnt Norton," in *Four Quartets* (New York: Harcourt Brace Jovanovich, 1971).

26. Giorgios Kallis et al. "Research on Degrowth," *Annual Review of Environment and Resources* 43 (2018): 291–316. See also Tim J. Garrett, "No Way Out? The Double-Bind in Seeking Global Prosperity Alongside Mitigated Climate Change," *Earth System Dynamics* 3, no. 1 (2012): 1–17.

27. The question of how perceptions of climate catastrophe affect the decision to have children is a recurring topic of concern. In addition to my own "Raising My Daughter in a Doomed World," *New York Times*, July 16, 2018, https://www.nytimes.com/2018/07/16/opinion/climate-change-parenting.html, see also Meehan Crist, "Is It OK to Have a Child?" *London Review of Books* 42, no. 5 (2020): 9–14.

28. See, for instance, William MacAskill, *What We Owe the Future* (New York: Basic Books, 2022). Longtermism holds that "the world's long-run fate depends in part on the choices we make in our lifetimes," and that we have a moral commitment to positively influence that fate (MacAskill, 6). While "future people count" may be a commonsense moral intuition, there are significant flaws in MacAskill's argument. First, the value of human life is not simply cumulative, as in MacAskill's account. Second, we cannot know what future people will hold to be of value. Finally, given the sheer unpredictability of events, planetary complexity, and the limits of human cognition and action, there can be no plausible basis for either individual or collective contemporary responsibility for the long-term fate of the planet. The entire "longtermist" project is of fantastic arrogance.

29. In considering what he identifies as the three types of theories of justice applied to climate change (distributive justice, corrective justice, and theories of equality), Stephen Gardiner finds that "as they have been applied to climate change, they all suffer from two faults," which he calls "climate change blinders" and basic infeasibility. Stephen Gardiner, "The Role of Claims of Justice in Climate Change Policy," in *Debating Climate Ethics*, ed. Stephen M. Gardiner and David A. Weisbach (New York: Oxford University Press, 2016), 201–240, 201.

30. The authoritative history of the environmental justice movement in the United States is Dorceta E. Taylor, *Toxic Communities: Environmental Racism, Industrial Pollution, and Residential Mobility* (New York: New York University Press, 2014). See also "The First National People of Color Environmental Leadership Summit: Principles of Environmental Justice," *Race, Poverty & the Environment* 2, no. 3/4 (1991): 32–31; and Robert Figueroa and Claudia Mills, "Environmental Justice," in *A Companion to Environmental Philosophy*, ed. Dale Jamieson (Malden, MA: Blackwell, 2001), 426–438. Environmental ethics generally is taken to have begun as a distinct philosophical discipline in the 1970s, though obviously the questions it takes up go back well beyond that, at least to Marx's engagement with Justus von Liebeg and Malthus, if not John Evelyn's *Sylva, or A Discourse of Forest-Trees and the Propagation of Timber in His Majesty's Dominions* (1664). See J. Baird Callicott, "Introduction to Environmental Ethics," in *Environmental Philosophy: From Animal Rights to Radical*

Ecology, ed. Michael E. Zimmerman et al. (Upper Saddle River, NJ: Prentice-Hall, 2004).

31. Eduardo S. Brondízio et al., eds., *Global Assessment Report on Biodiversity and Ecosystem Services of the Intergovernmental Science-Policy Platform on Biodiversity and Ecosystem Services* (Bonn: IPBES Secretariat, 2019), https://doi.org/10.5281/zenodo.3831673.

32. Eva Avila Martin et al., "Wild Meat Hunting and Use by Sedentarised Baka Pygmies in Southeastern Cameroon," *PeerJ* 8 (September 17, 2020): e9906, doi: 10.7717/peerj.9906; Romain Duda, Sandrine Gallois, and Victoria Reyes-García, "Ethnozoology of Bushmeat," *Revue d'Ethnoécologie* 14 (2018), https://doi.org/10.4000/ethnoecologie.3976; Jason MacLean et al., "Polar Bears and the Politics of Climate Change: A Response to Simpson," *Journal of International Wildlife Law & Policy* 23, no. 2 (2020): 141–150; Claire Jean Kim, "Makah Whaling and the (Non) Ecological Indian," in *Colonialism and Animality: Anti-Colonial Perspectives in Critical Animal Studies*, ed. Kelly Struthers Montford and Chloë Taylor (New York: Routledge, 2020), 50–103.

33. Robinson Jeffers, "Carmel Point," in *The Collected Poetry of Robinson Jeffers: Volume 3 1938–1962*, ed. Tim Hunt (Stanford: Stanford University Press, 1991).

34. Robinson Jeffers, preface to *The Double Axe and Other Poems* (New York: Random House, 1948), vii.

35. Charles Darwin, *On the Origin of Species by Means of Natural Selection, Or, The Preservation of Favoured Races in the Struggle for Life* (United Kingdom: J. Murray, 1859), 490.

36. Richard A. Watson, "A Critique of Anti-Anthropocentric Biocentrism," *Environmental Ethics* 5, no. 3 (1983): 245–256, doi: 10.5840/enviroethics19835325; George Bradford, "How Deep is Deep Ecology?" *Fifth Estate* 22, no. 3 (Fall 1987), https://jstor.org/stable/community.28036651; Keith Makato Woodhouse, *The Ecocentrists: A History of Radical Environmentalism* (New York: Columbia University Press, 2018); William Cronon, "The Trouble with Wilderness," in *Uncommon Ground: Rethinking the Human Place in Nature* (New York: W.W. Norton, 1996).

37. Watson, "A Critique of Anti-Anthropocentric Biocentrism," 253.

38. Aldo Leopold, "Arizona and New Mexico," in *A Sand County Almanac* (New York: Ballantine, 1966), 137–141; J. Baird Callicott, *Thinking Like a Planet: The Land Ethic and the Earth Ethic* (New York: Oxford University Press, 2013).

39. Gyan Chandra Acharya, "Those Who Contributed Least to Climate Change Are Now Fighting for Survival," *The Guardian*, December 2, 2015, https://www.theguardian.com/global-development/2015/dec/02/climate-change-paris-talks-those-who-contributed-least-fighting-for-survival.

40. In the words of Adolph Reed and Walter Benn Michaels, speaking generally about the use of racial disparities as an analytic framework: "As a diagnosis, identifying disparities is taxonomic and rhetorical, not etiological. Insisting that we understand those inequalities as evidence of racism is a demand about how we

should classify and feel about them, not an effort to understand their specific causes." Adolph Reed and Walter Benn Michaels, "The Trouble with Disparity," *nonsite.org* 32 (September 10, 2020), https://nonsite.org/the-trouble-with-disparity/.

41. Enzo Rossi and Olúfẹ́mi Táíwò, "What's New About Woke Racial Capitalism (and What Isn't): 'Wokewashing' and the Limits of Representation," *Spectre*, December 2020; Olúfẹ́mi O. Táíwò, *Elite Capture: How the Powerful Took Over Identity Politics (And Everything Else)* (Chicago: Haymarket Books, 2022).

42. Nikhil Singh writes, "In ideological terms, racism is knowable as a narrative structure of positions and habits of perception that corresponds with and responds to a preceding, preexisting regime of racial categorization and differentiation." Nikhil Singh, "Racial Formation in an Age of Permanent War," in *Racial Formation in the Twenty-First Century* (Berkeley: University of California Press, 2012), 285.

43. Amitav Ghosh, *The Nutmeg's Curse: Parables for a Planet in Crisis* (Chicago: University of Chicago Press, 2021), 170.

44. Carla Zoe Cremer and Luke Kemp, "Democratising Risk: In Search of a Methodology to Study Existential Risk," December 28, 2021, SSRN.com, https://ssrn.com/abstract=3995225.

45. Peter Burdon and Paul Alberts have both turned to Hans Jonas's *The Imperative of Responsibility* as a resource for beginning to think this problem. Paul Alberts, "Responsibility Towards Life in the Early Anthropocene," *Angelaki: Journal of Theoretical Humanities* 16, no. 4 (2011): 5–17, https://doi.org/10.1080/0969725X.2011.641341; Peter D. Burdon, "Obligations in the Anthropocene," *Law and Critique* 31, no. 3 (2020): 309–328, https://doi.org/10.1007/s10978-020-09273-9.

46. Henry Shue, "Subsistence Emissions and Luxury Emissions," *Law & Policy* 15, no. 1 (1993): 39–60.

47. Olúfẹ́mi O. Táíwò and Beba Cibralic, "The Case for Climate Reparations," *Foreign Policy*, October 10, 2020, https://foreignpolicy.com/2020/10/10/case-for-climate-reparations-crisis-migration-refugees-inequality/.

48. Olúfẹ́mi Táíwò, *Reconsidering Reparations* (New York: Oxford University Press, 2022), 190.

49. Matthew 20:16.

50. "Oh, to be sure, it is not for himself or his children that the capitalist works, but for the immortality of the system. A violence without purpose, a joy, a pure joy in feeling oneself a wheel in the machine, traversed by flows, broken by schizzes. Placing oneself in the position where one is thus traversed, broken, fucked by the socius, looking for the right place where, according to the aims and the interests assigned to us, one feels something moving that has neither an interest nor a purpose. A sort of art for art's sake in the libido, a taste for a job well done, each one in his own place, the banker, the cop, the soldier, the technocrat, the bureaucrat, and why not the worker, the trade-unionist. Desire is agape." Gilles Deleuze and Félix Guattari, *Anti-Oedipus: Capitalism and Schizophrenia*, vol. 1 (Minneapolis: University of Minnesota Press, 1983), 346–347.

51. Paul A. Grout, William L. Megginson, and Anna Zalewska, "One Half-Billion Shareholders and Counting—Determinants of Individual Share Ownership Around the World," 22nd Australasian Finance and Banking Conference (2009), https://ssrn.com/abstract=1457482. Lucas Chancel et al., *World Inequality Report 2022* (World Inequality Lab, 2022), https://wir2022.wid.world.

52. See, for instance, Kenneth J. Arrow et al., "Determining Benefits and Costs for Future Generations," *Science* 341, no. 6144 (2013): 349–350, https://doi.org/10.1126/science.1235665. Dale Jamieson talks about discounting the future in *Reason in a Dark Time*, and the topic is a common one in the philosophy of climate change.

53. This also helps us see how arguments for climate justice typically conflate class, wealth, and race in a way that obscures the function of all three social determinants.

54. Adela Cortina, *Aporophobia: Why We Now Fear the Poor Instead of Helping Them* (Princeton: Princeton University Press, 2022).

Chapter 4

1. See Richard Slotkin, *Regeneration Through Violence; The Mythology of the American Frontier, 1600–1860* (Middletown, CT: Wesleyan University Press, 1973); and Leslie Fiedler, "Come Back to the Raft Ag'in, Huck Honey!," *Partisan Review* 15, no. 6 (1948), reprinted in Leslie Fielder, *Love and Death in the American Novel* (Normal, IL: Dalkey Archive Press, 1997).

2. Achille Mbembe, *Necropolitics* (Durham: Duke University Press, 2019), 27–28.

3. Déborah Danowski and Eduardo Viveiros de Castro, *The Ends of the World* (Cambridge: Polity, 2016), 19.

4. Donna J. Haraway, *Staying with the Trouble: Making Kin in the Chthulucene* (Durham: Duke University Press, 2016).

5. Dipesh Chakrabarty, "The Planet: An Emergent Humanist Category," *Critical Inquiry* 46, no. 1 (2019): 1–31.

6. Martin Heidegger, "Building Dwelling Thinking," in *Poetry, Language, Thought*, trans. Albert Hofstadter (New York: Harper & Row, 1971), 149.

7. Bruno Latour, *Facing Gaia: Eight Lectures on the New Climatic Regime* (Cambridge: Polity, 2017).

8. This ontological world—the conceptual, structural, social, and spatio-temporal whole in which the existence of particular beings meaningfully emerges—the "collectively imagined global chronotope of the now"—becomes visible through two strains of phenomenological thought, the Hegelian and the Heideggerian, which achieve their most sophisticated contemporary development in postcolonial theory and Afropessimism. This is not the place to tell the story of how Hegelian Idealism, particularly the narrativization of conceptual transformation in dialectical thought as presented in *The Phenomenology of Spirit*, evolved in one strain through American transcendentalism and William James's psychology into the sociological work of James's student W. E. B. Du Bois and in another strain into surrealism, *négritude*,

and Lacanian psychoanalysis, both of which strains are synthesized in the work of Frantz Fanon, along with the phenomenological tradition that has its roots in the pioneering ethological work of biologist Felix von Uexküll and the psychological work of priest and philosopher Franz Brentano, not to speak of the pervasive and complex influence of Nietzsche's radical empiricism, and then develops through the monumental and politically compromised work of Martin Heidegger into the Hegelian-Kojèvean French phenomenology of Sartre and Merleau-Ponty. Nor can we here trace the complex entanglements of Hegelian, Nietzschean, and Heideggerian thought that commingle through Derrida, Foucault, and others to emerge in new ways in Gayatri Spivak's critique of postcolonialism, Dipesh Chakrabarty's groundbreaking *Provincializing Europe* (Princeton: Princeton University Press, 2000), Sylvia Wynter's essays, and the lacerating work of Achille Mbembe. These European and postcolonial lineages merge with the Anglo-American Black Radical tradition, including the Marxian lineage of Cedric Robinson, C. L. R. James, and Stuart Hall, as well as with the more recent emergence of Black Studies and African American satire, in Afropessimism.

9. Kermode, *The Sense of an Ending*, 7.

10. As Gertrude Stein put it: "The only thing that is different from one time to another is what is seen and what is seen depends upon how everybody is doing everything. This makes the thing we are looking at very different and this makes what those who describe it make of it, it makes a composition, it confuses, it shows, it is, it looks, it likes it as it is, and this makes what is seen as it is seen. Nothing changes from generation to generation except the thing seen and that makes a composition." Gertrude Stein, "Composition as Explanation," in *Selected Writings of Gertrude Stein*, ed. Carl Van Vechten (New York: Vintage Books, 1972), 513.

11. Anne Carson, *The Autobiography of Red* (New York: Vintage Books, 1998), 82.

12. Danowski and de Castro, *The Ends of the World*, 20

13. Faisal Mahmud, "Coronavirus: In Dense Bangladesh, Social Distancing a Tough Task," *Al Jazeera*, March 20, 2020, https://www.aljazeera.com/news/2020/3/20/coronavirus-in-dense-bangladesh-social-distancing-a-tough-task.

14. "Omnis determinatio est negatio." Benedict Spinoza, letter to Jarig Jelles, dated June 2, 1674, *The Letters*, trans. by Samuel Shirley (Indianapolis: Hackett, 1995).

15. See Eric Havelock, *Preface to Plato* (Cambridge: Harvard University Press, 1963); and Anne Carson, *Eros the Bittersweet* (Normal, IL: Dalkey Archive, 1998).

16. For more on the idea of losing a conceptual worldview, see Cora Diamond, "Losing your Concepts," *Ethics* 98, no. 2 (1988): 255–277; and Alasdair MacIntyre, *After Virtue: A Study in Moral Theory* (Notre Dame: University of Notre Dame Press, 2007).

17. In the words of Ghosh: "Uncanny indeed are the similarities between the current planetary crisis and the environmental disruptions that destroyed the lifeworlds of innumerable Amerindian and Australian peoples." Amitav Ghosh, *The*

Nutmeg's Curse: Parables for a Planet in Crisis (Chicago: University of Chicago Press, 2021), 165.

18. Gerald Robert Vizenor, *Survivance: Narratives of Native Presence* (Lincoln: University of Nebraska Press, 2008).

19. Dina Gilio-Whitaker, *As Long as Grass Grows: The Indigenous Fight for Environmental Justice, from Colonization to Standing Rock* (Boston: Beacon Press, 2019), 49.

20. Aníbal Quijano, "Coloniality and Modernity/Rationality," *Cultural Studies* 21, no. 2–3 (2007): 168–178, 170.

21. On the cultural and conceptual devastation wrought by Columbian exchange, see, among voluminous literature, Walter Mignolo, *The Darker Side of the Renaissance: Literacy, Territoriality, and Colonization* (Ann Arbor: University of Michigan Press, 2003).

22. In addition to those arguments cited later in this chapter, see Elaine Stratford et al., "Islands, the Anthropocene, and Decolonisation," *Antipode* 55, no. 4 (July 2023): 1255–1274; Meg Parsons, Karen Fisher, and Roa Petra Crease, eds., *Decolonising Blue Spaces in the Anthropocene: Freshwater Management in Aotearoa New Zealand* (London: Palgrave MacMillan, 2021); Matthew Adams, "Indigenizing the Anthropocene? Specifying and Situating Multi-Species Encounters," *International Journal of Sociology and Social Policy* 41, no. 3/4 (2021): 282–297, https://doi.org/10.1108/IJSSP-04-2019-0084; Christopher A. Kiahtipes, "Decolonizing the Anthropocene," *General Anthropology* 27, no. 1 (2020): 1–11, https://doi.org/10.1111/gena.12064; Weronika Łaszkiewicz, "Decolonizing the Anthropocene: Reading Charles de Lint's 'Widdershins,'" *Acta Neophilologica* 2, no. 22 (2020): 161–172, https://doi.org/10.31648/an.5593; Chiara Xausa, "Decolonizing the Anthropocene: 'Slow Violence' and Indigenous Resistance in Cherie Dimaline's *The Marrow Thieves*," *Il Tolomeo* 22, no. 1 (2020), https://doi.org/10.30687/Tol/2499-5975/2020/22/022; Andrew Curley and Majerle Lister, "Already Existing Dystopias: Tribal Sovereignty, Extraction, and Decolonizing the Anthropocene," in *Handbook on the Changing Geographies of the State*, ed. Sami Moisio et al. (Cheltenham: Edward Elgar Publishing, 2020); Lisa Woynarski, "Decolonised Ecologies: Performance Against the Anthropocene," in *Ecodramaturgies: Theatre, Performance and Climate Change* (London: Palgrave MacMillan, 2020), 179–211; Paulo Ilich Bacca, "Indigenizing International Law and Decolonizing the Anthropocene: Genocide by Ecological Means and Indigenous Nationhood in Contemporary Colombia," *Maguaré* 33, no. 2 (2019): 139–169, https://doi.org/10.15446/mag.v33n2.86199; Shelby E. Ward, "Decolonizing the Cosmopolitan Geospatial Imaginary of the Anthropocene: Beyond Collapsed and Exclusionary Politics of Climate Change," *Pivot: A Journal of Interdisciplinary Studies and Thought* 7, no. 1 (2019), https://doi.org/10.25071/2369-7326.40293; Karsten Schulz, "Decolonising the Anthropocene: The Mytho-Politics of Human Mastery," in *Critical Epistemologies of Global Politics,* ed. Marc Woons and Sebastian Weier (Bristol: E-International Relations Publishing, 2017), https://www.e-ir.info/2017/07/01/decolonising-the

-anthropocene-the-mytho-politics-of-human-mastery; Renzo Taddei, Karen Shiratori, Rodrigo C. Bulamah, "Decolonizing the Anthropocene," in *The International Encyclopedia of Anthropology*, ed. Hilar Callan et al., https://doi.org/10.1002/9781118924396.wbiea2519; Zoe Todd, "Indigenizing the Anthropocene," in *Art in the Anthropocene: Encounters Among Aesthetics, Politics, Environments and Epistemologies*, ed. Heather Davis and Etienne Turpin (London: Open Humanities Press, 2015), 241–254; Winona LaDuke, *Recovering the Sacred: The Power of Naming and Claiming* (Boston: South End Press, 2005).

23. Tyson Yunkaporta, *Sand Talk: How Indigenous Thinking Can Save the World* (New York: HarperOne, 2020), 3.

24. Kyle Whyte, "Indigenous Climate Change Studies: Indigenizing Futures, Decolonizing the Anthropocene," *English Language Notes* 55, no. 1–2 (2017): 153–162, https://doi.org/10.1215/00138282-55.1-2.153.

25. Mark Jackson, "On Decolonizing the Anthropocene: Disobedience via Plural Constitutions," *Annals of the American Association of Geographers* 111, no. 3 (2021): 698–708, doi: 10.1080/24694452.2020.1779645.

26. Heather Davis and Zoe Todd, "On the Importance of a Date, or, Decolonizing the Anthropocene," *ACME: An International Journal for Critical Geographies* 16, no. 4 (2017): 761–780, https://acme-journal.org/index.php/acme/article/view/1539.

27. Robin Wall Kimmerer, *Braiding Sweetgrass: Indigenous Wisdom, Scientific Knowledge and the Teachings of Plants* (Minneapolis: Milkweed Editions, 2013), 9, 48–59, 205–215, 371, 374–379.

28. Firkret Berkes, "Traditional Ecological Knowledge in Perspective," in *Traditional Ecological Knowledge: Concepts and Cases*, ed. Julian Inglis (Ottawa: IDRC, 1993), 2. See also Daniel Shilling and Melissa K. Nelson, eds., *Traditional Ecological Knowledge: Learning from Indigenous Practices for Environmental Sustainability* (Cambridge: Cambridge University Press, 2018).

29. Kyle Whyte, "On the Role of Traditional Ecological Knowledge as a Collaborative Concept: A Philosophical Study," *Ecological Process* 2, no. 7 (2013), https://doi.org/10.1186/2192-1709-2-7.

30. Joseph P. Gone, "Considering Indigenous Research Methodologies: Critical Reflections by an Indigenous Knower," *Qualitative Inquiry* 25, no. 1 (2019): 45–56; Sweeney Windchief and Jason Cummins, "Considering Indigenous Research Methodologies: Bicultural Accountability and the Protection of Community Held Knowledge," *Qualitative Inquiry* 28, no. 2 (2021): 151–163. See also Marshall Sahlins, *How "Natives" Think: About Captain Cook, for Example* (Chicago: University of Chicago Press, 1995).

31. Vanessa de Oliveira Andreotti et al., "Mapping Interpretations of Decolonization in the Context of Higher Education," *Decolonization: Indigeneity, Education & Society* 4, no. 1 (2015).

32. Eve Tuck and K. Wayne Yang, "Decolonization Is Not a Metaphor," *Decolonization: Indigeneity, Education & Society* 1, no. 1 (2012).

33. Frantz Fanon, *The Wretched of the Earth*, trans. Richard Philcox (New York: Grove, 2004), 1–2.

34. "Keele Manifesto for Decolonizing the Curriculum," *Journal of Global Faultlines* 5, no. 1–2 (2018): 97–99, https://doi.org/10.13169/jglobfaul.5.1-2.0097.

35. "The shift from hunting and foraging to agriculture—a shift that was slow, halting, reversible, and sometimes incomplete—carried at least as many costs as benefits." James Scott, *Against the Grain: A Deep History of the Earliest States* (New Haven: Yale University Press, 2017), 10.

36. David Graeber and David Wengrow, *The Dawn of Everything* (New York: Farrar, Straus and Giroux, 2021). This provocative argument builds on Gregory Bateson's idea of schismogenesis (56–57). See Gregory Bateson, "Culture Contact and Schismogenesis," *Man* 35 (December 1935): 178–183. Also, while this isn't the place to offer a criticism of Graeber and Wengrow's book, I might direct interested readers to David Bell, "A Flawed History of Humanity," *Persuasion*, November 19, 2021, https://www.persuasion.community/p/a-flawed-history-of-humanity; and Kwame Anthony Appiah, "Digging for Utopia." *New York Review*, December 16, 2021, among others.

37. Paul Shepard, *Coming Home to the Pleistocene* (Washington, D.C.: Island Press, 2004).

38. Shepard, *Coming Home to the Pleistocene*, 173.

39. John Zerzan, *Why Hope?: The Stand Against Civilization* (Port Townsend: Feral House, 2015), 9.

40. Paul Kingsnorth and Dougald Hine, *Uncivilization: The Dark Mountain Manifesto* (Dark Mountain Project Press, 2009), https://dark-mountain.net/about/manifesto/.

41. Graeber and Wengrow, *The Dawn of Everything*, 525.

42. Robin Wall Kimmerer, "Weaving Traditional Ecological Knowledge into Biological Education: A Call to Action," *Bioscience* 52, no. 5 (2002): 432–438.

43. Firkret Berkes, "Traditional Ecological Knowledge in Perspective"; Cedric J. Robinson, *Black Marxism: The Making of the Black Radical Tradition*, 3rd ed. (University of North Carolina Press, 2020), ch. 1. See also Christopher Hill. *The English Revolution 1640: An Essay* (London: Lawrence and Wishart, 1955).

44. Michel Foucault, *"Society Must Be Defended": Lectures at the College de France 1975–1976*, trans. David Macey (New York: Picador, 2003), 103.

45. As Silvia Federici has argued, the seventeenth-century European witch trials offer insight into how racial colonial violence and gendered domestic violence were concurrent and connected strategies for the centralized appropriation of social power and land by emerging states. Silvia Federici, *Caliban and the Witch* (New York: Autonomedia, 2004). On the connections between the emergence of modern legal property rights and colonial land claims, see Henry Jones, "Property, Territory, and Colonialism: An International Legal History of Enclosure," *Legal Studies (Society of Legal Scholars)* 39, no. 2 (2019): 187–203, https://doi.org/10.1017/lst.2018.22; Alain Pot-

tage, "The Measure of Land," *Modern Law Review* 57, no. 3 (1994): 361–384, https://doi.org/10.1111/j.1468-2230.1994.tb01946.x; and David J. Seipp, "The Concept of Property in the Early Common Law," *Law and History Review* 12, no. 1 (1994): 29–91, https://doi.org/10.1017/S073824800001124X.

46. Henri Pirenne, *Medieval Cities: Their Origins and the Revival of Trade* (Princeton: Princeton University Press, 1952), 81.

47. Paul Kingsnorth, *The Wake* (Minneapolis: Graywolf, 2015). On this reading of *The Wake*, see Christian Schmitt-Kilb, "A Case for a Green Brexit?: Paul Kingsnorth, John Berger and the Pros and Cons of a Sense of Place," in *The Road to Brexit: A Cultural Perspective on British Attitudes to Europe*, ed. Ina Habermann (Manchester: Manchester University Press, 2020), 162–178; Joe Kennedy, "The Brexit Novel?," *New Socialist*, October 29, 2017, https://newsocialist.org.uk/the-brexit-novel/; Chris Jones, *Fossil Poetry: Anglo-Saxon and Linguistic Nativism in Nineteenth-Century Poetry* (Oxford: Oxford University Press, 2018), 10–11; and Kingsnorth's own article, "Brexit and the Culture of Progress," *Resurgence* (2016), https://www.paulkingsnorth.net/brexit. As Sylvia Wynter argues, the ontological template for biological racism was the Christian distinction between believers and non-believers. Sylvia Wynter, "Unsettling the Coloniality of Being/Power/Truth/Freedom: Towards the Human, After Man, Its Overrepresentation—An Argument," *CR: The New Centennial Review* 3, no. 3 (2003): 257–337, https://doi.org/10.1353/ncr.2004.0015. See also Frank B. Wilderson, *Red, White and Black* (Durham: Duke University Press, 2010).

48. Joseph Conrad, *Heart of Darkness* (New York: W. W. Norton, 2017), 5.

49. Conrad, *The Heart of Darkness*, 5–6.

50. It's worth pointing out that while the novella is based on Conrad's own experience in the Congo, Marlow is not Conrad, nor is he even the narrator of the novella, who remains anonymous. The novella's deliberate distancing and framing stymies any easy conflation of author and character and forces us to engage critically with Marlow's sophisticated narrative representation of racialized imperialism.

51. Davis and Todd, "On the Importance of a Date."

52. Steven A. LeBlanc, with Katherine E. Register, *Constant Battles: The Myth of the Peaceful, Noble Savage* (New York: St. Martin's Press, 2003); Ernest S. Burch, "Rationality and Resource Use Among Hunters," in *Native Americans and the Environment: Perspectives on the Ecological Indian*, ed. Michael Eugene Harkin and David Rich Lewis (Lincoln: University of Nebraska Press, 2007).

53. For further helpful perspectives on the racialized figure of the "ecological Indian," see Gregory D. Smithers, "Beyond the 'Ecological Indian': Environmental Politics and Traditional Ecological Knowledge in Modern North America," *Environmental History* 20, no. 1 (2015): 83–111, https://doi.org/10.1093/envhis/emu125; James D. Rice, "Beyond 'The Ecological Indian' and 'Virgin Soil Epidemics': New Perspectives on Native Americans and the Environment," *History Compass* 12, no. 9 (2014): 745–757, https://doi.org/10.1111/hic3.12184; Michael Eugene Harkin and David Rich Lewis, *Native Americans and the Environment: Perspectives on the Ecological Indian*

(Lincoln: University of Nebraska Press, 2007); Shepard Krech, *The Ecological Indian: Myth and History* (New York: W.W. Norton, 1999).

54. Michael E. Zimmerman, *Contesting Earth's Future: Radical Ecology and Postmodernity* (Berkeley: University of California Press, 1994), 307.

55. Matthew R. Baker et al., "Integrated Research in the Arctic–Ecosystem Linkages and Shifts in the Northern Bering Sea and Eastern and Western Chukchi Sea," *Deep Sea Research Part II: Topical Studies in Oceanography* (2023): 105251; Thomas J. Ballinger and James E. Overland, "The Alaskan Arctic Regime Shift Since 2017: A Harbinger of Years to Come?," *Polar Science* 32 (2022): 100841; Donna D. W. Hauser et al., "Co-Production of Knowledge Reveals Loss of Indigenous Hunting Opportunities in the Face of Accelerating Arctic Climate Change," *Environmental Research Letters* 16, no. 9 (2021): 95003, https://doi.org/10.1088/1748-9326/ac1a36; Kristin L. Laidre et al., "Arctic Marine Mammal Population Status, Sea Ice Habitat Loss, and Conservation Recommendations for the 21st Century," *Conservation Biology* 29, no. 3 (2015): 724–737.

56. Abram relies especially on the groundbreaking work of Keith Basso, *Wisdom Sits in Places: Landscape and Language Among the Western Apache* (Albuquerque: University of New Mexico Press, 1996).

57. David Abram, *The Spell of the Sensuous: Perception and Language in a More-Than-Human World* (New York: Vintage 1997), 179.

58. Abram, *The Spell of the Sensuous*, 178.

59. Carson, *Eros the Bittersweet*, 41–42.

60. This argument relies on the literacy hypothesis advanced by Walter Ong, Eric Havelock, Jack Goody, and David Olson and developed by David Abram and Anne Carson, along with Claude Lévi-Strauss's structural anthropology, particularly the distinction between wild and domesticated thought put forward in *La Pensée Sauvage*, and the human self-domestication hypothesis advanced by Richard Wrangham, Robert Bednarik, and others. On the literacy hypothesis, see Havelock, Abram, and Carson cited in earlier notes in this chapter, as well as David Olson, *The World on Paper: The Conceptual and Cognitive Implications of Writing and Reading* (Cambridge: Cambridge University Press, 1994); Walter J. Ong, *Orality and Literacy: The Technologizing of the Word* (London: Routledge, 1991); Jack Goody, *The Interface Between the Written and the Oral* (Cambridge: Cambridge University Press, 1987); David Olson, "The Cognitive Consequences of Literacy," *Canadian Psychology/Psychologie canadienne* 27, no. 2 (1986); Jack Goody, *The Domestication of the Savage Mind* (Cambridge: Cambridge University Press, 1977); Alexander R. Luria, *Cognitive Development, Its Cultural and Social Foundations* (Cambridge: Harvard University Press, 1976); Jack Goody, *Literacy in Traditional Societies* (Cambridge: Cambridge University Press, 1968); Jack Goody and Ian Watt, "The Consequences of Literacy," *Comparative Studies in Society and History* 5, no. 3 (1963): 304–345; H. A. Innis, *Empire and Communication* (Oxford: Clarendon Press, 1950). For criticisms of the literacy hypothesis, see Mazama Ama, "The Eurocentric Discourse on Writing: An

Exercise in Self-Glorification," *Journal of Black Studies* 29, no. 1 (1998): 3–16; John Halverson, "Goody and the Implosion of the Literacy Thesis," *Man* 27, no. 2 (1992): 301–317. These criticisms are answered here and elsewhere: David R Olson, "Footnotes to Goody: On Goody and His Critics," presented at "Ecritures: sur le traces de Jack Goody," ENSSIB, Lyon, France, January 24–26, 2008. On human self-domestication, see Dor Shilton et al., "Human Social Evolution: Self-Domestication or Self-Control?," *Frontiers in Psychology* 11, no. 134 (February 14, 2020), doi: 10.3389/fpsyg.2020.00134; Robert G. Bednarik, *The Domestication of Humans* (London: Routledge, 2020); Richard Wrangham, *The Goodness Paradox: The Strange Relationship Between Virtue and Violence in Human Evolution* (New York: Pantheon, 2019); Constantina Theofanopoulou et al., "Self-Domestication in *Homo Sapiens*: Insights from Comparative Genomics," *PLOS ONE* 12, no. 10 (2017): e0185306, doi: 10.1371/journal.pone.0185306; Gary Clark and Maciej Henneberg, "The Life History of 'Ardipithecus Ramidus': A Heterochronic Model of Sexual and Social Maturation," *Anthropological Review* 78, no. 2 (2015): 109–132; Robert L. Cieri et al. "Craniofacial Feminization, Social Tolerance, and the Origins of Behavioral Modernity," *Current Anthropology* 55, no. 4 (2014): 419–443, https://doi.org/10.1086/677209; Christopher Boehm, *Moral Origins: the Evolution of Virtue, Altruism, and Shame* (New York: Basic Books, 2012); Martin Brüne, "On Human Self-Domestication, Psychiatry, and Eugenics," *Philosophy, Ethics, and Humanities in Medicine* 2 (2007): 1–9; Helen M. Leach, "Human Domestication Reconsidered." *Current Anthropology* 44, no. 3 (2003): 349–368, https://doi.org/10.1086/368119.

61. Stanislas Dehaene et al., "Illiterate to Literate: Behavioural and Cerebral Changes Induced by Reading Acquisition," *Nature Reviews Neuroscience* 16, no. 4 (2015): 234–244, https://doi.org/10.1038/nrn3924.

62. The central role the remediation of oral cultures played in Romanticism is described by Maureen McLane, *Balladeering, Minstrelsy, and the Making of British Romantic Poetry* (Cambridge: Cambridge University Press, 2008).

63. Recent work in cliodynamics on the Seshat database provides further support for a robust connection between the emergence of literacy and specific thresholds of social and political complexity. See Gary M. Feinman and David M. Carballo, "Communication, Computation, and Governance: A Multiscalar Vantage on the Prehispanic Mesoamerican World," *Journal of Social Computing* 3, no. 1 (2022): 91–118; Timothy Kohler, et al., "Social Scale and Collective Computation: Does Information Processing Limit Rate of Growth in Scale?," *Journal of Social Computing* 3, no. 1 (March 2022): 1–17, doi: 10.23919/JSC.2021.0020; Ian Morris, "Scale, Information-Processing, and Complementarities in Old-World Axial Age Societies," *Journal of Social Computing* 3, no. 2 (June 2022): 119–127, doi: 10.23919/JSC.2022.0001; Jaeweon Shin et al., "Scale and Information-Processing Thresholds in Holocene Social Evolution," *Nature Communications* 11, no. 1 (2020): 2394, https://doi.org/10.1038/s41467-020-16035-9; Daniel Austin Mullins et al., "A Systematic Assessment of 'Axial Age' Proposals Using Global Comparative Historical Evidence," *American Sociological Review* 83, no. 3 (2018): 596–626,

https://doi.org/10.1177/0003122418772567; Peiter François et al., "A Macroscope for Global History: Seshat Global History Data-bank, a Methodological Overview," *Digital Humanities Quarterly* 10, no. 4 (2016), http://www.digitalhumanities.org/dhq/vol/10/4/000272/000272.html; Peter Turchin et al., "The Equinox 2020 Seshat Data Release," *Cliodynamics* 11, no. 1 (2020), http://dx.doi.org/10.21237/C7clio11148620, retrieved from https://escholarship.org/uc/item/4wj1j1vb; Peter Turchin et al., "An Introduction to Seshat: Global History Databank," *Journal of Cognitive Historiography* 5 (2018): 115–123, https://doi.org/10.1558/jch.39395.

64. Claude Lévi-Strauss, *Tristes Tropiques*, trans. John and Doreen Weightman (New York: Penguin, 1973), 299.

65. Édouard Glissant, *Poetics of Relation* (Ann Arbor: University of Michigan Press, September 1997).

66. See, for instance, Achille Mbembe, "Meditation on the Second Creation," *e-flux journal* 114 (December 2020).

67. Bronislaw Szerszynski, "Gods of the Anthropocene: Geo-Spiritual Formations in the Earth's New Epoch," *Theory, Culture & Society* 34, no. 2–3 (2017): 253–275, 253.

68. Ong, *Orality and Literacy*, 31.

69. Abram, *The Spell of the Sensuous*, 124.

70. Similarly, Paul Gilroy writes, "Fanon's insights have been devoured and tamed by timid, often parochial fields like 'critical race theory' and 'postcolonial theory,' which are now moving on to new pastures from which US racial technologies can be exported worldwide as part of a modernizing enterprise measurable through the government of diversity.... Against any damning, simplifying judgement, I want to argue that, rather than Fanon's insights being redundant or anachronistic, the full impact of his political and philosophical writing has not so far been appreciated." Paul Gilroy, "Fanon and Améry: Theory, Torture and the Prospect of Humanism," *Theory, Culture & Society* 27, no. 7–8 (2010): 16–32, https://doi.org/10.1177/0263276410383716.

71. Fanon, *The Wretched of the Earth*, 3.

72. Davis and Todd, "On the Importance of a Date, or, Decolonizing the Anthropocene." Kyle Whyte, "Too Late for Indigenous Climate Justice: Ecological and Relational Tipping Points," *Wiley Interdisciplinary Reviews: Climate Change* 11, no. 1 (2020): e603–n/a, https://doi.org/10.1002/wcc.603.

73. Jonathan Lear, *Radical Hope: Ethics in the Face of Cultural Devastation* (Cambridge: Harvard University Press, 2008).

74. The dream and its reception are recounted in Frank Linderman, *Plenty-Coups: Chief of the Crows* (Lincoln: University of Nebraska Press, 1962), 60–67 and 68–76.

75. Linderman, *Plenty-Coups*, 65.

76. Lear, *Radical Hope*, 9.

77. Frederick E. Hoxie, *Parading Through History: The Making of the Crow Nation in America, 1805–1935* (Cambridge: Cambridge University Press, 1995), 92.

78. Lear, *Radical Hope*, 146–147.

79. Lear, *Radical Hope*, 100.

80. Lear, *Radical Hope*, 150–151. Louis Warren argues persuasively that the Ghost Dance should be understood as a kind of cultural dreamwork akin to Plenty-Coups's dream—that is, an attempt to merge traditional native beliefs with those of the white settlers and accommodate native cultures to their new reality. Louis Warren, *God's Red Son: The Ghost Dance Religion and the Making of Modern America* (New York City: Basic Books, 2017).

81. Rachel A. Leavitt et al., "Suicides Among American Indian/Alaska Natives—National Violent Death Reporting System, 18 States, 2003–2014," *Morbidity and Mortality Weekly Report* 67, no. 8 (March 2, 2018), US Centers for Disease Control and Prevention, 237–242, https://www.cdc.gov/mmwr/volumes/67/wr/mm6708a1.htm; Julian Brave NoiseCat, "13 Issues Facing Native People Beyond Mascots and Casinos," *HuffPost*, July 30, 2015, https://www.huffpost.com/entry/13-native-american-issues_n_55b7d801e4b0074ba5a6869c; Suzanne Macartney, Alemayehu Bishaw, and Kayla Fontenot, "Poverty Rates for Selected Detailed Race and Hispanic Groups by State and Place: 2007–2011," US Census Bureau, February 2013, https://www.census.gov/library/publications/2013/acs/acsbr11-17.html; US Census Bureau, *The American Community—American Indians and Alaska Natives: 2004 American Community Survey Reports* (May 2007), https://www2.census.gov/library/publications/2007/acs/acs-07.pdf.

82. Danowski and de Castro, *The Ends of the World*, 105–106. Authors' italics.

83. Tapio Schneider, Colleen M. Kaul, and Kyle G. Pressel, "Possible Climate Transitions from Breakup of Stratocumulus Decks Under Greenhouse Warming," *Nature Geoscience* 12, no. 3 (2019): 164, doi: 10.1038/s41561-019-0310-1; see also California Institute of Technology, "High CO_2 Levels Can Destabilize Marine Layer Clouds: Loss of Low-Level Clouds Can Lead to a Spike in Global Warming, New Study Shows," *Science Daily*, February 25, 2019, https://www.sciencedaily.com/releases/2019/02/190225123036.htm; Natalie Wolchover, "A World Without Clouds," *Quanta Magazine*, February 25, 2019, https://www.quantamagazine.org/cloud-loss-could-add-8-degrees-to-global-warming-20190225/.

84. Kermode, *The Sense of an Ending*, 179.

85. Walter Benjamin, "The Work of Art in the Age of Mechanical Reproduction," in *Illuminations*, trans. Harry Zohn (New York: Schocken, 1978), 242.

86. Jon Bridle and Alexandra van Rensburg, "Discovering the Limits of Ecological Resilience," *Science* 367, no. 6478 (February 7, 2020): 626.

87. For a good recent discussion of this, see Michael Strevens, *The Knowledge Machine: How Irrationality Created Modern Science* (New York: Liveright, 2020).

Chapter 5

1. Lauren B. Alloy and Lyn Y. Abramson, "Judgment of Contingency in Depressed and Nondepressed Students: Sadder but Wiser?," *Journal of Experimental Psychology.*

General 108, no. 4 (1979): 441–485, doi: 10.1037/0096-3445.108.4.441. Other researchers had identified positive biases before Alloy and Abramson, notably Margaret W. Matlin and David J. Stang, *The Pollyanna Principle : Selectivity in Language, Memory, and Thought* (Cambridge: Schenkman, 1978); and Jerry Boucher and Charles Osgood, "The Pollyanna Hypothesis," *Journal of Verbal and Learning Behavior* 8, no. 1 (1969): 1–8.

2. Steven F. Maier and Martin E. Seligman, "Learned Helplessness: Theory and Evidence," *Journal of Experimental Psychology: General* 105, no. 1 (1976): 3. Maier and Seligman later found that "the original theory got it backwards. Passivity in response to shock is not learned. It is the default, unlearned response to prolonged aversive events and it is mediated by the serotonergic activity of the dorsal raphe nucleus, which in turn inhibits escape." Steven F. Maier and Martin E. Seligman, "Learned Helplessness at Fifty: Insights from Neuroscience," *Psychological Review* 123, no. 4 (2016): 349.

3. "Subjects were assigned to a depressed or nondepressed group on the basis of their Beck Depression Inventory (BDI) scores," a twenty-one-item, self-report rating inventory that measures characteristic attitudes and symptoms of depression (Alloy and Abramson, "Judgment of Contingency," 448). See also Aaron Beck et al., "An Inventory for Measuring Depression," *Archives of General Psychiatry* 4 (1961): 561–571.

4. Alloy and Abramson, "Judgment of Contingency," 474.

5. Lauren B. Alloy and Lyn Y. Abramson, "Depressive Realism," in *Encyclopedia of Social Psychology*, vol. 1, ed. Roy Baumeister and Kathleen Vohs, 242–243 (Los Angeles: Sage Publications, 2007).

6. Christoph W. Korn et al., "Depression Is Related to an Absence of Optimistically Biased Belief Updating About Future Life Events," *Psychological Medicine* 44, no. 3 (2014): 579–592, doi: 10.1017/S0033291713001074; Michael T. Moore and David M. Fresco, "Depressive Realism: A Meta-Analytic Review," *Clinical Psychology Review* 32, no. 6 (2012): 496–509, doi: 10.1016/j.cpr.2012.05.004; Daniel R. Strunk, Howard Lopez, and Robert J. DeRubeis, "Depressive Symptoms Are Associated with Unrealistic Negative Predictions of Future Life Events," *Behaviour Research and Therapy* 44, no. 6 (2006): 861–882, https://doi.org/10.1016/j.brat.2005.07.001. See also C. Randall Colvin and Jack Block, "Do Positive Illusions Foster Mental Health? An Examination of the Taylor and Brown Formulation," *Psychological Bulletin* 116, no. 1 (1994): 3–20, https://doi.org/10.1037/0033-2909.116.1.3; Ben Hayden, "Depressive Realism May Not Be Real," *Psychology Today*, August 17, 2011, https://www.psychologytoday.com/us/blog/the-decision-tree/201108/depressive-realism-may-not-be-real.

7. Lauren B. Alloy and Lyn Y. Abramson, "Depressive Realism: Four Theoretical Perspectives," in *Cognitive Processes in Depression*, ed. Lauren B. Alloy (New York: Guilford Press, 1988), 224.

8. T. S. Eliot, "Burnt Norton," in *Four Quartets* (New York: Harcourt Brace Jovanovich, 1971). This line of thought has given rise to what we might call a philosophy

of depressive realism. See the 2016 special issue of *Self and Society* devoted to exploring the ramifications of depressive realism from the viewpoint of Humanistic Psychology, particularly the guest editor's introduction: Colin Feltham, "Depressive Realism: What It Is and Why It Matters to Humanistic Psychology," *Self and Society* 44, no. 2 (2016): 88–93, https://doi.org/10.1080/03060497.2016.1191831. See also Colin Feltham, *Depressive Realism: Interdisciplinary Perspectives* (London: Routledge, 2016).

9. "A rather consistent finding across many studies is that nondepressed individuals exhibit optimistic biases and illusions in their perceptions and judgments about themselves but not in their inferences about others. Likewise, depressed individuals are often (though not always) more realistic and less biased in their self-relevant perceptions and inferences but succumb to optimistic biases or distortions in their other-referent judgments." Alloy and Abramson, "Depressive Realism," 243.

10. Tiffany Szu-Ting Fu et al., "Confidence Judgment in Depression and Dysphoria: The Depressive Realism vs. Negativity Hypotheses," *Journal of Behavior Therapy and Experimental Psychiatry* 43, no. 2 (2011): 699–704, doi: 10.1016/j.jbtep.2011.09.014; Tiffany Fu et al., "Depression, Confidence, and Decision: Evidence Against Depressive Realism," *Journal of Psychopathology and Behavioral Assessment* 27, no. 4 (2005): 243–252, doi: 10.1007/s10862-005-2404-x.

11. Lisa Bortolotti and Magdalena Antrobus, "Costs and Benefits of Realism and Optimism," *Current Opinion in Psychiatry* 28, no. 2 (2015): 194; Moore and Fresco, "Depressive Realism: A Meta-Analytic Review."

12. Ravi Philip Rajkumar, "Depressive Realism and Functional Fear: An Alternative Perspective on Psychological Distress During the COVID-19 Pandemic," *Primary Care Companion for CNS Disorders* 22, no. 4 (2020), https://doi.org/10.4088/PCC.20com02714; Ravi Philip Rajkumar, "Contamination and Infection: What the Coronavirus Pandemic Could Reveal About the Evolutionary Origins of Obsessive-Compulsive Disorder," *Psychiatry Research* 289 (2020): 113062; Midori Tanaka, Dennis K. Kinney, and Sherry Anders, "Depression as an Evolutionary Strategy for Defense Against Infection," *Brain, Behavior, and Immunity* 31 (2013): 9–22; Melissa Bateson, Ben Brilot, and Daniel Nettle, "Anxiety: An Evolutionary Approach," *The Canadian Journal of Psychiatry* 56, no. 12 (2011): 707–715; Dennis K. Kinney and Midori Tanaka, "An Evolutionary Hypothesis of Depression and Its Symptoms, Adaptive Value, and Risk Factors," *The Journal of Nervous and Mental Disease* 197, no. 8 (2009): 561–567.

13. Ethan Zell et al., "The Better-Than-Average Effect in Comparative Self-Evaluation: A Comprehensive Review and Meta-Analysis," *Psychological Bulletin* 146, no. 2 (2020); Raffael Kalisch, Marianne B. Müller, and Oliver Tüscher, "A Conceptual Framework for the Neurobiological Study of Resilience," *Behavioral and Brain Sciences* 38 (2015); Christoph W. Korn et al., "Depression Is Related to an Absence of Optimistically Biased Belief Updating About Future Life Events," *Psychological Medicine* 44, no. 3 (2014): 579–592, doi: 10.1017/S0033291713001074; Helena

Matute et al., "Illusions of Causality at the Heart of Pseudoscience," *The British Journal of Psychology* 102, no. 3 (2011): 392–405, doi: 10.1348/000712610X532210; Manju Puri and David T. Robinson, "Optimism and Economic Choice," *Journal of Financial Economics* 86, no. 1 (2007): 71–99; Julie K. Norem, "Defensive Self-Deception and Social Adaptation Among Optimists," *Journal of Research in Personality* 36, no. 6 (2002): 549–555, doi: 10.1016/S0092-6566(02)00504-4; Shelley E. Taylor and Jonathon D. Brown, "Positive Illusions and Well-Being Revisited," *Psychological Bulletin* 116, no. 1 (1994): 21–27, doi: 10.1037/0033-2909.116.1.21; Shelley E. Taylor, *Positive Illusions: Creative Self-Deception and the Healthy Mind* (New York: Basic Books, 1989); Neil D. Weinstein, "Optimistic Biases About Personal Risks." *Science* 246, no. 4935 (1989): 1232–1233, https://doi.org/10.1126/science.2686031; Shelley E. Taylor and Jonathon D. Brown, "Illusion and Well-Being: A Social Psychological Perspective on Mental Health," *Psychological Bulletin* 103, no. 2 (1988): 193–210, doi: 10.1037//0033-2909.103.2.193. See also David Hume, *An Enquiry Concerning Human Understanding* (Oxford: Oxford University Press, 2007); and Friedrich Nietzsche, *Beyond Good and Evil: Prelude to a Philosophy of the Future*, trans. R. J. Hollingdale (New York: Penguin Books, 2003).

14. Peter Sheridan Dodds et al., "Human Language Reveals a Universal Positivity Bias," *Proceedings of the National Academy of Sciences* 112, no. 8 (2015): 2389–2394, doi: 10.1073/pnas.1411678112.

15. Tali Sharot, "The Optimism Bias," *Current Biology* 21, no. 23 (December 2011): R941–R945, doi: 10.1016/j.cub.2011.10.030. See also Tali Sharot, *The Optimism Bias: A Tour of the Irrationally Positive Brain* (New York: Pantheon, 2011), 192, where she cites data from Yale psychologist David Armor. Ronald Fischer and Anna Chalmers, "Is Optimism Universal? A Meta-Analytical Investigation of Optimism Levels Across 22 Nations," *Personality and Individual Differences* 45, no. 5 (2008): 378–382. On the other hand, Edward Chang finds evidence of cultural variations in levels of optimism and pessimism that demand further research. Edward C. Chang, "Cultural Influences on Optimism and Pessimism: Differences in Western and Eastern Construals of the Self," in *Optimism & Pessimism: Implications for Theory, Research, and Practice*, ed. Edward C. Chang (Washington, D.C.: APA, 2001), 257–280.

16. Tali Sharot et al., "How Unrealistic Optimism Is Maintained in the Face of Reality," *Nature Neuroscience* 14, no. 11 (2011): 1475–1479, https://doi.org/10.1038/nn.2949.

17. Samuel J. Gershman, "How to Never Be Wrong," *Psychonomic Bulletin & Review* 26, no. 1 (2019): 13–28, doi: 10.3758/s13423-018-1488-8.

18. Sharot et al., "How Unrealistic Optimism Is Maintained."

19. Dong Haur Phua and Nigel C. K. Tan, "Cognitive Aspect of Diagnostic Errors," *Annals of the Academy of Medicine, Singapore* 42, no. 1 (2013): 33–41; Deborah Kelemen et al., "Professional Physical Scientists Display Tenacious Teleological Tendencies: Purpose-Based Reasoning as a Cognitive Default," *Journal of Experimental Psychology: General* 142, no. 4 (2013): 1074–1083, doi: 10.1037/a0030399; Richard Wise-

man and Caroline Watt, "Belief in Psychic Ability and the Misattribution Hypothesis: A Qualitative Review," *The British Journal of Psychology* 97, no. 3 (2006): 323–338, doi: 10.1348/000712605X72523.

20. Susan T. Fiske and Shelley E. Taylor, *Social Cognition*, 2nd ed. (New York: McGraw-Hill, 1991), 88.

21. Timothy C. Bates, "The Glass Is Half Full and Half Empty: A Population-Representative Twin Study Testing If Optimism and Pessimism Are Distinct Systems," *The Journal of Positive Psychology* 10, no. 6 (February 25, 2015): 533–542, doi: 10.1080/17439760.2015.1015155; Dominic D. P. Johnson and James H. Fowler, "The Evolution of Overconfidence," *Nature* 477, no. 7364 (2011): 317–320; John M. McNamara et al., "Environmental Variability Can Select for Optimism or Pessimism," *Ecology Letters* 14, no. 1 (2011): 58–62, doi: 10.1111/j.1461-0248.2010.01556.x; Tali Sharot, "The Optimism Bias"; Marvin Zuckerman, "Optimism and Pessimism: Biological Foundations," in *Optimism and Pessimism: Implications for Theory, Research, and Practice*, ed. Edward C. Chang (Washington, D.C.: APA, 2001); Lionel Tiger, *Optimism: The Biology of Hope* (New York: Simon and Schuster, 1979).

22. Taylor and Brown, "Illusion and Well-Being," 194.

23. Roy F. Baumeister, "The Optimal Margin of Illusion," *Journal of Social and Clinical Psychology* 8, no. 2 (1989). See also Roy F. Baumeister et al., "Does High Self-Esteem Cause Better Performance, Interpersonal Success, Happiness, or Healthier Lifestyles?," *Psychological Science in the Public Interest*, May 2003, https://doi.org/10.1111/1529-1006.01431.

24. Roy F. Baumeister, Laura Smart, and Joseph M. Boden, "Relation of Threatened Egotism to Violence and Aggression: The Dark Side of High Self-Esteem," *Psychological Review* 103, no. 1 (1996).

25. C. Randall Colvin, Jack Block, and David C. Funder, "Overly Positive Self-Evaluations and Personality: Negative Implications for Mental Health," *Journal of Personality and Social Psychology* 68, no. 6 (1995). Baumeister et al., "Does High Self-Esteem Cause Better Performance?" See also Norem, "Defensive Self-Deception and Social Adaptation Among Optimists."

26. Hunter A. McAllister et al., "The Optimal Margin of Illusion Hypothesis: Evidence from the Self-Serving Bias and Personality Disorders," *Journal of Social and Clinical Psychology* 21, no. 4 (2002): 414–426, 422.

27. Baumeister, "The Optimal Margin of Illusion," 177; Pietro Ortoleva and Erik Snowberg, "Overconfidence in Political Behavior," *American Economic Review* 105, no. 2 (2015): 504–535.

28. D. Dunning, C. Heath, and J. M. Suls, "Flawed Self-Assessment: Implications for Health, Education, and the Workplace," *Psychological Science in the Public Interest* 5, no. 3 (2004): 69–106, https://doi.org/10.1111/j.1529-1006.2004.00018.x.

29. For examples illustrating the consequences of optimistic overconfidence, in addition to Naomi Oreskes and Erik Conway, *Merchants of Doubt: How a Handful of Scientists Obscured the Truth on Issues from Tobacco Smoke to Global Warming* (New

York: Bloomsbury Press, 2010); and Dale Jamieson, *Reason in a Dark Time: Why the Struggle Against Climate Change Failed—and What It Means for Our Future* (Oxford: Oxford University Press, 2014), see also Elizabeth Kolbert, *Under a White Sky: The Nature of the Future* (New York: Crown, 2021); Wenjie Chen, Mico Mrkaic, and Malhar S. Nabar, "The Global Economic Recovery 10 Years After the 2008 Financial Crisis," *IMF Working Paper 2019/083* (Washington, DC: International Monetary Fund, 2019); Nathaniel Rich, *Losing Earth: A Recent History* (New York: MCD, 2019); Paul Beaudry and Tim Willems, "On the Macroeconomic Consequences of Over-Optimism," *NBER Working Paper No. 24685* (Cambridge: National Bureau of Economic Research, 2018); Dominic D. P. Johnson, *Overconfidence and War: the Havoc and Glory of Positive Illusions* (Cambridge: Harvard University Press, 2004); Baumeister, "The Optimal Margin of Illusion"; Barbara Tuchman, *The March of Folly* (New York: Knopf, 1984); and Charles McKay, *Memoirs of Extraordinary Popular Delusions* (London: Richard Bentley, 1841).

30. See Kolbert, *Under a White Sky*, for an exploration of how optimistic technological solutions have led—and will almost certainly continue to lead—to unanticipated problems at least as serious as the ones that the interventions had been developed to solve.

31. Hans Jonas, *The Imperative of Responsibility* (Chicago: University of Chicago Press, 1985), 27–34; Luke Kemp et al., "Climate Endgame: Exploring Catastrophic Climate Change Scenarios," *Proceedings of the National Academy of Sciences* 119, no. 34 (2022): 1–e2108146119, https://doi.org/10.1073/pnas.2108146119; Daniel Steel et al., "Climate Change and the Threat to Civilization," *Proceedings of the National Academy of Sciences* 119, no. 42 (2022): 1–e2210525119, https://doi.org/10.1073/pnas.2210525119.

32. Joshua Foa Dienstag, *Pessimism: Philosophy, Ethic, Spirit* (Princeton: Princeton University Press, 2006), 161.

33. Carle Riley et al., "Trends and Variation in the Gap Between Current and Anticipated Life Satisfaction in the United States, 2008–2020," *American Journal of Public Health* 112, no. 3 (2022): 509–17, doi: 10.2105/AJPH.2021.306589; Courtney Johnson, "'Particularly Good Days' Are Common in Africa, Latin America and The U.S.," Pew Research, January 2, 2018, https://www.pewresearch.org/fact-tank/2018/01/02/particularly-good-days-are-common-in-africa-latin-america-and-the-u-s/.

34. Barbara Ehrenreich, *Bright-Sided: How the Relentless Promotion of Positive Thinking Has Undermined America* (New York: Metropolitan Books, 2009); Lauren Berlant, *Cruel Optimism* (Durham: Duke University Press, 2011); Susan Cain, *Bittersweet: How Sorrow and Longing Make Us Whole* (New York: Crown, 2022). See also William Davies, *The Happiness Industry: How the Government and Big Business Sold Us Well-Being* (London: Verso, 2015).

35. Ehrenreich. *Bright-Sided*, 45–46.

36. Chang quoted in Jared Keller, "What Makes Americans So Optimistic?" *The Atlantic*, March 25, 2015, https://www.theatlantic.com/politics/archive/2015/03/the-american-ethic-and-the-spirit-of-optimism/388538.

37. Mara van der Lugt, *Dark Matters: Pessimism and the Problem of Suffering* (Princeton: Princeton University Press, 2021), 10.

38. *Monty Python's Life of Brian* (1979), dir. Terry Jones; Megan Kelly, Alex DiBranco, and Dr. Julia R. DeCook, "Misogynist Incels and Male Supremacism: Overview and Recommendations for Addressing the Threat of Male Supremacist Violence," *New America*, February 18, 2021, https://www.newamerica.org/political-reform/reports/misogynist-incels-and-male-supremacism/; Urban Dictionary, "black pilled," https://www.urbandictionary.com/define.php?term=Black%20Pilled; Mike Pearl, "Climate Change Edgelords Are the New Climate Change Deniers," *Vice*, October 10, 2018, https://www.vice.com/en/article/pa9vg8/climate-change-edgelords-are-the-new-climate-change-deniers.

39. Eugene Thacker, *Cosmic Pessimism* (Minneapolis: Univocal Publishing, 2015), 56.

40. Joe Bailey, *Pessimism* (New York: Routledge, 1988), 159.

41. Jane Gillham et al., "Optimism, Pessimism, and Explanatory Style," in *Optimism and Pessimism: Implications for Theory, Research, and Practice*, ed. Edward C. Chang (Washington, D.C.: APA, 2001).

42. Edward C. Chang, "Introduction: Optimism and Pessimism and Moving Beyond the Most Fundamental Question," in *Optimism and Pessimism: Implications for Theory, Research, and Practice*, ed. Edward C. Chang (Washington, D.C.: APA, 2001), 5.

43. Heather Craig et al., "The Association of Optimism and Pessimism and All-Cause Mortality: A Systematic Review," *Personality and Individual Differences* 177 (2021): 110788; Chester Kam and John P. Meyer, "Do Optimism and Pessimism Have Different Relationships with Personality Dimensions? A Re-Examination," *Personality and Individual Differences* 52, no. 2 (2012): 123–127; William N. Dember, "The Optimism-Pessimism Instrument," in *Optimism and Pessimism: Implications for Theory, Research, and Practice*, ed. Edward C. Chang (Washington, D.C.: APA, 2001); Edward C. Chang, "Distinguishing Between Optimism and Pessimism: A Second Look at The Optimism-Neuroticism Hypothesis," in *Viewing Psychology as a Whole: The Integrative Science of William N. Dember*, ed. Robert R. Hoffman, Michael F. Sherrick, and Joel S. Warm (Washington, DC: APA, 1998), 415–432, https://doi.org/10.1037/10290-019; Edward C. Chang, Albert Maydeu-Olivares, and Thomas J. D'Zurilla, "Optimism and Pessimism as Partially Independent Constructs: Relationship to Positive and Negative Affectivity and Psychological Well-Being," *Personality and Individual Differences* 23, no. 3 (1997): 433–440; Grant N. Marshall et al., "Distinguishing Optimism from Pessimism: Relations to Fundamental Dimensions of Mood and Personality," *Journal of Personality and Social Psychology* 62, no. 6 (1992): 1067; Timothy C. Bates, "The Glass Is Half Full and Half Empty."

44. Julie K. Norem, *The Positive Power of Negative Thinking: Using Defensive Pessimism to Harness Anxiety and Perform at Your Peak* (New York: Basic Books, 2001), 25.

45. Christian Unkelbach, Hans Alves, and Alex Koch, "Negativity Bias, Positivity Bias, and Valence Asymmetries: Explaining the Differential Processing of Positive and Negative Information," in *Advances in Experimental Social Psychology*, vol. 62, ed. Bertram Gawronski (Cambridge: Academic Press, 2020), 115–187; Maria Lewicka et al., "Positive-Negative Asymmetry or When the Heart Needs a Reason," *European Journal of Social Psychology* 22, no. 5 (1992): 425–434; Guido Peeters and Janusz Czapinski, "Positive-Negative Asymmetry in Evaluations: The Distinction Between Affective and Informational Negativity Effects," *European Review of Social Psychology* 1, no. 1 (1990): 33–60, https://doi.org/10.1080/14792779108401856; Guido Peeters, "The Positive-Negative Asymmetry: On Cognitive Consistency and Positivity Bias," *European Journal of Social Psychology* 1, no. 4 (1971): 455–474, https://doi.org/10.1002/ejsp.2420010405.

46. Paul Rozin and Edward B. Royzman, "Negativity Bias, Negativity Dominance, and Contagion," *Personality and Social Psychology Review* 5, no. 4 (2001): 296–320, 297.

47. Anders Nordgren, "Pessimism and Optimism in the Debate on Climate Change: A Critical Analysis," *Journal of Agricultural and Environmental Ethics* 34, no. 22 (2021), https://doi.org/10.1007/s10806-021-09865-0.

48. Norem, *The Positive Power of Negative Thinking*.

49. Julie K. Norem and Nancy Cantor, "Defensive Pessimism," *Journal of Personality and Social Psychology* 51, no. 6 (1986): 1208–1217, https://doi.org/10.1037/0022-3514.51.6.1208; Julie K. Norem and Nancy Cantor, "Anticipatory and Post Hoc Cushioning Strategies: Optimism and Defensive Pessimism in 'Risky' Situations," *Cognitive Therapy and Research* 10, no. 3 (1986): 347–362, https://doi.org/10.1007/BF01173471.

50. Stacie M. Spencer and Julie K. Norem, "Reflection and Distraction Defensive Pessimism, Strategic Optimism, and Performance," *Personality & Social Psychology Bulletin* 22, no. 4 (1996): 354–365, https://doi.org/10.1177/0146167296224003.

51. Julie K. Norem, "Defensive Pessimism, Anxiety, and the Complexity of Evaluating Self-Regulation," *Social and Personality Psychology Compass* 2, no. 1 (2008): 121–34, https://doi.org/10.1111/j.1751-9004.2007.00053.x.

52. Julie K. Norem and Edward C. Chang, "The Positive Psychology of Negative Thinking," *Journal of Clinical Psychology* 58, no. 9 (2002): 993–1001.

53. Glenn Affleck, Howard Tennen, and Andrea Apter, "Optimism, Pessimism, and Daily Life with Chronic Illness," in *Optimism and Pessimism: Implications for Theory, Research, and Practice*, ed. Edward C. Chang (Washington, D.C.: APA, 2001); Gerhard Andersson, "The Benefits of Optimism: A Meta-Analytic Review of the Life Orientation Test," *Personality and Individual Differences* 21 (1996): 719–725; Martin E. P. Seligman, *Learned Optimism* (New York: Pocket Books, 1992).

54. Jonathon Brown and Margaret Marshall, "Great Expectations: Optimism and Pessimism in Achievement Settings," in *Optimism and Pessimism: Implications for Theory, Research, and Practice*, ed. Edward C. Chang (Washington, D.C.: APA, 2001).

55. Julia K. Boehm, Ying Chen, et al., "Unequally Distributed Psychological

Assets: Are There Social Disparities in Optimism, Life Satisfaction, and Positive Affect?," *PLoS ONE* 10, no. 2 (2015): e0118066, https://doi.org/10.1371/journal.pone.011 8066; Kathryn A. Robb et al., "Socioeconomic Disparities in Optimism and Pessimism," *International Journal of Behavioral Medicine* 16, no. 4 (2009): 331–338, https://doi.org/10.1007/s12529-008-9018-0.

56. Sanaz Talaifar et al., "Asymmetries in Mutual Understanding: People with Low Status, Power, and Self-Esteem Understand Better Than They Are Understood," *Perspectives on Psychological Science* 16, no. 2 (2021), https://doi.org/10.1177/1745691620958003.

57. Baumeister et al., "Does High Self-Esteem Cause Better Performance?"

58. Kati Heinonen et al., "Socioeconomic Status in Childhood and Adulthood: Associations with Dispositional Optimism and Pessimism Over a 21-Year Follow-Up," *Journal of Personality* 74, no. 4 (2006): 1111–1126, https://doi.org/10.1111/j.1467-6494.2006.00404.x. See also Dorsa Amir et al., "An Uncertainty Management Perspective on Long-Run Impacts of Adversity: The Influence of Childhood Socioeconomic Status on Risk, Time, and Social Preferences," *Journal of Experimental Social Psychology* 79 (2018): 217–226; as well as other more recent research on long-term impacts of childhood SES.

59. Daniel Kahneman, *Thinking, Fast and Slow* (New York: Farrar, Straus and Giroux, 2011), ch. 31; Ian Bateman et al., "Testing Competing Models of Loss Aversion: An Adversarial Collaboration," *Journal of Public Economics* 89, no. 8 (2005): 1561–1580, https://doi.org/10.1016/j.jpubeco.2004.06.013.

60. Marianne Bertrand et al., "Behavioral Economics and Marketing in Aid of Decision Making Among the Poor," *Journal of Public Policy & Marketing* 25, no. 1 (2006): 8–23, https://doi.org/10.1509/jppm.25.1.8.

61. Karen A. Cerulo, *Never Saw It Coming: Cultural Challenges to Envisioning the Worst* (Chicago: University of Chicago Press, 2006). See also Richard H. Thaler, *Nudge: Improving Decisions About Health, Wealth, and Happiness* (New York: Penguin Books, 2009); and Kahneman, *Thinking Fast and Slow*.

62. Stephanie Mertens et al., "The Effectiveness of Nudging: A Meta-Analysis of Choice Architecture Interventions Across Behavioral Domains," *Proceedings of the National Academy of Sciences* 119, no. 1 (2022), https://doi.org/10.1073/pnas.2107346 118; Stefano DellaVigna and Elizabeth Linos, "RCTs to Scale: Comprehensive Evidence from Two Nudge Units," *Econometrica* 90 (2022): 81–116, https://doi.org/10.3982/ECTA18709; Maximilian Maier et al., "No Evidence for Nudging After Adjusting for Publication Bias," *Proceedings of the National Academy of Sciences* 119, no. 31 (2022), doi: e2200300119; Barnabas Szaszi et al., "No Reason to Expect Large and Consistent Effects of Nudge Interventions," *Proceedings of the National Academy of Sciences* 119, no. 31 (2022), doi: e2200732119; Jonathan Z. Bakdash and Laura R. Marusich, "Left-Truncated Effects and Overestimated Meta-Analytic Means," *Proceedings of the National Academy of Sciences* 119, no. 31 (2022): e2203616119; Stephanie Mertens et al., "Reply to Maier et al., Szaszi et al., and Bakdash and Marusich: The

Present and Future of Choice Architecture Research," *Proceedings of the National Academy of Sciences* 119, no. 31 (2022), doi: e2202928119; Mark Kosters and Jeroen van der Heijden, "From Mechanism to Virtue: Evaluating Nudge Theory," *Evaluation* 21, no. 3 (2015): 276–291, https://doi.org/10.1177/1356389015590218.

63. Bailey, *Pessimism*, 26. See also Henry Vyverberg, *Historical Pessimism in the French Enlightenment* (Cambridge: Harvard University Press, 1958).

64. Hernán D. Caro, *The Best of All Possible Worlds?: Leibniz's Philosophical Optimism and Its Critics 1710–1755* (Leiden: Brill, 2020), ch. 3.

65. Louis Bertrand Castel, review of the second edition of the *Theodicy*, in *Mémoires pour l'histoire des Sciences et des beaux arts = Mémoires de Trévoux* 37 (1737; repr., Geneva: Slatkine Reprints, 1968), 6–36, 198–241, 444–471 and 954–991, https://babel.hathitrust.org/cgi/pt?id=msu.31293025270350&seq=11. Cited in Caro, *The Best of All Possible Worlds?*

66. Bayle criticized Leibniz's arguments in his "Rosarius" article, a version of which was published in the *Journal des Savants* (June 27, 1695), to which Leibniz first replied in the *Histoire des Ouvrages des Savants* in July 1698, then again in a private letter in 1702, which, according to Leibniz scholar Austin Farrer, "reads almost like a sketch for the *Theodicy*." Farrer notes, "The point of connection between Rosarius and Leibniz was no more than this, that both held views about the souls of beasts." Austin Farrer, "Editor's Introduction," in *Theodicy: Essays on the Goodness of God, the Freedom of Man, and the Origin of Evil*, by G. W. Leibniz, trans. E. M. Huggard (New Haven: Yale University Press, 1952), 46, 34. Leibniz also discusses the origins of the *Theodicy* in that work's preface. See also Steven Nadler, *The Best of all Possible Worlds: A Story of Philosophers, God, and Evil* (New York: Farrar, Straus and Giroux, 2008), esp. 78–106.

67. See Van der Lugt, *Dark Matters*. The entries on the Manichians and Paulitians can be found in the standard translation of Bayle still in print, by Popkin. For the entry on Xenophanes, readers must go to editions from the early eighteenth century, today accessible in digital collections. Pierre Bayle, *Historical and Critical Dictionary: Selections*, trans. Richard Popkin (Indianapolis: Hackett, 1991); Pierre Bayle, "Xenophanes," *An Historical and Critical Dictionary By Monsieur Bayle*, vol. 4 (London: C. Harper et al., 1710), 3047–3059, Gale ECCO, Gale document no. GALE |CW0114129616.

68. In addition to van der Lugt, *Dark Matters*, see Dienstag, *Pessimism*, 9; Richard Henry Popkin, *The History of Scepticism: From Savonarola to Bayle*, rev. and expanded ed. (Oxford: Oxford University Press, 2003), ch. 18; and Brian Domino and Daniel W. Conway, "Optimism and Pessimism from a Historical Perspective," in *Optimism and Pessimism: Implications for Theory, Research, and Practice*, ed. Edward Chang (Washington, D.C.: APA, 2001), 13–30.

69. Gottfried Wilhelm Leibniz, *Theodicy: Essays on the Goodness of God, the Freedom of Man, and the Origin of Evil*, trans. E. M. Huggard (New Haven: Yale University Press, 1952), 61–62.

70. Van der Lugt, *Dark Matters*, 72. See also 42–66.

71. Van der Lugt, *Dark Matters*.

72. Van der Lugt, *Dark Matters*, 73.

73. It should be noted that most Europeans of the seventeenth and eighteenth century weren't bothered by such puzzles and incongruities and went along happily murdering each other on the basis of faith, syncretizing contradictory beliefs, and living their lives according to received wisdom and the dictates of earthly authorities. It was only a tiny caste of privileged bourgeois and aristocratic malcontents who thought about such questions.

74. In the words of Austin Farrer, "Leibniz . . . like all the philosophers of the seventeenth century, was reforming scholasticism in the light of a new physical science." Farrer, "Editor's Introduction to Leibniz's *Theodicy*," 12.

75. Leibniz, *Theodicy*, 128.

76. Leibniz, *Theodicy*, 442.

77. Although Pope denied ever having read Leibniz, his *Essay* was read as a verse *Theodicy*. Alexander Pope, *An Essay on Man* (Princeton: Princeton University Press, 2016).

78. Pope, *An Essay on Man*.

79. Voltaire, *Candide*, in *Candide, Zadig, and Selected Stories*, trans. Daniel Frame (New York: Signet Classics, 1961), 2.

80. Voltaire, *Candide*, 90.

81. Voltaire, *Candide*, 96.

82. Ira Wade, *Voltaire and Candide* (Princeton: Princeton University Press, 1959), 87. See also George R. Havens, "The Composition of Voltaire's Candide," *Modern Language Notes* 47, no. 4 (1932): 225–234, https://doi.org/10.2307/2913581; George R. Havens, "Some Notes on Candide," *MLN* 88, no. 4 (1973): 841–847, https://doi.org/10.2307/2907412.

83. Will Durant and Ariel Durant, *The Age of Voltaire* (New York: Simon & Schuster, 1965), 462–463.

84. Voltaire, "The Lisbon Earthquake," in *Portable Voltaire*, ed. Ben Redman (New York: Viking, 1949), 569.

85. Susan Neiman, *Evil in Modern Thought: An Alternative History of Philosophy*, First Princeton Classic Edition, with a new afterword by the author (Princeton: Princeton University Press, 2015), 2–3, 8.

86. Neiman, *Evil in Modern Thought*, 8.

87. Neiman, *Evil in Modern Thought*, 115.

88. Theodor Adorno, *Negative Dialectics* (New York: Continuum, 1973), 361.

89. Voltaire, *Works*, XVIa, 144. Quoted in Durant and Durant, *The Age of Voltaire*, 487.

90. Voltaire, *Memoirs of Voltaire*, from Jean-Antoine-Nicolas de Caritat, *The Life of Voltaire, by the Marquis de Condorcet. To which are added, Memoirs of Voltaire, written by himself*, vol. 2 (G. G. J. and J. Robinson, Pater-Noster-Row, 1790), 166, Gale ECCO, Gale document no. GALE|CW0110000462.

91. Ian Davidson, *Voltaire: A Life* (New York: Pegasus Books, 2010), 291.

92. Elie Fréron, review of *Candide, Année Littéraire*, vol. 2 (Amsterdam: Michel Lambert, 1759), 204–205, repr. in vol. 6 (Genève: Slatkine Reprints, 1966), 147. See Aurélien Demars, "Le Pessimisme Jubilatoire de Cioran: Enquête sur un Paradigme Métaphysique Négatif," thèse pour obtenir le grade de docteur en Philosophie de l'Université Jean Moulin Lyon 3, 15 Octobre 2007, 37–38, https://scd-resnum.univ-lyon3.fr/out/theses/2007_out_demars_a.pdf.

93. Roger Pearson draws an explicit connection between Candide's circumstances at the end of the novel and Voltaire's in Ferney. Roger Pearson, *Voltaire Almighty: A Life in Pursuit of Freedom* (New York: Bloomsbury, 2005), 261.

94. K. Rockett, "An Optimistic Streak in Voltaire's Thought," *The Modern Language Review* 39, no. 1 (1944): 24–27, https://doi.org/10.2307/3716455.

95. Will Durant and Ariel Durant, *Rousseau and Revolution* (New York: Simon and Schuster, 1967), 145.

96. Karl Löwith, *Meaning in History: Theological Implications of the Philosophy of History* (Chicago: University of Chicago Press, 1949), 111.

97. See Larry M. Jorgensen and Samuel Newlands, "Introduction," in *New Essays on Leibniz's Theodicy*, ed. Larry M. Jorgensen and Samuel Newlands (Oxford: Oxford University Press, 2014), 4.

98. Brian Domino and Daniel Conway trace the origins of philosophical optimism to the conviction of René Descartes that "a methodical application of human reason can unlock the mysteries of the natural world." Domino and Conway, "Optimism and Pessimism from a Historical Perspective," 14.

99. Plato, *The Apology of Socrates*, trans. Benjamin Jowett, 30e–31a.

Chapter 6

1. William Hazlitt, *The Spirit of the Age*, in *Lectures on the English Poets & Spirit of the Age: Or Contemporary Portraits* (London: JP Dent & Sons 1910), 182–183.

2. William Godwin, *An Enquiry Concerning Political Justice*, book 1, iv (Oxford: Oxford University Press, 2013), 20.

3. Godwin, *Enquiry*, 1, vi, 33.

4. Godwin, *Enquiry*, 2, vi, 72.

5. Godwin, *Enquiry*, 1, iii, 16, and 1, vii, 41. See also the quotation he attributes to Benjamin Franklin, "Mind will one day become omnipotent over matter" (8, vii, 453).

6. Godwin, *Enquiry*, 6, viii, 350–354.

7. Godwin, *Enquiry*, 8, vii, 458.

8. Godwin, *Enquiry*, 8, viii, 468–469.

9. Alison Bashford and Joyce E. Chaplin, *The New Worlds of Thomas Robert Malthus: Rereading the Principle of Population* (Princeton: Princeton University Press, 2016), 56, 57–63.

10. Robert Mayhew, *Malthus: The Life and Legacies of an Untimely Prophet* (Cambridge: Harvard University Press, 2014), 62–63; Bashford and Chaplin, *The New Worlds*, 57–58.

11. Mayhew, *Malthus*, 15–19.

12. Mayhew, *Malthus*, 64–65. See also Walter M. Stern, "The Bread Crisis in Britain, 1795–96." *Economica* 31, no. 122 (1964): 168–187. https://doi.org/10.2307/2551353; and John Stevenson, "Food Riots in England, 1792–1818," in *Popular Protest and Public Order: Six Studies in British History, 1790–1920*, ed. John Stevenson and Roland Quinault (London: George Allen and Unwin, 1974), 33–74.

13. Stevenson, "Food Riots in England," 51.

14. Mayhew, *Malthus*, 64–65.

15. Thomas R. Malthus, Preface, in *An Essay on the Principle of Population*, 1st ed., ed. Joyce E. Chaplin (New York: W. W. Norton, 2018), 33; Thomas R. Malthus, *An Essay on the Principle of Population*, variorum ed. based on 1803 text, ed. Donald Winch (Cambridge: Cambridge University Press, 1992), I. I, 13.

16. Malthus, *Essay*, 1st ed., I, 38–39; Malthus, *Essay*, var. ed., I. I, 15, et passim.

17. Bashford and Chaplin, *The New Worlds*, 66. Indeed, as sociologist and demographer William Petersen argues, "by our measure of politics, Godwin was more often the reactionary, Malthus more often the progressive." William Petersen, "The Malthus-Godwin Debate, Then and Now." *Demography* 8, no. 1 (1971): 13–26, 13.

18. Malthus, *Essay*, var. ed., I. I, 14.

19. Maureen N. McLane, "Malthus Our Contemporary?: Toward a Political Economy of Sex." *Studies in Romanticism* 52, no. 3 (2013): 337–362, https://doi.org/10.1353/srm.2013.0026.

20. Malthus, *Essay*, var. ed., III. I, 53.

21. Malthus, *Essay*, var. ed., IV. III, 229.

22. Malthus, *Essay*, 1st ed., XIX, 153.

23. Malthus, *Essay*, var. ed., IV. XIV, 328, and Appendix, 1806, 362–364. Petersen, "The Malthus-Godwin Debate," 20.

24. Giorgos Kallis offers a sophisticated but idiosyncratic and fundamentally one-sided interpretation of Malthus in *Limits: Why Malthus Was Wrong and Why Environmentalists Should Care* (Stanford: Stanford University Press, 2019). According to Kallis, Malthus has been misread: "Contrary to his iconic status as a prophet of limits," he argues, "Malthus was in fact a prophet of growth" (15). Kallis's version of Malthus—at once puritanical killjoy and proto-capitalist free-market ideologue—relies heavily on a single, out-of-print sociological interpretation that Kallis admits not having read in full, that does not support his interpretation, and that explicitly ignores every edition of the *Essay* but the first, which was the most religious and least empirical, thus conveniently eliding the ways Malthus revised and developed his argument in response to critics.

25. Michael Shermer, "Why Malthus Is Still Wrong," *Scientific American*, May 1, 2016, doi:10.1038/scientificamerican0516-72.

26. Malthus, Preface, *Essay*, 1st ed., 34.

27. The word *ecopessimism* appears no earlier than the 1990s, as far as I can tell.

28. Desrochers and Szurmak, *Population Bombed!: Exploding the Link Between Overpopulation and Climate Change* (London: GWPF Books, 2018), 14–15.

29. Desrochers and Szurmak, *Population Bombed!*, xvi.

30. Desrochers and Szurmak, *Population Bombed!*, 14–15.

31. Pierre Desrochers and Joanna Szurmak, "Eco-Pessimism versus Techno-Optimism," *Aero*, June 8, 2019, https://areomagazine.com/2019/08/06/eco-pessimism-versus-techno-optimism/. See also Pierre Desrochers and Joanna Szurmak, "The Long History of Eco-Pessimism," *Spiked.com*, October 25, 2019, https://www.spiked-online.com/2019/10/25/the-long-history-of-eco-pessimism/.

32. Desrochers and Szurmak, *Population Bombed!*, 175.

33. Pierre Desrochers and Joanna Szurmak, "Seven Billion Solutions Strong: Why Markets, Growth, and Innovation Are the Antidote to Eco-Pessimism," The Breakthrough Institute, February 7, 2020, https://thebreakthrough.org/journal/no-12-winter-2020/seven-billion-solutions.

34. Desrochers and Szurmak, *Population Bombed!*, 175.

35. Desrochers and Szurmak, *Population Bombed!*, 145–169 et passim.

36. Jared Sexton, "Affirmation in the Dark: Racial Slavery and Philosophical Pessimism," *The Comparatist* 43 (2019): 90–111, doi: 10.1353/com.2019.0005; W. G. Lambert, *Babylonian Wisdom Literature* (Clarendon Press, 1960), 139–149; E. A. Speiser, "The Case of the Obliging Servant," *Journal of Cuneiform Studies* 8, no. 3 (1954): 98–105, https://doi.org/10.2307/1359094.

37. Sophocles, *The Oedipus at Colonus of Sophocles*, ed. Richard Jebb (Cambridge: Cambridge University Press, 1889), line 1225. J. C. Opstelten also is insightful on the complex relationship between pessimism and tragedy. Johannes Cornelis Opstelten, *Sophocles and Greek Pessimism* (Amsterdam: North-Holland Publishing, 1952), 24–41.

38. Ecclesiastes 4:2–3 (New International Version).

39. Sexton, "Affirmation in the Dark."

40. Techno-optimism, for instance, ably defined if not persuasively defended by John Danaher, is the view holding "that technology plays a key role in insuring that the good prevails over the bad." John Danaher, "Techno-Optimism: An Analysis, an Evaluation and a Modest Defence," *Philosophy & Technology* 35, no. 2 (2022): 54, https://doi.org/10.1007/s13347-022-00550-2. For a nuanced account of the complex relationship between optimism and progress in the eighteenth century, see Haydn Mason, "Optimism, Progress, and Philosophical History," in *The Cambridge History of Eighteenth-Century Political Thought* (Cambridge: Cambridge University Press, 2006), 195–217.

41. Although Mark Migotti argues that "philosophical pessimism . . . is an inherently reactive phenomenon; it arose in opposition to an established tradition of philosophical optimism," I think the case van der Lugt makes for the roughly simultaneous emergence of the two perspectives in response to the challenges posed to a theistic universe by Baconian-Cartesian rationality is more persuasive. Mark Migotti, "Schopenhauer's Pessimism in Context," in *The Oxford Handbook of Schopenhauer* (Oxford: Oxford University Press, 2020).

42. As Frederick Beiser puts it, "pessimism was the rediscovery of the problem

of evil after the collapse of theism." Frederick C. Beiser, *Weltschmerz: Pessimism in German Philosophy, 1860–1900* (Oxford: Oxford University Press, 2016), 7.

43. As does George W. Harris: "A secular version of the problem of evil is the problem of pessimism, which asserts that, given the fact of human suffering, it would have been better that human life as a whole never evolved." George W. Harris, "Pessimism," *Ethical Theory and Moral Practice* 5, no. 3 (2002): 271–286, https://doi.org/10.1023/A:1019621509461.

44. Joshua Foa Dienstag, *Pessimism: Philosophy, Ethic, Spirit* (Princeton: Princeton University Press, 2006), 18.

45. Mara van der Lugt, *Dark Matters: Pessimism and the Problem of Suffering* (Princeton: Princeton University Press, 2021), 69.

46. Pierre Bayle, "Acosta, Uriel," *Dictionary*, quoted in Richard Popkin, *The History of Skepticism* (Oxford: Oxford University Press, 2003), 288.

47. Karl Löwith, *Meaning in History: Theological Implications of the Philosophy of History* (Chicago: University of Chicago Press, 1949), 191.

48. Eugene Thacker, *Cosmic Pessimism* (Minneapolis: Univocal Publishing, 2015), 3.

49. Dienstag, *Pessimism*, 19.

50. See Beiser, *Weltschmerz*. Wittgenstein's pessimism and its connections to Schopenhauer have been noted but not fully integrated into the narrative of philosophical pessimism. Dale Jacquette, "Wittgenstein and Schopenhauer," in *A Companion to Wittgenstein*, ed. Hans-Johan Glock (Newark: John Wiley & Sons, 2016), 57–73; William J. DeAngelis, *Ludwig Wittgenstein—A Cultural Point of View: Philosophy in the Darkness of This Time* (London: Routledge, 2016); Rupert Read, "Wittgenstein and the Illusion of 'Progress': On Real Politics and Real Philosophy in a World of Technocracy," *Royal Institute of Philosophy Supplement* 78 (2016): 265–284, https://doi.org/10.1017/S1358246116000321; Constantine Sandis, "'If Some People Looked Like Elephants and Others Like Cats': Wittgenstein on Understanding Others and Forms of Life," *Nordic Wittgenstein Review* (2015): 131–153, https://doi.org/10.15845/nwr.v4i0.337; Cengiz Cakmak, "Schopenhauer & Wittgenstein," *Philosophical Inquiry* 25, no. 1 (2003): 115–124, https://doi.org/10.5840/philinquiry2003251/29; Michael Hymers, "Wittgenstein, Pessimism and Politics," *The Dalhousie Review* 80, no. 2 (2000): 187–216; David Avraham Weiner, *Genius and Talent: Schopenhauer's Influence on Wittgenstein's Early Philosophy* (Rutherford: Fairleigh Dickinson University Press, 1992); Russell B. Goodman, "Schopenhauer and Wittgenstein on Ethics," *Journal of the History of Philosophy* 17, no. 4 (1979): 437–447; A. Phillips Griffiths, "Wittgenstein, Schopenhauer, and Ethics," *Royal Institute of Philosophy Supplement* 7 (1973): 96–116, https://doi.org/10.1017/S0080443600000297.

51. Dienstag, *Pessimism*, 6, 45. Susan Neiman, *Evil in Modern Thought: An Alternative History of Philosophy*, First Princeton Classic Edition, with a new afterword by the author (Princeton: Princeton University Press, 2015), 8.

52. Richard Rorty, *Philosophy and the Mirror of Nature* (Princeton: Princeton University Press, 1979), 367.

53. Rorty, *Philosophy and the Mirror*, 369.

54. Rorty, *Philosophy and the Mirror*, 371.

55. Dienstag, *Pessimism*, 6.

56. Raihan Kadri, *Reimagining Life: Philosophical Pessimism and the Revolution of Surrealism* (Rutherford: Farleigh Dickinson University Press, 2011), 20. See also Walter Benjamin, "Surrealism: The Last Snapshot of the European Intelligentsia," *New Left Review*, no. 108 (1978): 47–56.

57. Mark Schmitt's insightful *Spectres of Pessimism: A Cultural Logic of the Worst* (Cham: Palgrave Macmillan, 2023) also identifies a strain of queer pessimism in the work of Sara Ahmed, Jack Halberstam, and Lee Edelman, as well as a Marxian pessimism in the work of Slavoj Žižek, Jacques Derrida, Mark Fisher, and Stuart Hall.

58. John F. Kennedy, Address upon Accepting the Liberal Party Nomination for President, New York, New York, September 14, 1960, https://www.jfklibrary.org/archives/other-resources/john-f-kennedy-speeches/liberal-party-nomination-nyc-19600914.

59. Amanda Anderson, *Bleak Liberalism* (Chicago: University of Chicago Press, 2016), 22.

60. Anderson, *Bleak Liberalism*, 23.

61. See my own "How John Hersey Bore Witness," *The New Republic*, July 27, 2019; Jeremy Treglown, *Mr. Straight Arrow: The Career of John Hersey, Author of Hiroshima* (New York: Farrar, Straus and Giroux, 2019).

62. Dillon S. Tatum, "A Pessimistic Liberalism: Jacob Talmon's Suspicion and the Birth of Contemporary Political Thought," *The British Journal of Politics and International Relations* 21, no. 4 (2019): 650–666, https://doi.org/10.1177/1369148119866086.

63. Sara Marcus, *Political Disappointment: A Cultural History from Reconstruction to the AIDS Crisis* (Cambridge: Harvard University Press, 2023).

64. Arthur Schlesinger, "Forgetting Reinhold Niebuhr," *New York Times*, September 18, 2005, https://www.nytimes.com/2005/09/18/books/review/forgetting-reinhold-niebuhr.html.

65. Reinhold Niebuhr, *Beyond Tragedy* (New York: Charles Scribner's Sons, 1937), 18. See also, Reinhold Niebuhr, "Augustine's Political Realism," in *Christian Realism and Political Problems* (New York: Faber and Faber, 1954).

66. Andrew J. Bacevich, Introduction, in *American Conservatism: Reclaiming an Intellectual Tradition* (New York: Library of America, 2020), xi.

67. William F. Buckley Jr., "Our Mission Statement," *National Review* (November 19, 1955), https://www.nationalreview.com/1955/11/our-mission-statement-william-f-buckley-jr/.

68. Lionel Trilling, *The Liberal Imagination* (New York: Viking, 1950), xv; Bertrand Russell, *The History of Western Philosophy* (New York: Simon & Schuster, 1945), 759.

69. See, for instance, the example of Whittaker Chambers, as dramatized in Lionel Trilling's novel *The Middle of the Journey*. I no more have space here to explore

the complex history of conservatism than I do to explore the history of liberalism, but the following may offer a start: Bacevich, *American Conservatism*; Richard Bourke, "What Is Conservatism? History, Ideology and Party," *European Journal of Political Theory* 17, no. 4 (2018): 449–475, https://doi.org/10.1177/1474885118782384; Samuel P. Huntington, "Conservatism as an Ideology," *The American Political Science Review* 51, no. 2 (1957): 454–473, https://doi.org/10.2307/1952202.

70. See, for instance, Roger Scruton, *The Uses of Pessimism and the Danger of False Hope* (Oxford: Oxford University Press, 2010); and John Gray, *Straw Dogs: Thoughts on Humans and Other Animals* (London: Granta, 2002).

71. Oliver Bennett, *Cultural Pessimism: Narratives of Decline in the Postmodern World* (Edinburgh: Edinburgh University Press, 2001), 1. See also John Derbyshire, *We Are Doomed: Reclaiming Conservative Pessimism* (New York: Crown, 2009).

72. Paul Prescott, "What Pessimism Is," *Journal of Philosophical Research* 37 (2012): 337–356, https://doi.org/10.5840/jpr20123716.

73. See, for example, Mads Rosendahl Thomsen and Jacob Wamberg, eds., *The Bloomsbury Handbook of Posthumanism* (New York: Bloomsbury Publishing, 2020); Alastair Hunt and Stephanie Youngblood, eds., *Against Life* (Evanston: Northwestern University Press, 2016); Cary Wolfe, *What Is Posthumanism?* (Minneapolis: University of Minnesota Press, 2010); Eugene Thacker, *After Life* (Chicago: University of Chicago, 2010); Lee Edelman, *No Future: Queer Theory and the Death Drive* (Durham: Duke University Press, 2004); N. Katherine Hayles, *How We Became Posthuman: Virtual Bodies in Cybernetics, Literature, and Informatics* (Chicago: University of Chicago Press, 1999). See also Richard Grusin, ed. *The Nonhuman Turn* (Minneapolis: University of Minnesota Press, 2015); and Dana Luciano and Mel Y. Chen, eds., "Dossier: Theorizing Queer Inhumanisms," *GLQ: A Journal of Lesbian and Gay Studies* 21, no. 2–3 (June 2015): 209–248.

74. Francesca Ferrando, "Posthumanism, Transhumanism, Antihumanism, Metahumanism, and New Materialisms: Differences and Relations," *Existenz* 8, no. 2 (2013): 26–32.

75. Wolfe, *What Is Posthumanism?*, xvi.

76. Frank Wilderson III, *Afropessimism* (New York: Liveright, 2020), 174. A note on capitalization and typography: I capitalize *Afropessimism* following Wilderson, although I don't capitalize other forms of pessimism, mainly because Afropessimism is a distinct movement, like Surrealism, rather than a general category. There is no standard as to whether *Afropessimism* is hyphenated; again, I follow Wilderson in not hyphenating. On the capitalization of *black*, *white*, *human*, and *being* in this chapter and elsewhere, I capitalize these terms only when directly or indirectly quoting someone who does so or when referring to North American Black or African-American cultural identity.

77. Patrice Douglass, Selamawit D. Terrefe, and Frank B. Wilderson III, "Afro-Pessimism," *Oxford Bibliographies Online*, August 28, 2018, doi: 10.1093/OBO/9780190280024-0056.

78. For a suggestive analysis of the geopolitical aspects of this context, see Law-

rence Jackson, *Hold It Real Still: Clint Eastwood, Race, and the Cinema of the American West* (Baltimore: Johns Hopkins, 2022); and Nikhil Singh, "Racial Formation in an Age of Permanent War," in *Racial Formation in the Twenty-First Century* (Berkeley: University of California Press, 2012).

79. Wilderson, *Afropessimism*, 41–42.

80. *Qui Parle* 13, no. 2 (2003), http://www.jstor.org/stable/20686150.

81. Jared Sexton and Huey Copeland, "Raw Life: An Introduction," *Qui Parle* 13, no. 2 (2003): 53–62, http://www.jstor.org/stable/20686150.

82. Gayatri Chakravorty Spivak, "Can the Subaltern Speak?," in *Marxism and the Interpretation of Culture*, ed. Cary Nelson and Lawrence Grossberg (Urbana: University of Illinois Press, 1987), 271–313.

83. Orlando Patterson, *Slavery and Social Death: A Comparative Study* (Cambridge: Harvard University Press, 1982), 46.

84. Calvin Warren, "Black Nihilism and the Politics of Hope," *CR: The New Centennial Review* 15, no. 1 (Spring 2015): 215–248, 217–218. My emphasis.

85. Jared Sexton, "Afro-Pessimism: The Unclear Word," *Rhizomes* 29 (2016).

86. Annie Olaloku-Teriba, "Afro-Pessimism and the (Un)Logic of Anti-Blackness," *Historical Materialism* 26, no. 2 (2018): 96–122, 118–119.

87. Kevin Ochieng Okoth, "The Flatness of Blackness: Afro-Pessimism and the Erasure of Anti-Colonial Thought," *Salvage*, January 16, 2020.

88. Greg Thomas, "Afro-Blue Notes: The Death of Afro-Pessimism (2.0)?" *Theory & Event* 21, no. 1 (January 2018): 282–317, 284.

89. Jesse McCarthy, "On Afropessimism," in *Who Will Pay Reparations on My Soul?* (New York: Liveright, 2021), 201–202.

90. McCarthy, "On Afropessimism," 204.

91. McCarthy, "On Afropessimism," 216.

92. McCarthy, "On Afropessimism," 209.

93. I owe this insight to Lawrence Jackson.

94. See Danielle Fuentes Morgan, *Laughing to Keep from Dying* (Urbana: University of Illinois Press, 2020); Terrence T. Tucker, *Furiously Funny: Comic Rage in Late 20th Century African-American Literature* (Gainesville: University Press of Florida, 2017); Glenda Carpio, *Laughing Fit to Kill: Black Humor in the Fictions of Slavery* (Oxford: Oxford University Press, 2008); and Darryl Dickson-Carr, *African American Satire: The Sacredly Profane Novel* (Columbia: University of Missouri Press, 2001).

95. Jonathan Daniel Greenberg, *The Cambridge Introduction to Satire* (Cambridge: Cambridge University Press, 2018), 10.

96. Dickson-Carr, *African American Satire*, 32.

97. Northrop Frye, *Anatomy of Criticism: Four Essays* (Princeton: Princeton University Press, 2020), 222–223. On the difficulty of defining satire, see Greenberg, *The Cambridge Introduction to Satire*, ch. 1; and John T. Gilmore, *Satire* (New York: Routledge, 2018), ch. 1. On comic theory more generally, see Peter C. Kunze and Jared N.

Champion, *Taking a Stand: Contemporary US Stand-Up Comedians as Public Intellectuals* (Jackson: University Press of Mississippi, 2021); Sianne Ngai, *Our Aesthetic Categories: Zany, Cute, Interesting* (Cambridge: Harvard University Press, 2012); Alenka Zupančič, *The Odd One In: On Comedy* (Cambridge: MIT Press, 2008); Sigmund Freud, *The Joke and Its Relation to the Unconscious* (New York: Penguin Books, 2003); John Limon, *Stand-up Comedy in Theory, or, Abjection in America* (Durham: Duke University Press, 2000). See also my own *Total Mobilization: World War II and American Literature* (Chicago: University of Chicago Press, 2019), ch. 3.

98. Wilderson, *Afropessimism*, 3.
99. Shakespeare, *Hamlet*.
100. There are other resonances between *Afropessimism* and *Hamlet* that might be productively explored.
101. Wilderson, *Afropessimism*, 3.
102. Wilderson, *Afropessimism*, 5.
103. Wilderson, *Afropessimism*, 6.
104. Wilderson, *Afropessimism*, 17.
105. Frye, *Anatomy of Criticism*, 235.
106. Wilderson, *Afropessimism*, 31.
107. McCarthy, "On Afropessimism," 221. Italics in original.
108. McCarthy, "On Afropessimism," 222.
109. Patrick Farnsworth, "#286 | Afropessimism: Blackness, at the End of this World w/ Frank B. Wilderson III," *Last Born in the Wilderness*, January 28, 2021, https://www.lastborninthewilderness.com/episodes/frank-wilderson.

Chapter 7

1. Ursula K. Le Guin, "The Ones Who Walk Away from Omelas," *The Wind's Twelve Quarters: Short Stories* (New York: Harper & Row, 1975), 278.
2. Le Guin, "Omelas," 281.
3. Le Guin, "Omelas," 282.
4. See Robert Silverberg, ed., *New Dimensions 3* (New York: Nelson Doubleday/SFBC, 1973).
5. World Bank, *Poverty and Shared Prosperity 2022: Correcting Course* (Washington, DC: World Bank., 2022), doi:10.1596/978-1-4648-1893-6. 3.
6. Emily A. Shrider and John Creamer, *Poverty in the United States: 2022* (Report P60–280), US Census Bureau, Current Population Reports (Washington, D.C.: US Government Publishing Office, 2023), https://www.census.gov/library/publications/2023/demo/p60-280.html.
7. Le Guin, "Omelas," 283.
8. Le Guin, "Omelas," 284.
9. Paul Firenze, "'[T]hey, Like the Child, Are Not Free': An Ethical Defense of the Ones Who Remain in Omelas," *Response: The Journal of American and Popular Culture* 2, no. 1 (November 2017), https://responsejournal.net/issue/2017-11/article/%E2

%80%98they-child-are-not-free%E2%80%99-ethical-defense-ones-who-remain-omelas.

10. Firenze, "'[T]hey, Like the Child, Are Not Free.'"

11. Frederick C. Beiser, *Weltschmerz: Pessimism in German Philosophy, 1860–1900* (Oxford: Oxford University Press, 2016), 163.

12. Paul Prescott, "What Pessimism Is," *Journal of Philosophical Research* 37 (2012): 337–356, https://doi.org/10.5840/jpr20123716.

13. Joshua Foa Dienstag, *Pessimism: Philosophy, Ethic, Spirit* (Princeton: Princeton University Press, 2006), 22.

14. Beiser, *Weltschmerz*, 44.

15. Beiser, *Weltschmerz*, 163.

16. Prescott, "What Pessimism Is," 340.

17. Julie K. Norem, *The Positive Power of Negative Thinking: Using Defensive Pessimism to Harness Anxiety and Perform at Your Peak* (New York: Basic Books, 2001), et passim.

18. Julie K. Norem et al., "Defensive Pessimism and Precautionary Action During the COVID Pandemic," *Social and Personality Psychology Compass* 17, no. 11 (2023), https://doi.org/10.1111/spc3.12853.

19. Prescott, "What Pessimism Is," 341.

20. Prescott, "What Pessimism Is," 347–348.

21. *Oxford English Dictionary Online*, s.v. "despair, n.," June 2024, www.oed.com/view/Entry/50935.

22. See William James, "Is Life Worth Living?" *The International Journal of Ethics* 6, no. 1 (1895): 1–24; and Céline Leboeuf, "Fearing the Future: Is Life Worth Living in the Anthropocene?" *The Journal of Speculative Philosophy* 35, no. 3 (2021): 273–288.

23. Albert Camus, "The Myth of Sisyphus," *The Myth of Sisyphus and Other Essays*, trans. Justin O'Brien (New York: Vintage, 1955), 3.

24. This point is made by Schopenhauer, as discussed by Mark Migotti, "Schopenhauer's Pessimism in Context," in *The Oxford Handbook of Schopenhauer* (Oxford: Oxford University Press, 2020).

25. See Henri Hubert and Marcel Mauss, *Sacrifice: Its Nature and Function* (Chicago: University of Chicago Press, 1964); Claude Lévi-Strauss, *Wild Thought: A New Translation of "La Pensée Sauvage"* (Chicago: University of Chicago Press, 2021); René Girard, *The Scapegoat* (Baltimore: Johns Hopkins University Press, 1986); Roy Scranton, *Total Mobilization* (Chicago: University of Chicago Press, 2019); et passim.

26. Samuel Scheffler, *Death and the Afterlife* (Oxford: Oxford University Press, 2014).

27. Arthur Schopenhauer, *The World as Will and Representation*, vol. 1 (Mineola, NY: Dover, 1969), 293, §55.

28. Aldo Leopold, "The Land Ethic," in *A Sand County Almanac* (New York: Ballantine, 1966), 238.

29. William James, *The Varieties of Religious Experience* (New York: Penguin, 1982), 163

30. Frank Kermode, *The Sense of an Ending: Studies in the Theory of Fiction* (Oxford: Oxford University Press, 2000), 164.

31. Schopenhauer, *The World as Will and Representation*, 196, §38. See also 311–326, §57–59.

32. Pierre Bayle, "Xenophanes," *An Historical and Critical Dictionary By Monsieur Bayle*, vol. 4 (London: C. Harper et al., 1710), 3047–3059, Gale ECCO, Gale document no. GALE|CW0114129616; Mara van der Lugt, *Dark Matters: Pessimism and the Problem of Suffering* (Princeton: Princeton University Press, 2021), 38–39, 46–66.

33. Virginia Woolf, *On Being Ill* (Ashfield, MA: Paris Press, 2002).

34. Elizabeth Stone, quoted in Ellen Cantorow, "No Kids." *The Village Voice*, January 15, 1985. Verified by personal correspondence with Stone.

35. Ernst Jünger, *On Pain* (Candor, NY: Telos Press Publishing, 2008), 1.

36. David Benatar, *Better Never to Have Been: The Harm of Coming into Existence* (Oxford: Oxford University Press, 2006), 28.

37. Benatar, *Better Never to Have Been*, 28–59.

38. See Roy Baumeister et al., "Bad Is Stronger Than Good," *Review of General Psychology* 5, no. 4 (2001): 323–370, https://doi.org/10.1037//1089-2680.5.4.323"; and Paul Rozin and Edward B. Royzman, "Negativity Bias, Negativity Dominance, and Contagion," *Personality and Social Psychology Review* 5, no. 4 (2001): 296–320, as discussed in ch. 2; as well as Daniel Kahneman and Amos Tversky, "Prospect Theory: An Analysis of Decision Under Risk," *Econometrica* 47 (1979): 263–291.

39. Rozin and Royzman, "Negativity Bias, Negativity Dominance, and Contagion," 297.

40. Kahneman and Tversky, "Prospect Theory"; David Gal et al., "The Loss of Loss Aversion: Will It Loom Larger Than Its Gain?" *Journal of Consumer Psychology* 28, no. 3 (2018): 497–516, https://doi.org/10.1002/jcpy.104; Itamar Simonson et al., "Bringing (Contingent) Loss Aversion Down to Earth—A Comment on Gal and Rucker's Rejection of 'Losses Loom Larger Than Gains,'" *Journal of Consumer Psychology* 28, no. 3 (2018): 517–522, https://doi.org/10.1002/jcpy.1046; Richard H. Thaler, "From Homo Economicus to Homo Sapiens," *The Journal of Economic Perspectives* 14, no. 1 (2000): 133–141, https://doi.org/10.1257/jep.14.1.133.

41. Daniel Kahneman et al., "Anomalies: The Endowment Effect, Loss Aversion, and Status Quo Bias," *The Journal of Economic Perspectives* 5, no. 1 (1991): 193–206, https://doi.org/10.1257/jep.5.1.193.

42. Benatar, *Better Never to Have Been*, 58.

43. Benatar, *Better Never to Have Been*, 60–92.

44. The discussion of consciousness is so vast, dubious, and tendentious I hesitate even to suggest a place to begin, for fear of signaling partisanship on an issue about which I remain skeptical. I merely suggest a few works readers might find useful: Anil K. Seth and Tim Bayne, "Theories of Consciousness," *Nature Reviews Neuroscience* 23, no. 7 (2022): 439–452; Galen Strawson, *Selves: An Essay in Revisionary Metaphysics* (Oxford: Oxford University Press, 2009); Andy Clark and David Chalmers, "The Extended Mind," *Analysis* 58, no. 1 (1998): 7–19; William James,

"Does 'Consciousness' Exist?," *The Journal of Philosophy, Psychology and Scientific Methods* 1, no. 18 (1904): 477–491.

45. George Eliot, *Middlemarch* (Oxford: Oxford University Press, 2019), 182.

46. Timothy D. Wilson et al., "Just Think: The Challenges of The Disengaged Mind," *Science* 345, no. 6192 (2014): 75–77.

47. Kōun Yamada, *The Gateless Gate: The Classic Book of Zen Koans*, case 41 (Boston: Wisdom Publications, 2005).

48. Peter Wessel Zapffe, "The Last Messiah," in *Wisdom in the Open Air: The Norwegian Roots of Deep Ecology*, ed. Peter Reed and David Rothenberg (Minneapolis: University of Minnesota Press, 1993), 40–52.

49. Thomas Ligotti, *The Conspiracy Against the Human Race: A Contrivance of Horror* (New York: Penguin, 2018).

50. Zapffe, "The Last Messiah," 41.

51. Zapffe, "The Last Messiah." 42

52. Zapffe, "The Last Messiah." 43.

53. Zapffe, "The Last Messiah," 44.

54. Hans Jonas, *The Imperative of Responsibility* (Chicago: University of Chicago Press, 1985), 29.

55. Theodor Adorno, *Negative Dialectics* (New York: Continuum, 1973), 320.

56. Karl Löwith, *Meaning in History: Theological Implications of the Philosophy of History* (Chicago: University of Chicago Press, 1949), 4.

57. I'm sympathetic to environmental virtue ethics, but given the profoundly technocratic and anti-natural material affordances of the modern human lifeworld, I don't see how environmental ethics add up to anything more than an aesthetics of consumption signaling belonging to a loosely identified sociopolitical moiety (i.e., "environmentalists"). Any truly environmental ethics would be indistinguishable from ethics as such, since the necessary precondition for an environmental ethics would need to be the genuine integration of the human lifeworld with the nonhuman "environment," as one sees in some accounts of indigenous cultures—for instance, Keith Basso, *Wisdom Sits in Places: Landscape and Language Among the Western Apache* (Albuquerque: University of New Mexico Press, 1996). Such a transformation seems impossible to imagine sitting at my desk typing on this keyboard, looking out across the roofs of my neighborhood, listening to the mournful hoots of a distant train. One cannot, at present, escape the human landscape.

58. Clarence Darrow, *Facing Life Fearlessly: The Pessimistic Versus the Optimistic View of Life* (Girard, KS: Haldeman-Julius Publications, 1929), 30.

59. Kathryn J. Norlock, "Perpetual Struggle," *Hypatia* 34, no. 1 (2019): 6–19, https://doi.org/10.1111/hypa.12452, 7.

60. Ecclesiastes 11:10 (Revised Standard Version).

61. The Serenity Prayer was probably composed by Reinhold Niebuhr in the early 1930s. Fred Shapiro, "How I Discovered I Was Wrong about the Origin of the Serenity Prayer," US Catholic, May 15, 2014, https://uscatholic.org/news_item/commentary-how-i-discovered-i-was-wrong-about-the-origin-of-the-serenity-prayer/.

62. Anne Roe, *The Making of a Scientist* (New York: Dodd, Mead, 1953), 46–47.

63. Nassim Nicholas Taleb et al., "The Precautionary Principle (with Application to the Genetic Modification of Organisms)," Extreme Risk Initiative—NYU School of Engineering Working Paper Series, arXiv.org, 2014, https://arxiv.org/abs/1410.5787.

64. Jonas, *The Imperative of Responsibility*, 26–27.

65. Arthur Schopenhauer, *Studies in Pessimism* (New York: MacMillan, 1890), 29.

66. van der Lugt, *Dark Matters*, 27.

67. For instance, Kaida and Kaida's constructive pessimism hypothesis argues that pessimism facilitates pro-environmental behavior. See Naoko Kaida and Kosuke Kaida, "Facilitating Pro-Environmental Behavior: The Role of Pessimism and Anthropocentric Environmental Values," *Social Indicators Research* 126, no. 3 (2016): 1243–1260, https://doi.org/10.1007/s11205-015-0943-4; and Naoko Kaida and Kosuke Kaida, "Pro-Environmental Behavior Correlates with Present and Future Subjective Well-Being," *Environment, Development and Sustainability* 18 (2016): 111–127, https://doi.org/10.1007/s10668-015-9629-y.

68. Rosie Warren, "Some Final Thoughts on Pessimism," *Salvage,* January 4, 2016, https://salvage.zone/some-last-words-on-pessimism/.

69. Dienstag, *Pessimism*, 192.

70. Georg Henrik von Wright, "The Myth of Progress," in *The Tree of Knowledge and Other Essays* (New York: E. J. Brill, 1993), 202–228, 227.

Afterword

1. *First Reformed*, dir. Paul Schrader (New York: A24, 2018).

2. He is named for the Jewish-German expressionist playwright, poet, and anarchist, imprisoned for his part in the revolutionary 1919 Bavarian Soviet Republic, an insight I owe to Kristen Sieranski.

3. Commonwealth Club, "About," https://www.commonwealthclub.org/about.

4. See Roy Scranton, "Learning How to Die in the Anthropocene," *New York Times,* November 10, 2013, https://archive.nytimes.com/opinionator.blogs.nytimes.com/2013/11/10/learning-how-to-die-in-the-anthropocene/.

5. James Blue, "Excerpts from an Interview with Robert Bresson," June 1965, Los Angeles, p. 2, box 6, folder 41, James Blue Papers, Special Collections and University Archives, University of Oregon Libraries, quoted in Paul Schrader, *Transcendental Style in Film: Ozu, Bresson, Dreyer* (Berkeley: University of California Press, 2018), 87.

6. Schrader, *Transcendental Style in Film*, 88.

7. Schrader, *Transcendental Style, in Film* 40.

8. Schrader, *Transcendental Style in Film*, 42.

9. Alfred North Whitehead, *Process and Reality*, corr. ed. (New York: Free Press, 1978), 337–338.

10. John Keats, "Letter to George and Tom Keats, 21, ?27 December 1817," *Selected Letters of John Keats*, ed. Grant F. Scott (Cambridge: Harvard University Press, 2002).

11. T. S. Eliot, "East Coker," in *Four Quartets* (New York: Harcourt Brace Jovanovich, 1971).

12. Though I developed this idea independently of China Miéville's "apophatic Marxism," I was pleased to see the resonance between the two, as was pointed out to me in conversation by Meehan Crist. China Miéville, "Silence in Debris: Towards an Apophatic Marxism," *Salvage*, April 2, 2019.

13. Achille Mbembe, *Necropolitics* (Durham: Duke University Press, 2019), 28.

SELECTED REFERENCES

I have included a limited selection of works cited, organized into broad categories that loosely correspond to the book's chapters, to be of use to those who might like to follow up with their own research. I have tried to limit my focus to key references, especially sources that were important for my thinking, but I have also tried to include notable sources on controversial topics. For full references, please see the endnotes.

Societal Collapse and Chaco Canyon

Brozović, Danilo. "Societal Collapse: A Literature Review." *Futures: The Journal of Policy, Planning and Futures Studies* 145 (2023). https://doi.org/10.1016/j.futures.2022.103075.

Centeno, Miguel A., Peter W. Callahan, Paul Larcey, and Thayer S. Patterson, eds. *How Worlds Collapse: What History, Systems, and Complexity Can Teach Us About Our Modern World and Fragile Future.* New York: Routledge Press, 2023.

Costanza, Robert, Lisa J. Graumlich, and Will Steffen, eds. *Sustainability or Collapse?: An Integrated History and Future of People on Earth.* Cambridge: MIT Press, 2007.

Faulseit, Ronald K., ed. *Beyond Collapse: Archaeological Perspectives on Resilience, Revitalization, and Transformation in Complex Societies.* Carbondale: Southern Illinois University Press, 2016.

Fischer-Kowalski, Marina, and Helmut Haberl. "Sustainable Development: Socio-Economic Metabolism and Colonization of Nature." *International Social Science Journal*, English ed., 50, no. 158 (1998): 573–587. https://doi.org/10.1111/1468-2451.00169.

Frazier, Kendrick. *People of Chaco: A Canyon and Its Culture.* New York: W. W. Norton, 2005.

Glowacki, Donna. *Living and Leaving: A Social History of Regional Depopulation in Thirteenth-Century Mesa Verde.* Tucson: University of Arizona Press, 2015.

Haberl, Helmut. "The Global Socioeconomic Energetic Metabolism as A Sustainability Problem." *Energy* 31, no. 1 (2006): 87–99.

Haberl, Helmut, Dominik Wiedenhofer, Doris Virág, Gerald Kalt, Barbara Plank, Paul Brockway, Tomer Fishman, et al. "A Systematic Review of the Evidence on Decoupling of GDP, Resource Use and GHG Emissions, Part II: Synthesizing the Insights." *Environmental Research Letters* 15, no. 6 (2020): 65003. https://doi.org/10.1088/1748-9326/ab842a.

Homer-Dixon, Thomas, Brian Walker, Reinette Biggs, Anne-Sophie Crépin, Carl Folke, Eric F. Lambin, Garry D. Peterson, et al. "Synchronous Failure." *Ecology and Society* 20, no. 3 (2015): 6. https://doi.org/10.5751/ES-07681-200306.

Lawrence, Michael, Thomas Homer-Dixon, Scott Janzwood, Johan Rockström, Ortwin Renn, and Jonathan F. Donges. "Global Polycrisis: The Causal Mechanisms of Crisis Entanglement." *Global Sustainability* 7 (2024): e6. https://doi.org/10.1017/sus.2024.1.

Lekson, Steven. *The Chaco Meridian: One Thousand Years of Religious Power in the Ancient Southwest.* 2nd ed. Lanham, MD: Rowman and Littlefield, 2015.

McAnany, Patricia Ann, and Norman Yoffee, eds. *Questioning Collapse: Human Resilience, Ecological Vulnerability, and the Aftermath of Empire.* Cambridge: Cambridge University Press, 2010.

Michaux, Simon P. *The Mining of Minerals and the Limits to Growth.* Espoo: Geological Survey of Finland, 2021.

Smil, Vaclav. *Energy and Civilization: A History.* Cambridge: MIT Press, 2017.

———. *Energy in Nature and Society: General Energetics of Complex Systems.* Cambridge: MIT Press, 2008.

———. "What We Need to Know About the Pace of Decarbonization." *Substantia*, no. 2 (2019): 13–28. https://doi.org/10.13128/Substantia-XXX.

Steel, Daniel, C. Tyler Des Roches, and Kian Mintz-Woo. "Climate Change and the Threat to Civilization." *Proceedings of the National Academy of Sciences* 119, no. 42 (2022). https://doi.org/10.1073/pnas.2210525119.

Tainter, Joseph A. "Archaeology of Overshoot and Collapse." *Annual Review of Anthropology* 35, no. 1 (2006): 59–74. https://doi.org/10.1146/annurev.anthro.35.081705.123136.

———. "Collapse and Sustainability: Rome, the Maya, and the Modern World." *Archaeological Papers of the American Anthropological Association* 24 (2014): 201–214. https://doi.org/10.1111/apaa.12038.

———. *The Collapse of Complex Societies.* Cambridge: Cambridge University Press, 1988.

Turchin, Peter, Thomas E. Currie, Harvey Whitehouse, Pieter François, Kevin Feeney,

Daniel Mullins, Daniel Hoyer, et al. "Quantitative Historical Analysis Uncovers a Single Dimension of Complexity That Structures Global Variation in Human Social Organization." *Proceedings of the National Academy of Sciences* 115, no. 2 (2018): E144–E151. https://doi.org/10.1073/pnas.1708800115.

Wiedenhofer, Dominik, Doris Virág, Gerald Kalt, Barbara Plank, and Jan Streeck. "A Systematic Review of the Evidence on Decoupling of GDP, Resource Use and GHG Emissions, Part I: Bibliometric and Conceptual Mapping." *Environmental Research Letters* 15, no. 6 (2020). https://doi.org/10.1088/17489326/ab8429.

Yoffee, Norman, and George L. Cowgill, eds. *The Collapse of Ancient States and Civilizations*. Tucson: University of Arizona Press, 1988.

Progress and Its Critics

Bashford, Alison, and Joyce E. Chaplin. *The New Worlds of Thomas Robert Malthus: Rereading the Principle of Population*. Princeton: Princeton University Press, 2016.

Bauman, Zygmunt. *Legislators and Interpreters: On Modernity, Post-modernity, and Intellectuals*. Ithaca: Cornell University Press, 1987.

Burgen, Arnold, Peter McLaughlin, and Jürgen Mittelstraß. *The Idea of Progress*. Berlin: Walter de Gruyter, 1997.

Condorcet, Jean-Antoine-Nicolas de Caritat. *Outlines of an Historical View of the Progress of the Human Mind: Being a Posthumous Work of the Late M. de Condorcet*. London: Printed for J. Johnson, 1795. HathiTrust.

Federici, Silvia. *Caliban and the Witch*. New York: Autonomedia, 2004.

Godwin, William. *An Enquiry Concerning Political Justice*. Oxford: Oxford University Press, 2013.

Lasch, Christopher. *The True and Only Heaven*. New York: W.W. Norton, 1991.

Lévi-Strauss, Claude. *Tristes Tropiques*. New York: Penguin Classics, 2012.

Löwith, Karl. *Meaning in History: Theological Implications of the Philosophy of History*. Chicago: University of Chicago Press, 1949.

Malthus, Thomas R. *An Essay on the Principle of Population*. Var. ed. Edited by Donald Winch. Cambridge: Cambridge University Press, 1992.

Mayhew, Robert. *Malthus: The Life and Legacies of an Untimely Prophet*. Cambridge: Harvard University Press, 2014.

McNeill, J. R., and Peter Engelke. *The Great Acceleration*. Cambridge: Belknap Press, 2016.

Nisbet, Robert A. *History of the Idea of Progress*. New York: Basic Books, 1980.

Ong, Walter. *Orality and Literacy: The Technologizing of the Word*. New York: Routledge, 1982.

Oreskes, Naomi. *Why Trust Science?* Princeton: Princeton University Press, 2019.

Quijano, Aníbal. "Coloniality and Modernity/Rationality." *Cultural Studies* 21, no. 2–3 (2007): 168–178.

Rorty, Richard. *Philosophy and the Mirror of Nature*. Princeton: Princeton University Press, 1979.

Scott, James. *Against the Grain: A Deep History of the Earliest States.* New Haven: Yale University Press, 2017.
Scott, Joan Wallach. *On the Judgment of History.* New York: Columbia University Press, 2020.
Sloterdijk, Peter. "Rules for the Human Zoo: A Response to the Letter on Humanism." *Environment and Planning. D, Society & Space* 27, no. 1 (2009): 12–28. https://doi.org/10.1068/dst3.
Voegelin, Eric. *The New Science of Politics, an Introduction.* Chicago: University of Chicago Press, 1952.
Wittgenstein, Ludwig. *Culture and Value.* Translated by Peter Winch. Chicago: University of Chicago Press, 1980.
Wrangham, Richard. *The Goodness Paradox: The Strange Relationship between Virtue and Violence in Human Evolution.* New York: Pantheon Books, 2019.
Wright, Georg Henrik von. *The Tree of Knowledge and Other Essays.* New York: E. J. Brill, 1993.
Wynter, Sylvia. "The Ceremony Must Be Found: After Humanism." *Boundary 2,* no. 3 (1984): 19–70. https://doi.org/10.2307/302808.
———. "Unsettling the Coloniality of Being/Power/Truth/Freedom: Towards the Human, After Man, Its Overrepresentation—An Argument." *The New Centennial Review* 3, no. 3, 2003, 257–337, https://doi.org/10.1353/ncr.2004.0015.

Fictionalism, Biases, and Worldhood

Abram, David. *The Spell of the Sensuous: Perception and Language in a More-Than-Human World.* New York: Vintage, 1997.
Bakhtin, Mikhail M. *The Dialogic Imagination: Four Essays.* Translated by Caryl Emerson and Michael Holquist. Austin: University of Texas Press, 2020.
Basso, Keith. *Wisdom Sits in Places: Landscape and Language Among the Western Apache.* Albuquerque: University of New Mexico Press, 1996.
Carson, Anne. *Eros the Bittersweet.* Normal, IL: Dalkey Archive Press, 1998.
Diamond, Cora. "Losing your Concepts." *Ethics* 98, no. 2 (1988): 255–277.
Giere, Ronald N. *Scientific Perspectivism.* Chicago: University of Chicago Press, 2006.
Kahneman, Daniel. *Thinking, Fast and Slow.* New York: Farrar, Straus, and Giroux, 2013.
Kahneman, Daniel, and Amos Tversky. "Choices, Values, and Frames." *American Psychologist* 39, no. 4 (1984): 341–350. https://doi.org/10.1037/0003-066X.39.4.341.
———. "Prospect Theory: An Analysis of Decision Under Risk." *Econometrica* 47 (1979): 263–291. https://doi.org/10.2307/1914185.
Kahneman, Daniel, Jack L. Knetsch, and Richard H. Thaler. "Anomalies: The Endowment Effect, Loss Aversion, and Status Quo Bias." *The Journal of Economic Perspectives* 5, no. 1 (1991): 193–206. https://doi.org/10.1257/jep.5.1.193.
Kermode, Frank. *The Sense of an Ending: Studies in the Theory of Fiction.* New York: Oxford University Press, 2000.

Kuhn, Thomas S. *The Structure of Scientific Revolutions*. 50th anniv. ed. Chicago: University of Chicago Press, 2012.

MacIntyre, Alasdair. *After Virtue: A Study in Moral Theory*. Notre Dame: University of Notre Dame Press, 2007.

Quine, W. V. "Main Trends in Recent Philosophy: Two Dogmas of Empiricism." *The Philosophical Review* 60, no. 1 (1951): 20–43. https://doi.org/10.2307/2181906.

Shu Li, Yan-Ling Bi, and Li-Lin Rao. "Every Science/Nature Potter Praises His Own Pot—Can We Believe What He Says Based on His Mother Tongue?" *Journal of Cross-Cultural Psychology* 42, no. 1 (2011): 125–130. https://doi.org/10.1177/0022022110383425.

Tversky, Amos, and Daniel Kahneman. "Judgment Under Uncertainty: Heuristics and Biases." *Science* 185, no. 4157 (1974): 1124–1131. https://doi.org/10.1126/science.185.4157.1124.

Viveiros De Castro, Eduardo. "Cosmological Deixis and Amerindian Perspectivism." *Journal of the Royal Anthropological Institute* 4, no. 3 (1998): 469–488.

Whitehead, Alfred North. *Process and Reality*. Corr. ed. New York: Free Press, 1978.

Climate Change (General)

Anderson, Kevin, and Alice Bows. "Beyond 'Dangerous' Climate Change: Emission Scenarios for a New World." *Philosophical Transactions of the Royal Society A: Mathematical, Physical and Engineering Sciences* 369, no. 1934 (2011): 20–44.

Armstrong McKay, David I., et al. "Exceeding 1.5°C Global Warming Could Trigger Multiple Climate Tipping Points." *Science* 377, no. 6611 (2022): eabn7950.

Balaji, V., Fleur Couvreux, Julie Deshayes, Jacques Gautrais, Frédéric Hourdin, and Catherine Rio. "Are General Circulation Models Obsolete?" *Proceedings of the National Academy of Sciences* 119, no. 47 (2022): e2202075119. https://doi.org/10.1073/pnas.2202075119.

Ballinger, Thomas J., and James E. Overland. "The Alaskan Arctic Regime Shift Since 2017: A Harbinger of Years to Come?" *Polar Science* 32 (2022): 100841. https://doi.org/10.1016/j.polar.2022.100841.

Betts, Richard A., Matthew Collins, Deborah L. Hemming, Chris D. Jones, Jason A. Lowe, and Michael G. Sanderson. "When Could Global Warming Reach 4°C?" *Philosophical Transactions of the Royal Society A: Mathematical, Physical and Engineering Sciences* 369, no. 1934 (2011): 67–84.

Biermann, Frank, and Rakhyun E. Kim. "The Boundaries of the Planetary Boundary Framework: A Critical Appraisal of Approaches to Define a 'Safe Operating Space' for Humanity." *Annual Review of Environment and Resources* 45, no. 1 (2020): 497–521.

Brovkin, Victor, Edward Brook, John W. Williams, Sebastian Bathiany, Timothy M. Lenton, Michael Barton, Robert M. DeConto, et al. "Past Abrupt Changes, Tipping Points and Cascading Impacts in the Earth System." *Nature Geoscience* 14, no. 8 (2021): 550–558.

Committee on Abrupt Climate Change. *Abrupt Climate Change: Inevitable Surprises.* Washington, D.C.: National Academy Press, 2002.

Davis, W. Jackson. "Mass Extinctions and Their Relationship with Atmospheric Carbon Dioxide Concentration: Implications for Earth's Future." *Advancing Earth and Space Sciences* 11, no. 6 (2023). https://doi.org/10.1029/2022EF003336.

Deser, Clara. "Certain Uncertainty: The Role of Internal Climate Variability in Projections of Regional Climate Change and Risk Management." *Earth's Future* 8 (2020): e2020EF001854. https://doi.org/10.1029/2020EF001854.

Hansen, James E., Makiko Sato, Leon Simons, Larissa Nazarenko, Karina von Schuckmann, Norman Loeb, Matthew Osman, et al. "Global Warming in the Pipeline." *Oxford Open Climate Change*, December 8, 2022. https://doi.org/10.48550/arXiv.2212.04474.

Hansen, James E., Makiko Sato, Paul Hearty, Reto Ruedy, Maxwell Kelley, Valerie Masson-Delmotte, Gary Russell, et al. "Ice Melt, Sea Level Rise and Superstorms: Evidence from Paleoclimate Data, Climate Modeling, and Modern Observations That 2°C Global Warming Could Be Dangerous." *Atmospheric Chemistry and Physics* 16 (2016): 3761–3812. https://doi.org/10.5194/acp-16-3761-2016.

Hausfather, Zeke, Henri F. Drake, Tristan Abbott, and Gavin A. Schmidt. "Evaluating the Performance of Past Climate Model Projections." *Geophysical Research Letters* 47, no. 1 (2020). https://doi.org/10.1029/2019GL085378.

Hausfather, Zeke, Kate Marvel, Gavin A. Schmidt, John W. Nielsen-Gammon, and Mark Zelinka. "Climate Simulations: Recognize the 'Hot Model' Problem." *Nature* 605, no. 7908 (2022): 26–29. https://doi.org/10.1038/d41586-022-01192-2.

Hsiang, Solomon M., Kyle C. Meng, and Mark A. Cane. "Civil Conflicts Are Associated with the Global Climate." *Nature* 476, no. 7361 (2011): 438–441. https://doi.org/10.1038/nature10311.

Intergovernmental Panel on Climate Change. *Climate Change 2021: The Physical Science Basis.* Contribution of Working Group I to the Sixth Assessment Report of the Intergovernmental Panel on Climate Change. Cambridge: Cambridge University Press, 2021.

Lenton, Timothy M., Johan Rockström, Owen Gaffney, Stefan Rahmstorf, Katherine Richardson, Will Steffen, and Hans Joachim Schellnhuber. "Climate Tipping Points—Too Risky to Bet Against." *Nature* 575, no. 7784 (2019): 592–595. https://doi.org/10.1038/d41586-019-03595-0.

Le Roy Ladurie, Emmanuel. *Times of Feast, Times of Famine: A History of Climate Since the Year 1000.* New York: Farrar, Straus and Giroux, 1988.

Maslin, Mark, and Patrick Austin. "Climate Models at Their Limit?" *Nature* 486 (2012): 183–184. https://doi.org/10.1038/486183a.

Matthews, H. Damon, and Seth Wynes. "Current Global Efforts Are Insufficient to Limit Warming to 1.5°C." *Science* 376, no. 6600 (2022): 1404–1409.

Moraga, Jorge Sebastián, Nadav Peleg, Peter Molnar, Simone Fatichi, and Paolo Burlando. "Uncertainty in High-Resolution Hydrological Projections: Partitioning the Influence of Climate Models and Natural Climate Variability." *Hydrological Processes* 36, no. 10 (2022). https://doi.org/10.1002/hyp.14695.

New, Mark, Diana Liverman, Heike Schroder, and Kevin Anderson. "Four Degrees and Beyond: The Potential for a Global Temperature Increase of Four Degrees and Its Implications." *Philosophical Transactions of the Royal Society A: Mathematical, Physical and Engineering Sciences* 369, no. 1934 (2011): 6–19.

Nisbet, Euan G., M. R. Manning, E. J. Dlugokencky, R. E. Fisher, David Lowry, S. E. Michel, C. Lund Myhre, et al. "Very Strong Atmospheric Methane Growth in the 4 Years 2014–2017: Implications for the Paris Agreement." *Global Biogeochemical Cycles* 33 (2019): 318–342. https://doi.org/10.1029/2018GB006009.

Pindyck, Robert S. *Climate Future: Averting and Adapting to Climate Change*. New York: Oxford University Press, 2022.

———. "The Use and Misuse of Models for Climate Policy." *Review of Environmental Economics and Policy* 11, no. 1 (2017): 100–114. https://doi.org/10.1093/reep/rew012.

Raftery, Adrian E., Alec Zimmer, Dargan M. W. Frierson, Richard Startz, and Peiran Liu. "Less Than 2°C Warming by 2100 Unlikely." *Nature Climate Change* 7, no. 9 (2017): 637–641. https://doi.org/10.1038/nclimate3352.

Reckien, Diana, Alexandre K. Magnan, Chandni Singh, Megan Lukas-Sithole, Ben Orlove, E. Lisa F. Schipper, and Erin Coughlan de Perez. "Navigating the Continuum Between Adaptation and Maladaptation." *Nature Climate Change* 13, no. 9 (2023): 907–918.

Schneider, Stephen. "Climate Change and the World Predicament: A Case Study for Interdisciplinary Research." *Climatic Change* 1, no. 1 (1977): 21–43. https://doi.org/10.1007/BF00162775.

———. *Global Warming: Are We Entering the Greenhouse Century?* Oakland: Sierra Club Books, 1989.

Schneider, Tapio. "Uncertainty in Climate-Sensitivity Estimates." *Nature* 446, no. 7131 (2007): E1. https://doi.org/10.1038/nature05707.

Schneider, Tapio, Colleen M. Kaul, and Kyle G. Pressel. "Possible Climate Transitions from Breakup of Stratocumulus Decks Under Greenhouse Warming." *Nature Geoscience* 12, no. 3 (2019): 164. https://doi.org/10.1038/s41561-019-0310-1.

Sherwood, Steven C., Mark J. Webb, James D. Annan, Kyle C. Armour, Piers M. Forster, Julia C. Hargreaves, Gabriele Hegerl, et al. "An Assessment of Earth's Climate Sensitivity Using Multiple Lines of Evidence." *Reviews of Geophysics* 58, no. 4 (2020). https://doi.org/10.1029/2019RG000678.

Smith, Mark Stafford, Lisa Horrocks, Alex Harvey, and Clive Hamilton. "Rethinking Adaptation for a 4°C World." *Philosophical Transactions of the Royal Society A: Mathematical, Physical and Engineering Sciences* 369, no. 1934 (2011): 196–216.

Stainforth, Thorfinn, and Bartosz Brzezinski. "More Than Half of All CO_2 Emissions Since 1751 Emitted in the Last 30 Years." Institute for European Environmental Policy. April 29, 2020. https://ieep.eu/news/more-than-half-of-all-co2-emissions-since-1751-emitted-in-the-last-30-years.

Stern, Nicholas. "Economics: Current Climate Models Are Grossly Misleading." *Nature* 530 (2016): 407–409. https://doi.org//10.1038/530407a.
Tollefson, Jeff. "Top Climate Scientists Are Sceptical That Nations Will Rein In Global Warming." *Nature* 599 (2021): 22–24. https://doi.org/10.1038/d41586-021-02990-w.
Trisos, Christopher H., Cory Merow, and Alex L. Pigot. "The Projected Timing of Abrupt Ecological Disruption from Climate Change." *Nature* 580, no. 7804 (2020): 496–501.
Turetsky, Merritt R., Benjamin W. Abbott, Miriam C. Jones, Katey Walter Anthony, David Olefeldt, et al. "Permafrost Collapse Is Accelerating Carbon Release." *Nature* 569, no. 7754 (2019): 32–34.
Utterström, Gustaf. "Climatic Fluctuations and Population Problems in Early Modern History." *Scandinavian Economic History Review* 3, no. 1 (1955): 3–47.
Wagner, Gernot, and Richard J. Zeckhauser. "Climate Policy: Hard Problem, Soft Thinking." *Climatic Change* 110, no. 3–4 (2012): 507–521.
Wall, Casey J., Joel R. Norris, Anna Possner, Daniel T. McCoy, Isabel L. McCoy, and Nicholas J. Lutsko. "Assessing Effective Radiative Forcing from Aerosol-Cloud Interactions over the Global Ocean." *Proceedings of the National Academy of Sciences* 119, no. 46 (2022): 1-e2210481119. https://doi.org/10.1073/pnas.2210481119.
World Bank. *Turn Down the Heat: Why a 4°C Warmer World Must Be Avoided*. Washington, D.C.: World Bank Group, 2012.

Climate Change Communication and Fear Appeals

Anderson, Josh, and Anthony Dudo, "A View from the Trenches: Interviews with Journalists About Reporting Science News." *Science Communication* 45, no. 1 (2023): 39–64. https://doi.org/10.1177/10755470221149156.
Atkin, Emily. "The Power and Peril of 'Climate Disaster Porn.'" *The New Republic*, July 10, 2017. https://newrepublic.com/article/143788/power-peril-climate-disaster-porn.
Awagu, Chrysantus, and Debra Z. Basil. "Fear Appeals: The Influence of Threat Orientations." *Journal of Social Marketing* 6, no. 4 (2016): 361–376. https://doi.org/10.1108/JSOCM-12-2014-0089.
Bendell, Jem. "The Biggest Mistakes in Climate Communications, Part 1: Looking Back at the 'Incomparably Average.'" *Brave New Europe*, September 8, 2022. https://braveneweurope.com/jem-bendell-the-biggest-mistakes-in-climate-communications-part-1-looking-back-at-the-incomparably-average.
———. "The Biggest Mistakes in Climate Communications, Part 2: Climate Brightsiding." *Brave New Europe*, September 15, 2022. https://braveneweurope.com/jem-bendell-the-biggest-mistakes-in-climate-communications-part-2-climate-brightsiding.
Bernauer, Thomas, and Liam F. McGrath. "Simple Reframing Unlikely to Boost Public Support for Climate Policy." *Nature Climate Change* 6, no. 7 (2016): 680–683. https://doi.org/10.1038/nclimate2948.

Bowen, Mark Stander. *Censoring Science: Inside the Political Attack on Dr. James Hansen and the Truth of Global Warming.* New York: Penguin, 2008.

Brosch, Tobias. "Affect and Emotions as Drivers of Climate Change Perception and Action: A Review." *Current Opinion in Behavioral Sciences* 42 (2021): 15–21. https://doi.org/10.1016/j.cobeha.2021.02.001.

Bruine de Bruin, Wändi, Lila Rabinovich, Kate Weber, Marianna Babboni, Monica Dean, and Lance Ignon. "Public Understanding of Climate Change Terminology." *Climatic Change* 167, no. 3–4 (2021): 37. https://doi.org/10.1007/s10584-021-03183-0.

Brulle, Robert J. "The Climate Lobby: A Sectoral Analysis of Lobbying Spending on Climate Change in the USA, 2000 to 2016." *Climatic Change* 149, no. 3–4 (2018): 289–303. https://doi.org/10.1007/s10584-018-2241-z.

———. "Critical Reflections on the March for Science." *Sociological Forum* 33, no. 1 (2018): 255–258. https://doi.org/10.1111/socf.12398.

———. "Institutionalizing Delay: Foundation Funding and the Creation of U.S. Climate Change Counter-Movement Organizations." *Climatic Change* 122, no. 4 (2014): 681–694. https://doi.org/10.1007/s10584-013-1018-7.

Brulle, Robert J., and Kari Marie Norgaard. "Avoiding Cultural Trauma: Climate Change and Social Inertia." *Environmental Politics* 28, no. 5 (2019): 886–908. https://doi.org/10.1080/09644016.2018.1562138.

Bugden, Dylan. "Does Climate Protest Work? Partisanship, Protest, and Sentiment Pools." *Socius* 6 (2020). https://doi.org/2378023120925949.

Chinn, Sedona, and P. Sol Hart. "Climate Change Consensus Messages Cause Reactance." *Environmental Communication* 17, no. 1 (2023): 51–59. https://doi.org/10.1080/17524032.2021.1910530.

Dunwoody, Sharon. "Science Journalism: Prospects in the Digital Age." In *Routledge Handbook of Public Communication of Science and Technology*, edited by Massimiano Bucchi and Brian Trench. London: Routledge, 2021. 14–32.

Farrell, Justin. "Corporate Funding and Ideological Polarization About Climate Change." *Proceedings of the National Academy of Sciences* 113, no. 1 (2016): 92–97. https://doi.org/10.1073/pnas.1509433112.

Feldman, Lauren, and P. Sol Hart. "Is There Any Hope? How Climate Change News Imagery and Text Influence Audience Emotions and Support for Climate Mitigation Policies." *Risk Analysis* 38, no. 3 (2018): 585–602. https://doi.org/10.1111/risa.12868.

———. "Using Political Efficacy Messages to Increase Climate Activism: The Mediating Role of Emotions." *Science Communication* 38, no. 1 (2016): 99–127. https://doi.org/10.1177/1075547015617941.

Fesenfeld, Lukas P., Yixian Sun, Michael Wicki, and Thomas Bernauer. "The Role and Limits of Strategic Framing for Promoting Sustainable Consumption and Policy." *Global Environmental Change* 68 (2021): 102266. https://doi.org/10.1016/j.gloenvcha.2021.102266.

Gauchat, Gordon. "Politicization of Science in the Public Sphere: A Study of Public

Trust in the United States, 1974 to 2010." *American Sociological Review* 77, no. 2 (2012). https://doi.org/10.1177/0003122412438225.

Gifford, Robert. "The Dragons of Inaction: Psychological Barriers That Limit Climate Change Mitigation and Adaptation." *American Psychologist* 66, no. 4 (2011): 290–302. https://doi.org/10.1037/a0023566.

Gjalt-Jorn Ygram Peters, Robert A. C. Ruiter, and Gerjo Kok. "Threatening Communication: A Critical Re-Analysis and a Revised Meta-Analytic Test of Fear Appeal Theory." *Health Psychology Review* 7, Suppl 1 (2013): S8–S31. https://doi.org/10.1080/17437199.2012.703527.

Glavovic, Bruce C., Timothy F. Smith, and Iain White. "The Tragedy of Climate Change Science." *Climate and Development* 14, no. 9 (2022): 829–833. https://doi.org/0.1080/17565529.2021.2008855.

Goldberg, Matthew H., Abel Gustafson, Seth A. Rosenthal, and Anthony Leiserowitz. "Shifting Republican Views on Climate Change Through Targeted Advertising." *Nature Climate Change* 11 (2021): 573–577. https://doi.org/10.1038/s41558-021-01070-1.

Grundmann, Reiner. "'Climategate' and The Scientific Ethos." *Science, Technology, & Human Values* 38, no. 1 (2013): 67–93. https://doi.org/10.1177/0162243911432318.

Hamilton, Clive. *Requiem for a Species: Why We Resist the Truth About Climate Change.* Washington, D.C.: Earthscan, 2010.

Hamilton, Lawrence C., Joel Hartter, Barry D. Keim, Angela E. Boag, Michael W. Palace, Forrest R. Stevens, and Mark J. Ducey. "Wildfire, Climate, and Perceptions in Northeast Oregon." *Regional Environmental Change* 16, no. 6 (2016): 1819–1832. https://doi.org/10.1007/s10113-015-0914-y.

Hamilton, Lawrence C., Cameron P. Wake, Joel Hartter, Thomas G. Safford, and Alli J. Puchlopek. "Flood Realities, Perceptions and the Depth of Divisions on Climate." *Sociology* 50, no. 5 (2016): 913–933. https://doi.org/10.1177/0038038516648547.

Hart, P. Sol, and Erik C. Nisbet. "Boomerang Effects in Science Communication: How Motivated Reasoning and Identity Cues Amplify Opinion Polarization About Climate Mitigation Policies." *Communication Research* 39, no. 6 (2012): 701–723. https://doi.org/10.1177/0093650211416646.

Hastings, Gerard, and Martine Stead. "Fear Appeals in Social Marketing: Strategic and Ethical Reasons for Concern." *Psychology and Marketing* 21, no. 11 (2004): 961–986. https://doi.org/10.1002/mar.20043.

Hibbing, John R., Kevin B. Smith, and John R. Alford. "Differences in Negativity Bias Underlie Variations in Political Ideology." *Behavioral and Brain Sciences* 37, no. 3 (2014): 297–307. https://doi.org/10.1017/S0140525X13001192.

Hornsey, Matthew J. "The Role of Worldviews in Shaping How People Appraise Climate Change." *Current Opinion in Behavioral Sciences* 42 (2021): 36–41. https://doi.org/10.1016/j.cobeha.2021.02.021.

Hornsey, Matthew J., and Kelly S. Fielding. "A Cautionary Note About Messages of

Hope: Focusing on Progress in Reducing Carbon Emissions Weakens Mitigation Motivation." *Global Environmental Change* 39 (2016): 26–34. https://doi.org/10.1016/j.gloenvcha.2016.04.003.

———. "Understanding (and Reducing) Inaction on Climate Change." *Social Issues and Policy Review* 14, no. 1 (2020): 3–35. https://doi.org/10.1111/sipr.12058.

Hornsey, Matthew J., Kelly S. Fielding, Ryan McStay, Joseph P. Reser, Graham L. Bradley, and Katharine H. Greenaway. "Evidence for Motivated Control: Understanding the Paradoxical Link Between Threat and Efficacy Beliefs About Climate Change." *Journal of Environmental Psychology* 42 (2015): 57–65. https://doi.org/10.1016/j.jenvp.2015.02.003.

Hornsey, Matthew J., Emily Harris, Paul Bain, and Kelly Fielding. "Meta-Analyses of the Determinants and Outcomes of Belief in Climate Change." *Nature Climate Change* 6, no. 6 (2016): 622–626. https://doi.org/10.1038/nclimate2943.

Huber, Robert A. "The Role of Populist Attitudes in Explaining Climate Change Skepticism and Support for Environmental Protection." *Environmental Politics* 29, no. 6 (2020): 959–982.

Jehn, Florian, Luke Kemp, Ekaterina Ilin, Christoph Funk, Jason Wang, and Lutz Breuer. "Focus of the IPCC Assessment Reports Has Shifted to Lower Temperatures." *Earth's Future* 10 (2022). https://doi.org/10.1029/2022EF002876.

Jehn, Florian U., Marie Schneider, Jason R. Wang, Luke Kemp, and Lutz Breuer. "Betting on the Best Case: Higher End Warming Is Underrepresented in Research." *Environmental Research Letters* 16, no. 8 (2021): 084036.

Kahan, Dan M. "Climate-Science Communication and the Measurement Problem." *Political Psychology* 36 (2015): 1–43.

Kahan, Dan M., and Jonathan C. Corbin. "A Note on the Perverse Effects of Actively Open-Minded Thinking on Climate-Change Polarization." *Research & Politics* 3, no. 4 (October 2016): 2053168016676705. https://doi.org/10.1177/2053168016676705.

Kahan, Dan M., Ellen Peters, Erica Cantrell Dawson, and Paul Slovic. "Motivated Numeracy and Enlightened Self-Government." *Behavioral Public Policy* 1, no. 1 (2017): 54–86. https://doi.org/10.1017/bpp.2016.2.

Kaida, Naoko, and Kosuke Kaida. "Facilitating Pro-Environmental Behavior: The Role of Pessimism and Anthropocentric Environmental Values." *Social Indicators Research* 126, no. 3 (2016): 1243–1260. https://doi.org/10.1007/s11205-015-0943-4.

———. "Pro-environmental Behavior Correlates with Present and Future Subjective Well-being." *Environmental Development and Sustainability* 18 (2016): 111–127. https://doi.org/10.1007/s10668-015-9629-y.

Kemp, Luke, Chi Xu, Joanna Depledge, Kristie L. Ebi, Goodwin Gibbins, Timothy A. Kohler, Johan Rockström, et al. "Climate Endgame: Exploring Catastrophic Climate Change Scenarios." *Proceedings of the National Academy of Sciences* 119, no. 34 (2022). https://doi.org/10.1073/pnas.2108146119.

Krause, Nicole M., Dominique Brossard, Dietram A. Scheufele, Michael A. Xenos, and Keith Franke. "Trends—Americans' Trust in Science and Scientists." *Public Opinion Quarterly* 83, no. 4 (2019): 817–836. https://doi.org/10.1093/poq/nfz041.

Lahsen, Myanna, and Esther Turnhout. "How Norms, Needs, and Power in Science Obstruct Transformations Towards Sustainability." *Environmental Research Letters* 16, no. 2 (2021): 025008. https://doi.org/10.1088/1748-9326/abdcfo.

Lamb, William F., Giulio Mattioli, Sebastian Levi, J. Timmons Roberts, Stuart Capstick, Felix Creutzig, et al. "Discourses of Climate Delay." *Global Sustainability*, vol. 3 (2020). https://doi.org/10.1017/sus.2020.13.

Leiserowitz, Anthony A., Edward W. Maibach, Connie Roser-Renouf, Nicholas Smith, and Erica Dawson. "Climategate, Public Opinion, and the Loss of Trust." *The American Behavioral Scientist* 57, no. 6 (2013): 818–837. https://doi.org/10.1177/0002764212458272.

Loeb, Zachary. "Be Afraid! But Not That Afraid?—On Climate Doom." *Librarian Shipwreck*, August 16, 2019. https://librarianshipwreck.wordpress.com/2019/08/16/be-afraid-but-not-that-afraid-on-climate-doom/.

Ma, Xin, Yin Yang, and Liang Chen. "Promoting Behaviors to Mitigate the Effects of Climate Change: Using the Extended Parallel Process Model at the Personal and Collective Level in China." *Environmental Communication* 17, no. 4 (2023): 353–369. https://doi.org/10.1080/17524032.2023.2181134.

Ma, Yanni, Graham Dixon, and Jay D. Hmielowski. "Psychological Reactance from Reading Basic Facts on Climate Change: The Role of Prior Views and Political Identification." *Environmental Communication* 13, no. 1 (2019): 71–86. https://doi.org/10.1080/17524032.2018.1548369.

Maibach, Edward, Anothony Leiserowitz, Sara Cobb, and Michael Shank. "The Legacy of Climategate: Undermining or Revitalizing Climate Science and Policy?" *WIREs Climate Change*, 3 (2012): 289–295. https://doi.org/10.1002/wcc.168.

Maibach, Edward W., Connie Roser-Renouf, and Anthony Leiserowitz. "Communication and Marketing as Climate Change-Intervention Assets." *American Journal of Preventive Medicine* 35, no. 5 (2008): 488–500. https://doi.org/10.1016/j.amepre.2008.08.016.

Maloney, Erin K., Maria K. Lapinski, and Kim Witte. "Fear Appeals and Persuasion: A Review and Update of the Extended Parallel Process Model." *Social and Personality Psychology Compass* 5, no. 4 (2011): 206–19. https://doi.org/10.1111/j.1751-9004.2011.00341.x.

Mann, Christopher B., Kevin Arceneaux, and David W. Nickerson. "Do Negatively Framed Messages Motivate Political Participation? Evidence from Four Field Experiments." *American Politics Research* 48, no. 1 (2020): 3–21. https://doi.org/10.1177/1532673X19840732.

Marshall, George. *Don't Even Think About It*. London: Bloomsbury Publishing, 2015.

Martel-Morin, Marjolaine, and Erick Lachapelle. "Code Red for Humanity or Time for Broad Collective Action? Exploring the Role of Positive and Negative Messaging in (De)motivating Climate Action." *Frontiers in Communication* 7 (2022): 252. https://doi.org/10.3389/fcomm.2022.968335.

Martin, Paul S. "Inside the Black Box of Negative Campaign Effects: Three Reasons Why Negative Campaigns Mobilize." *Political Psychology* 25, no. 4 (2004): 545–562. https://doi.org/10.1111/j.1467-9221.2004.00386.x.

McCright, Aaron M., Meghan Charters, Katherine Dentzman, and Thomas Dietz. "Examining the Effectiveness of Climate Change Frames in the Face of a Climate Change Denial Counter-Frame." *Topics in Cognitive Science* 8, no. 1 (2016): 76–97. https://doi.org/10.1111/tops.12171.

———. "The Influence of Political Ideology on Trust in Science." *Environmental Research Letters* 8, no. 4 (2013): 044029. https://doi.org/10.1088/1748-9326/8/4/044029.

Merkel, Sonia Hélène, Angela M. Person, Randy A. Peppler, and Sarah M. Melcher. "Climate Change Communication: Examining the Social and Cognitive Barriers to Productive Environmental Communication." *Social Science Quarterly* 101, no. 5 (2020): 2085–2100. https://doi.org/10.1111/ssqu.12843.

Moore, Frances C., Nick Obradovich, Flavio Lehner, and Patrick Baylis. "Rapidly Declining Remarkability of Temperature Anomalies May Obscure Public Perception of Climate Change." *Proceedings of the National Academy of Sciences* 116, no. 11 (2019): 4905–4910.

Morris, Brandi S., Polymeros Chrysochou, Simon T. Karg, and Panagiotis Mitkidis. "Optimistic vs. Pessimistic Endings in Climate Change Appeals." *Humanities and Social Sciences Communications* 7, no. 1 (2020). https://doi.org/10.1057/s41599-020-00574-z.

Moser, Deyshawn, Peter Steiglechner, and Achim Schlueter. "Facing Global Environmental Change: The Role of Culturally Embedded Cognitive Biases." *Environmental Development* 44 (2022): 100735.

Moser, Suzanne C. "More Bad News: The Risk of Neglecting Emotional Responses to Climate Change Information." In *Creating a Climate for Change*, edited by Susanne Moser and Lisa Dilling. Cambridge: Cambridge University Press, 2007.

———. "The Work After 'It's Too Late' (to Prevent Dangerous Climate Change)." *Wiley Interdisciplinary Reviews: Climate Change* 11, no. 1 (2020): e606.

Motta, Matthew. "The Polarizing Effect of the March for Science on Attitudes Toward Scientists." *PS: Political Science & Politics* 51, no. 4 (2018): 782–788. https://doi.org/10.1017/S1049096518000938.

Myrick, Jessica Gall, and Suzannah Evans Comfort. "The Pope May Not Be Enough: How Emotions, Populist Beliefs, and Perceptions of an Elite Messenger Interact to Influence Responses to Climate Change Messaging." *Mass Communication & Society* 23, no. 1 (2020): 1–21. https://doi.org/10.1080/15205436.2019.1639758.

Nerlich, Brigitte, Nelya Koteyko, and Brian Brown. "Theory and Language of Climate Change Communication." *Wiley Interdisciplinary Reviews: Climate Change* 1, no. 1 (2010): 97–110. https://doi.org/10.1002/wcc.2.

Nisbet, Erik C., Kathryn E. Cooper, and R. Kelly Garrett. "The Partisan Brain: How Dissonant Science Messages Lead Conservatives and Liberals to (Dis)Trust Science." *The Annals of the American Academy of Political and Social Science* 658, no. 1 (2015). https://doi.org/10.1177/0002716214555474.

Norgaard, Kari. *Living in Denial*. Cambridge: MIT Press, 2011.

Nurse, Matthew S., and Will J. Grant. "I'll See It When I Believe it: Motivated Numeracy in Perceptions of Climate Change Risk." *Environmental Communication* 14, no. 2 (2020): 184–201. https://doi.org/10.1080/17524032.2019.1618364.

Ogunbode, Charles A., Rouven Doran, Daniel Hanss, Maria Ojala, Katariina Salmela-Aro, Karlijn L. van den Broek, Navjot Bhullar, et al. "Climate Anxiety, Wellbeing and Pro-Environmental Action: Correlates of Negative Emotional Responses to Climate Change in 32 Countries." *Journal of Environmental Psychology* 84 (2022). https://doi.org/10.1016/j.jenvp.2022.101887.

Oppenheimer, Michael, Keynyn Brysse, Dale Jamieson, Naomi Oreskes, Jessica O'Reilly, Matthew Shindell, and Milena Wazeck. *Discerning Experts: The Practices of Scientific Assessment for Environmental Policy*. Chicago: University of Chicago Press, 2019.

Oreskes, Naomi, and Erik M. Conway. *Merchants of Doubt: How a Handful of Scientists Obscured the Truth on Issues from Tobacco Smoke to Global Warming*. New York: Bloomsbury Press, 2010.

Palm, Risa, Toby Bolsen, and Justin T. Kingsland. "'Don't Tell Me What to Do': Resistance to Climate Change Messages Suggesting Behavior Changes." *Weather, Climate, and Society* 12, no. 4 (2020). https://doi-org./10.1175/WCAS-D-19-0141.1.

Reddy, Christopher. *Science Communication in a Crisis: An Insider's Guide*. New York: Routledge, 2023.

Reser, Joseph P., and Graham L. Bradley. "Fear Appeals in Climate Change Communication." In *The Oxford Encyclopedia of Climate Change Communication*, edited by Matthew Nisbet. New York: Oxford University Press, 2018.

Rode, Jacob B., Amy L. Dent, Caitlin N. Benedict, Daniel B. Brosnahan, Ramona L. Martinez, and Peter H. Ditto. "Influencing Climate Change Attitudes in the United States: A Systematic Review and Meta-Analysis." *Journal of Environmental Psychology* 76 (2021): 101623. https://doi.org/10.1016/j.jenvp.2021.101623.

Rode, Jacob B., and Peter H. Ditto. "Can the Partisan Divide in Climate Change Attitudes Be Bridged? A Review of Experimental Interventions." In *The Psychology of Political Polarization*, edited by Jan-Willem van Prooijen. Oxfordshire: Taylor & Francis Group, 2021.

Romm, Joseph. "Apocalypse Not: Oscars, Media, and the Myths of Climate Message." *Energy Central*, February 25, 2013. https://energycentral.com/c/ec/apocalypse-not-oscars-media-and-myths-climate-message.

Safford, Thomas G., Emily H. Whitmore, and Lawrence C. Hamilton. "Questioning Scientific Practice: Linking Beliefs About Scientists, Science Agencies, and Climate Change." *Environmental Sociology* 6, no. 2 (2019). https://doi.org/10.1080/23251042.2019.1696008.

Smith, Nicholas, and Anthony Leiserowitz. "The Role of Emotion in Global Warming Policy Support and Opposition." *Risk Analysis* 34, no. 5 (2014): 937–948. https://doi.org/10.1111%2Frisa.12140.

Stanley, Samantha K., Anna Klas, Edward J. R. Clarke, and Iain Walker. "The Effects of a Temporal Framing Manipulation on Environmentalism: A Replication and Extension." *PLOS ONE* 16, no. 2 (2021). https://doi.org/10.1371/journal.pone.0246058.

Stern, Paul C. "Fear and Hope in Climate Messages." *Nature Climate Change* 2, no. 8 (2012): 572–573. https://doi.org/10.1038/nclimate1610.

Stevenson, Kathryn, and Nils Peterson. "Motivating Action Through Fostering Climate Change Hope and Concern and Avoiding Despair Among Adolescents." *Sustainability* 8, no. 1 (2016): 6. https://doi.org/10.3390/su8010006.

Stoknes, Per Espen. *What We Think About When We Try Not to Think About Global Warming*. White River Junction, VT: Chelsea Green Publishing, 2015.

Supran, Geoffrey, and Naomi Oreskes. "Assessing ExxonMobil's Climate Change Communications (1977–2014)." *Environmental Research Letters* 12, no. 8 (2017). https://doi.org/10.1088/1748-9326/aa815f.

———. "Rhetoric and Frame Analysis of ExxonMobil's Climate Change Communications." *One Earth* 4, no. 5 (2021): 696–719. https://doi.org/10.1016/j.oneear.2021.04.014.

Tannenbaum, Melanie B., Justin Hepler, Rick S. Zimmerman, Lindsey Saul, Samantha Jacobs, Kristina Wilson, and Dolores Albarracín. "Appealing to Fear: A Meta-Analysis of Fear Appeal Effectiveness and Theories." *Psychological Bulletin* 141, no. 6 (2015): 1178–204. https://doi.org/10.1037/a0039729.

Van der Linden, Sander, Anthony Leiserowitz, and Edward Maibach. "The Gateway Belief Model: A Large-Scale Replication." *Journal of Environmental Psychology* 62 (2019): 49–58. https://doi.org/10.1016/j.jenvp.2019.01.009.

Van der Linden, Sander, Edward Maibach, and Anthony Leiserowitz. "Improving Public Engagement with Climate Change: Five 'Best Practice' Insights from Psychological Science." *Perspectives on Psychological Science* 10, no. 6 (2015): 758–763. https://doi.org/0.1177/1745691615598516.

Van Valkengoed, Anne M., and Linda Steg. "Meta-Analyses of Factors Motivating Climate Change Adaptation Behaviour." *Nature Climate Change* 9, no. 2 (2019): 158–163. https://doi.org/10.1038/s41558-018-0371-y.

Van Zomeren, Martijn, Russell Spears, and Colin Wayne Leach. "Experimental Evidence for a Dual Pathway Model Analysis of Coping with the Climate Crisis." *Journal of Environmental Psychology* 30, no. 4 (2010): 339–346. https://doi.org/10.1016/j.jenvp.2010.02.006.

Vlasceanu, Madalina, Kimberly C. Doell, Joseph B. Bak-Coleman, Boryana Todorova, Michael M. Berkebile-Weinberg, Samantha J. Grayson, Yash Patel, et al. "Addressing Climate Change with Behavioral Science: A Global Intervention Tournament in 63 Countries." *Science Advances* 10, no. 6 (2024). https://doi.org/10.1126/sciadv.adj5778.

Voelkel, Jan G., Adina Abeles, and Robb Willer. "The Effects of Dire and Solvable Messages on Belief in Climate Change: A Replication Study." Preprint. *PsyArXiv*, 2021. https://doi.org/10.31234/osf.io/prk9f.

Voelkel, Jan G., Ashwini Ashokkumar, Neil Malhotra, Adina Abeles, Jarret T. Crawford, Jane K. Willenbring, and Robb Willer. "A Large Experimental Test of the Effectiveness of Highly Cited Climate Change Messages." Registered Report Stage 1 Protocol. figshare, May 13, 2024. https://doi.org/10.6084/m9.figshare.25807429.v1.

Vu, Linh, Ivan Soraperra, Margarita Leib, Joël van der Weele, and Shaul Shalvi. "Ignorance by Choice: A Meta-Analytic Review of the Underlying Motives of Willful Ignorance and Its Consequences." *Psychological Bulletin* 149, no. 9–10 (2023): 611.

Wallace-Wells, David. "The Uninhabitable Earth." *New York Magazine*, July 9, 2017. https://nymag.com/intelligencer/2017/07/climate-change-earth-too-hot-for-humans.html.

Whitmarsh, Lorraine, Lois Player, Angelica Jiongco, Melissa James, Marc Williams, Elizabeth Marks, and Patrick Kennedy-Williams. "Climate Anxiety: What Predicts It and How Is It Related to Climate Action?" *Journal of Environmental Psychology* 83 (2022): 101866. https://doi.org/10.1016/j.jenvp.2022.101866.

Wilson, Robyn S., Atar Herziger, Matthew Hamilton, and Jeremy S. Brooks. "From Incremental to Transformative Adaptation in Individual Responses to Climate-Exacerbated Hazards." *Nature Climate Change* 10, no. 3 (2020): 200–208. https://doi.org/10.1038/s41558-020-0691-6.

Witte, Kim. "Putting the Fear Back into Fear Appeals: The Extended Parallel Process Model." *Communication Monographs* 59, no. 4 (1992): 329–349. https://doi.org/10.1080/03637759209376276.

Witte, Kim, and Mike Allen. "A Meta-Analysis of Fear Appeals: Implications for Effective Public Health Campaigns." *Health Education & Behavior* 27, no. 5 (2000): 591–615. https://doi.org/10.1177/109019810002700506.

Wolsko, Christopher, Hector Ariceaga, and Jesse Seiden. "Red, White, and Blue Enough to Be Green: Effects of Moral Framing on Climate Change Attitudes and Conservation Behaviors." *Journal of Experimental Social Psychology* 65 (2016): 7–19.

Xue, Wen, Donald W. Hine, Anthony D. G. Marks, Wendy J. Phillips, Patrick Nunn, and Shouying Zhao. "Combining Threat and Efficacy Messaging to Increase Public Engagement with Climate Change in Beijing, China." *Climatic Change* 137, no. 1–2 (2016): 43–55. https://doi.org/10.1007/s10584-016-1678-1.

Climate Politics

Antonio, Robert J., and Robert J. Brulle. "The Unbearable Lightness of Politics: Climate Change Denial and Political Polarization." *Sociological Quarterly* 52, no. 2 (2011): 195–202. https://doi.org/10.1111/j.1533-8525.2011.01199.x.

Arrow, Kenneth, Maureen Cropper, Christian Gollier, Ben Groom, Geoffrey Heal, Richard Newell, William Nordhaus, et al. "Determining Benefits and Costs for Future Generations." *Science* 341, no. 6144 (2013): 349–50. https://doi.org/10.1126/science.1235665.

Biehl, Janet, and Peter Staudenmaier. *Ecofascism Revisited: Lessons from the German Experience*. New York: New Compass Press, 2011.

Cohen, Daniel Aldana. "New York City as 'Fortress of Solitude' After Hurricane

Sandy: A Relational Sociology of Extreme Weather's Relationship to Climate Politics." *Environmental Politics* 30, no. 5 (2021): 687–707. https://doi.org/10.1080/09644016.2020.1816380.

Dembicki, Geoff. *The Petroleum Papers: Inside the Far-Right Conspiracy to Cover Up Climate Change*. Vancouver: Greystone Books, 2022.

Edwards, Paul N. "Is Climate Change Ungovernable?" Proceedings of the 2023 Stanford Existential Risks Conference, September 14, 2023, 133–146. https://doi.org/10.25740/yc096zw4572.

Faye, Guillaume. *Convergence of Catastrophes*. Translated by E. Christian Kopff. London: Arktos, 2012.

Fleming, Sean. "The Unabomber and the Origins of Anti-tech Radicalism." *Journal of Political Ideologies* 27, no. 2 (2022): 207–225. https://doi.org/10.1080/13569317.2021.1921940.

Forchtner, Bernhard. "Climate Change and the Far Right." *Wiley Interdisciplinary Reviews. Climate Change* 10, no. 5 (2019). https://doi.org/10.1002/wcc.604.

Gilens, Martin, and Benjamin I. Page. "Testing Theories of American Politics: Elites, Interest Groups, and Average Citizens." *Perspectives on Politics* 12, no. 3 (2014): 564–581. https://doi.org/10.1017/S1537592714001595.

Hickel, Jason, and Giorgos Kallis. "Is Green Growth Possible?" *New Political Economy* 25, no. 4 (2020): 469–486. https://doi.org/10.1080/13563467.2019.1598964.

Hughes, Brian, Dave Jones, and Amarnath Amarasingam. "Ecofascism: An Examination of the Far-Right/Ecology Nexus in the Online Space." *Terrorism and Political Violence* 34, no. 5 (2022): 997–1023. https://doi.org/10.1080/09546553.2022.2069932.

Javeline, Debra, and Gregory Shufeldt. "Scientific Opinion in Policymaking: The Case of Climate Change Adaptation." *Policy Sciences* 47, no. 2 (2014): 121–139. https://doi.org/10.1007/s11077-013-9187-9.

Javeline, Debra, and Sophia N. Chau. "The Unexplored Politics of Climate Change Adaptation." In *Handbook of US Environmental Policy*, edited by David Konisky, 357–372. Cheltenham: Edward Elgar, 2020. https://doi.org/10.4337/9781788972840.00035.

Javeline, Debra, Jessica J. Hellmann, Rodrigo Castro Cornejo, and Gregory Shufeldt. "Expert Opinion on Climate Change and Threats to Biodiversity." *Bioscience* 63, no. 8 (2013): 666–673. https://doi.org/10.1525/bio.2013.63.8.9.

Joosse, Paul. "Elves, Environmentalism, and 'Eco-Terror': Leaderless Resistance and Media Coverage of the Earth Liberation Front." *Crime, Media, Culture* 8, no. 1 (2012): 75–93. https://doi.org/10.1177/1741659011433366.

Kehler, Sarah, and S. Jeff Birchall. "Climate Change Adaptation: How Short-Term Political Priorities Trample Public Well-Being." *Environmental Science & Policy* 146 (2023): 144–150.

Koubi, Vally. "Climate Change and Conflict." *Annual Review of Political Science* 22 (2019): 343–360. https://doi.org/10.1146/annurev-polisci-050317-070830.

Levin, Kelly, Benjamin Cashore, Steven Bernstein, and Graeme Auld. "Playing It Forward: Path Dependency, Progressive Incrementalism, and the 'Super Wicked' Problem of Global Climate Change." *Earth and Environmental Science* 6, no. 50 (January 2007). https://doi.org/10.1088/1755-1307/6/0/502002.

Linkola, Pentti. *Can Life Prevail? A Revolutionary Approach to the Environmental Crisis*. London: Arktos Media, 2011.

Lockwood, Matthew. "Right-Wing Populism and the Climate Change Agenda: Exploring the Linkages." *Environmental Politics* 27, no. 4 (2018): 712–732.

Lubarda, Balša. "Beyond Ecofascism? Far-Right Ecologism (FRE) as a Framework for Future Inquiries." *Environmental Values* 29, no. 6 (2020): 713–732. https://doi.org/10.3197/096327120X15752810323922.

Macklin, Graham. "The Extreme Right, Climate Change and Terrorism." *Terrorism and Political Violence* 34, no. 5 (2022): 979–996. https://doi.org/10.1080/09546553.2022.2069928.

MacLean, Jason, Susan G. Clark, Lee Foote, Thomas S. Jung, David S. Lee, and Douglas A. Clark. "Polar Bears and the Politics of Climate Change: A Response to Simpson." *Journal of International Wildlife Law & Policy* 23, no. 2 (2020): 141–150.

Malm, Andreas. *How to Blow Up a Pipeline*. New York: Verso Books, 2020.

Mann, Geoff, and Joel Wainwright. *Climate Leviathan*. New York: Verso Books, 2018.

Moore, Sam, and Alex Roberts. *The Rise of Ecofascism*. Cambridge: Polity Books, 2022.

Parenti, Christian. *Tropic of Chaos*. New York: Hachette Book Group, 2012.

Parker, Geoffrey. *Global Crisis: War, Climate Change and Catastrophe in the Seventeenth Century*. New Haven: Yale University Press, 2013.

Purdy, Jedediah. *After Nature: A Politics for the Anthropocene*. Cambridge: Harvard University Press, 2015.

Richards, Imogen, Callum Jones, and Gearóid Brinn. "Eco-Fascism Online: Conceptualizing Far-Right Actors' Response to Climate Change on Stormfront." *Studies in Conflict and Terrorism*, July 25, 2022. https://doi.org/10.1080/1057610X.2022.2156036.

Riofrancos, Thea. "The Security–Sustainability Nexus: Lithium Onshoring in the Global North." *Global Environmental Politics* 23, no. 1 (2023): 20–41. https://doi.org/10.1162/glep_a_00668.

Sageman, Marc. *Turning to Political Violence: The Emergence of Terrorism*. Philadelphia: University of Pennsylvania Press, 2017.

Saito, Kohei. *Karl Marx's Ecosocialism: Capital, Nature, and the Unfinished Critique of Political Economy*. New York: NYU Press, 2017.

Táíwò, Olúfẹ́mi O. *Elite Capture: How the Powerful Took Over Identity Politics (And Everything Else)*. Chicago: Haymarket Books, 2022.

———. *Reconsidering Reparations*. New York: Oxford University Press, 2022.

Taylor, Blair. "Alt-Right Ecology: Ecofascism and Far-Right Environmentalism in the United States." In *The Far Right and the Environment*, edited by Bernard Forchtner. New York: Routledge, 2019.

Taylor, Bron. "Religion, Violence and Radical Environmentalism: From Earth First! to the Unabomber to the Earth Liberation Front." *Terrorism and Political Violence* 10, no. 4 (1998): 1–42. https://doi.org/10.1080/09546559808427480.

Zacher, Sam. "Polarization of the Rich: The New Democratic Allegiance of Affluent Americans and the Politics of Redistribution." *Perspectives on Politics* 22, no. 2 (2024): 338–356. https://doi.org/10.1017/S1537592722003310.

Environmental Philosophy and Philosophy of Climate Change

Alberts, Paul. "Responsibility Towards Life in the Early Anthropocene." *Angelaki: Journal of Theoretical Humanities* 16, no. 4 (2011): 5–17. https://doi.org/10.1080/0969725X.2011.641341.

Bookchin, Murray. "Social Ecology versus Deep Ecology: A Challenge for the Ecology Movement." *Green Perspectives: Newsletter of the Green Program Project*, nos. 4–5 (1987). http://dwardmac.pitzer.edu/Anarchist_Archives/bookchin/socecovdeepeco.html.

Boulding, Kenneth. "The Economics of the Coming Spaceship Earth." In *Environmental Quality in a Growing Economy: Essays from the Sixth RFF Forum*, edited by Henry Jarrett, 3–14. Baltimore: Johns Hopkins University Press, 1966.

Bradford, George. "How Deep Is Deep Ecology?" *Fifth Estate* 22, no. 3 (1987): 327. JSTOR, https://jstor.org/stable/community.28036651.

Bridge, Damian J. "The Ethics of Climate Change: A Systematic Literature Review." *Account Finance* 62 (2022): 2651–2665. https://doi-org.proxy.library.nd.edu/10.1111/acfi.12877.

Bridle, Jon, and Alexandra van Rensburg. "Discovering the Limits of Ecological Resilience." *Science* 367, no. 6478 (2020): 626. https://doi.org/10.1126/science.aba6432.

Broome, John. *Counting the Cost of Global Warming: A Report to the Economic and Social Research Council on Research*. Cambridge: White Horse Press, 1992.

Budolfson, Mark, Tristam McPherson, and David Plunkett, eds. *Philosophy and Climate Change*. Oxford: Oxford University Press, 2021.

Burdon, Peter D. "Obligations in the Anthropocene." *Law and Critique* 31, no. 3 (2020): 309–328. https://doi.org/10.1007/s10978-020-09273-9.

Callicott, J. Baird. "Animal Liberation: A Triangular Affair." *Environmental Ethics* 2, no. 4 (1980): 311–338.

———. *Thinking like a Planet : The Land Ethic and the Earth Ethic*. New York: Oxford University Press, 2013.

Chakrabarty, Dipesh. *The Climate of History in a Planetary Age*. Chicago: University of Chicago Press, 2021.

———. "Postcolonial Studies and the Challenge of Climate Change." *New Literary History* 43, no. 1 (2012): 1–18.

Chase, Steve, ed. *Defending the Earth: A Dialogue Between Murray Bookchin and Dave Foreman*. Boston: South End Press, 1991.

Clark, Timothy. *Ecocriticism on the Edge: The Anthropocene as a Threshold Concept*. London: Bloomsbury, 2015.

Colebrook, Claire. "We Have Always Been Post-Anthropocene: The Anthropocene Counterfactual." In *Anthropocene Feminism*, edited by Richard Grusin, 1–20. Minneapolis: University of Minnesota Press, 2017.

———. "What Is the Anthropo-political?" In *Twilight of the Anthropocene Idols*, edited by J. Hillis Miller. London: Open Humanities Press, 2016. https://doi.org/10.26530/OAPEN_588463.

Crist, Meehan. "Is It OK to Have a Child?" *London Review of Books* 42, no. 5 (2020): 9–14.

Danowski, Déborah, and Eduardo Viveiros de Castro. *The Ends of the World*. New York: Polity, 2016.

Davis, Heather, and Zoe Todd. "On the Importance of a Date, or Decolonizing the Anthropocene." *ACME: An International Journal for Critical Geographies* 16, no. 4 (2017): 761–780. https://acme-journal.org/index.php/acme/article/view/1539. Top of Form

Desroches, Pierre, and Joanna Szurmak. *Population Bombed! Exploding the Link Between Overpopulation and Climate Change*. London: GWPF Books, 2018.

Di Paola, Marcello, and Sven Nyholm. "Climate Change and Anti-Meaning." *Ethical Theory and Moral Practice* 26, no. 5 (2023): 709–724.

"The First National People of Color Environmental Leadership Summit: Principles of Environmental Justice." *Race, Poverty & the Environment* 2, no. 3/4 (1991): 32–31.

Foster, John Bellamy. *Marx's Ecology: Materialism and Nature*. New York: Monthly Review Press, 2000.

Fuller, R. Buckminster. *Operating Manual for Spaceship Earth*. New York: Simon and Schuster, 1969.

Gardiner, Stephen M. "Ethics and Global Climate Change." *Ethics* 114, no. 3 (2004): 555–600. https://doi.org/10.1086/382247.

———. *A Perfect Moral Storm: The Ethical Tragedy of Climate Change*. New York: Oxford University Press, 2011.

Gardiner, Stephen M., Simon Caney, Dale Jamieson, and Henry Shue, eds. *Climate Ethics: Essential Readings*. New York: Oxford University Press, 2010.

Gardiner, Stephen M., and David A. Weisbach. *Debating Climate Ethics*. New York: Oxford University Press, 2016.

Garrett, Tim J. "No Way Out? The Double-Bind in Seeking Global Prosperity Alongside Mitigated Climate Change." *Earth System Dynamics* 3, no. 1 (2012): 1–17.

Garvey, James. "Climate Change and Causal Inefficacy: Why Go Green When It Makes No Difference?" *Royal Institute of Philosophy Supplements* 69 (2011): 157–174.

Ghosh, Amitav. *The Great Derangement: Climate Change and the Unthinkable*. Chicago: University of Chicago Press, 2016.
———. *The Nutmeg's Curse: Parables for a Planet in Crisis*. Chicago: University of Chicago Press, 2021.
Gilio-Whitaker, Dina. *As Long as Grass Grows: The Indigenous Fight for Environmental Justice, from Colonization to Standing Rock*. Boston: Beacon Press, 2019.
Haraway, Donna J. *Staying with the Trouble: Making Kin in the Chthulucene*. Durham: Duke University Press, 2016.
Hardin, Garrett. "Living on a Lifeboat." *Bioscience* 24, no. 10 (1974): 561–568. https://doi.org/10.2307/1296629.
———. "The Tragedy of the Commons." *Science* 162, no. 3859 (1968): 1243–1248. http://www.jstor.org/stable/1724745.
Jamieson, Dale, ed. *A Companion to Environmental Philosophy*. Malden, MA: Blackwell, 2001.
Jamieson, Dale. *Reason in a Dark Time: Why the Struggle Against Climate Change Failed—and What It Means for Our Future*. New York: Oxford University Press, 2014.
Jonas, Hans. *The Imperative of Responsibility: In Search of an Ethics for the Technological Age*. Chicago: University of Chicago Press, 1979.
Katz, Eric. *Nature as Subject: Human Obligation and Natural Community*. Lanham, MD: Rowman & Littlefield, 1997.
Kekes, John. "On the Supposed Obligation to Relieve Famine." *Philosophy* 77, no. 4 (2002): 503–517.
Kimmerer, Robin Wall. *Braiding Sweetgrass: Indigenous Wisdom, Scientific Knowledge and the Teachings of Plants*. Minneapolis: Milkweed Editions, 2013.
Kingsnorth, Paul, and Dougald Hine. "Uncivilization: The Dark Mountain Manifesto." London: Dark Mountain Project Press, 2009. https://dark-mountain.net/about/manifesto/.
Krech, Shepard. *The Ecological Indian: Myth and History*. New York: W.W. Norton, 1999.
Latour, Bruno. *Facing Gaia: Eight Lectures on the New Climatic Regime*. Cambridge: Polity, 2017.
Lear, Jonathan. *Radical Hope: Ethics in the Face of Cultural Devastation*. Cambridge: Harvard University Press, 2008.
Leboeuf, Céline. "Fearing the Future: Is Life Worth Living in the Anthropocene?" *The Journal of Speculative Philosophy* 35, no. 3 (2021): 273–288.
Leopold, Aldo. *A Sand County Almanac*. New York: Ballantine Books, 1966.
Mbembe, Achille. *Necropolitics*. Durham: Duke University Press Books, 2019.
Merchant, Carolyn. *The Death of Nature: Women, Ecology, and the Scientific Revolution*. New York: HarperCollins, 1980.
Mitchell, Timothy. *Carbon Democracy*. New York: Verso Books, 2013.
Moen, Ole Martin. "The Unabomber's Ethics." *Bioethics* 33, no. 2 (2019): 223–229.

Nordgren, Anders. "Pessimism and Optimism in the Debate on Climate Change: A Critical Analysis." *Journal of Agricultural and Environmental Ethics* 34, no. 22 (2021). https://doi.org/10.1007/s10806-021-09865-0.
Rodman, John. "The Liberation of Nature?" *Inquiry* 20, no. 1–4 (1977): 83–131.
Rolston, Holmes, III. "Respect for Life: Counting what Singer Finds of No Account." In *Singer and His Critics*. Dale Jamieson, editor. Malden, MA: Blackwell Publishers, 1999.
Scavenius, Theresa. "Climate Change and Moral Excuse: The Difficulty of Assigning Responsibility to Individuals." *Journal of Agricultural and Environmental Ethics* 31 (2018): 1–15.
Sessions, George. *Deep Ecology for the Twenty-First Century*. New York: Random House, 1994.
Sessions, George, and Bill Devall. *Deep Ecology*. Salt Lake City: Peregrine Smith Books, 1985.
Shue, Henry. "Subsistence Emissions and Luxury Emissions." In *The Climate Change Crisis: An Introductory Guide*, 200–214. New York: Oxford University Press, 2010. https://doi.org/10.1093/oso/9780195399622.003.0021.
Singer, Peter. "Famine, Affluence, and Morality." *Philosophy & Public Affairs* 1, no. 3 (1972): 229–243. https://www.jstor.org/stable/2265052.
Smithers, Gregory D. "Beyond the 'Ecological Indian': Environmental Politics and Traditional Ecological Knowledge in Modern North America." *Environmental History* 20, no. 1 (2015): 83–111. https://doi.org/10.1093/envhis/emu125.
Szerszynski, Bronislaw. "Gods of the Anthropocene: Geo-Spiritual Formations in the Earth's New Epoch." *Theory, Culture & Society* 34, no. 2–3 (2017): 253–275.
Taylor, Dorceta E. *Toxic Communities: Environmental Racism, Industrial Pollution, and Residential Mobility*. New York: New York University Press, 2014.
Uexküll, Jakob von. *A Foray into the Worlds of Animals and Humans, with a Theory of Meaning*. Translated by Joseph D. O'Neil. Minneapolis: University of Minnesota Press, 2010.
Watson, Richard A. "A Critique of Anti-Anthropocentric Biocentrism." *Environmental Ethics* 5, no. 3 (1983): 245–256. https://doi.org/10.5840/enviroethics19835325.
Whyte, Kyle. "Indigenous Climate Change Studies: Indigenizing Futures, Decolonizing the Anthropocene." *English Language Notes* 55, no. 1–2 (2017): 153–162. https://doi.org/10.1215/00138282-55.1-2.153.
———. "Too Late for Indigenous Climate Justice: Ecological and Relational Tipping Points." *Wiley Interdisciplinary Reviews. Climate Change* 11, no. 1 (2020): e603. https://doi.org/10.1002/wcc.603.
Wilson, Edward O. *The Social Conquest of Earth*. New York: Liveright, 2012.
———. *Sociobiology*. Cambridge: Belknap Press of Harvard University Press, 1980.
Woodhouse, Keith Makato. *The Ecocentrists: A History of Radical Environmentalism*. New York: Columbia University Press, 2018.
Zimmerman, Michael E. *Contesting Earth's Future: Radical Ecology and Postmodernity*. Berkeley: University of California Press, 1994.

Zimmerman, Michael E., J. Baird Callicott, Karen Warren, Irene Klaver, and John Clark. *Environmental Philosophy: From Animal Rights to Radical Ecology*. 4th ed. Upper Saddle River, NJ : Pearson/Prentice Hall, 2004.

Optimism and Pessimism (General)

Alloy, Lauren B., and Lyn Y. Abramson. "Judgment of Contingency in Depressed and Nondepressed Students: Sadder but Wiser?" *Journal of Experimental Psychology: General* 108, no. 4 (1979): 441–485. https://doi.org/10.1037/0096-3445.108.4.441.

Andersson, Gerhard. "The Benefits of Optimism: A Meta-Analytic Review of the Life Orientation Test." *Personality and Individual Differences* 21 (1996): 719–725.

Bates, Timothy C. "The Glass Is Half Full and Half Empty: A Population-Representative Twin Study Testing If Optimism and Pessimism Are Distinct Systems." *The Journal of Positive Psychology* 10, no. 6 (2015): 533–542. https://doi.org/10.1080/17439760.2015.1015155.

Baumeister, Roy F. "The Optimal Margin of Illusion." *Journal of Social and Clinical Psychology* 8, no. 2 (1989). https://doi.org/10.1521/jscp.1989.8.2.176.

Baumeister, Roy F, Ellen Bratslavsky, Catrin Finkenauer, and Kathleen D. Vohs. "Bad Is Stronger Than Good." *Review of General Psychology* 5, no. 4 (2001): 323–370. https://doi.org/10.1037//1089-2680.5.4.323.

Baumeister, Roy F., Laura Smart, and Joseph M. Boden. "Relation of Threatened Egotism to Violence and Aggression: The Dark Side of High Self-Esteem." *Psychological Review* 103, no. 1 (1996): 5. https://doi.org/10.1037/0033-295x.103.1.5.

Baumeister, Roy, and John Tierney. *The Power of Bad: How the Negativity Effect Rules Us and How We Can Rule It*. London: Penguin Classics, 2019.

Bortolotti, Lisa, and Magdalena Antrobus. "Costs and Benefits of Realism and Optimism." *Current Opinion in Psychiatry* 28, no. 2 (2015): 194. https://doi.org/10.1097/YCO.0000000000000143.

Cerulo, Karen A. *Never Saw It Coming: Cultural Challenges to Envisioning the Worst*. Chicago: University of Chicago Press, 2006.

Chang, Edward C. "Distinguishing Between Optimism and Pessimism: A Second Look at the Optimism-Neuroticism Hypothesis." In *Viewing Psychology as a Whole: The Integrative Science of William N. Dember*, edited by Robert Hoffmann, Michael Sherrick, and Joel S. Warm, 415–432. American Psychological Association, 1998.

Chang, Edward C., ed. *Optimism and Pessimism: Implications for Theory, Research, and Practice*. Washington, D.C.: American Psychological Association, 2001.

Chang, Edward C., Albert Maydeu-Olivares, and Thomas J. D'Zurilla. "Optimism and Pessimism as Partially Independent Constructs: Relationship to Positive and Negative Affectivity and Psychological Well-Being." *Personality and Individual Differences* 23, no. 3 (1997): 433–440. https://doi.org/10.1016/S0191-8869(97)80009-8.

Colvin, C. Randall, Jack Block, and David C. Funder. "Overly Positive Self-

Evaluations and Personality: Negative Implications for Mental Health." *Journal of Personality and Social Psychology* 68, no. 6 (1995).

Corns, Jennifer. "Rethinking the Negativity Bias." *Review of Philosophy and Psychology* 9, no. 3 (2018): 607–625.

Craig, Heather, Rosanne Freak-Poli, Aung Zaw Zaw Phyo, Joanne Ryan, and Danijela Gasevic. "The Association of Optimism and Pessimism and All-Cause Mortality: A Systematic Review." *Personality and Individual Differences* 177 (2021): 110788. https://doi.org/10.1016/j.paid.2021.110788.

Dodds, Peter Sheridan, Eric M. Clark, Suma Desu, Morgan R. Frank, Andrew J. Reagan, Jake Ryland Williams, Lewis Mitchell, et al. "Human Language Reveals a Universal Positivity Bias." *Proceedings of the National Academy of Sciences* 112, no. 8 (2015): 2389–2394. https://doi.org/10.1073/pnas.1411678112.

Dunning, David, Chip Heath, and Jerry M. Suls. "Flawed Self-Assessment: Implications for Health, Education, and the Workplace." *Psychological Science in the Public Interest* 5, no. 3 (2004): 69–106. https://doi.org/10.1111/j.1529-1006.2004.00018.x.

Ehrenreich, Barbara. *Bright-Sided: How the Relentless Promotion of Positive Thinking Has Undermined America.* New York: Metropolitan Books, 2009.

Fischer, Ronald, and Anna Chalmers. "Is Optimism Universal? A Meta-Analytical Investigation of Optimism Levels Across 22 Nations." *Personality and Individual Differences* 45, no. 5 (2008): 378–382.

Gershman, Samuel J. "How to Never Be Wrong." *Psychonomic Bulletin & Review* 26, no. 1 (2019): 13–28. https://doi.org/10.3758/s13423-018-1488-8.

Hilbig, Benjamin E. "Sad, Thus True: Negativity Bias in Judgments of Truth." *Journal of Experimental Social Psychology* 45, no. 4 (2009): 983–986. https://doi.org/10.1016/j.jesp.2009.04.012.

Johnson, Dominic D. P., and James H. Fowler. "The Evolution of Overconfidence." *Nature* 477, no. 7364 (2011): 317–320.

Kam, Chester, and John P. Meyer. "Do Optimism and Pessimism Have Different Relationships with Personality Dimensions? A Re-examination." *Personality and Individual Differences* 52, no. 2 (2012): 123–127. https://doi.org/10.1016/j.paid.2011.09.011.

Kinney, Dennis K., and Midori Tanaka. "An Evolutionary Hypothesis of Depression and Its Symptoms, Adaptive Value, and Risk Factors." *The Journal of Nervous and Mental Disease* 197, no. 8 (2009): 561–567. https://doi.org/10.1097/NMD.0b013e3181b05fa8.

Korn, Christoph W., Tali Sharot, Hendrik Walter, Hauke R. Heekeren, and Raymond J. Dolan. "Depression Is Related to an Absence of Optimistically Biased Belief Updating about Future Life Events." *Psychological Medicine* 44, no. 3 (2014): 579–592. https://doi.org/10.1017/S0033291713001074.

Lewicka, Maria, Janusz Czapinski, and Guido Peeters. "Positive-Negative Asymmetry or When the Heart Needs a Reason." *European Journal of Social Psychology* 22, no. 5 (1992): 425–434.

Marshall, Grant N., Camille Wortman, Jeffrey W. Kusulas, and Linda K. Hervig. "Distinguishing Optimism from Pessimism: Relations to Fundamental Dimensions of Mood and Personality." *Journal of Personality and Social Psychology* 62, no. 6 (1992): 1067. https://doi.org/10.1037/0022-3514.62.6.1067.

McAllister, Hunter A., Jeffrey D. Baker, and Catherine Mannes. "The Optimal Margin of Illusion Hypothesis: Evidence from the Self-Serving Bias and Personality Disorders." *Journal of Social and Clinical Psychology* 21, no. 4 (2002): 414–426. https://doi.org/10.1521/jscp.21.4.414.22593.

McNamara, John M, Pete C. Trimmer, Anders Eriksson, James A. R. Marshall, and Alasdair I. Houston. "Environmental Variability Can Select for Optimism or Pessimism." *Ecology Letters* 14, no. 1 (2011): 58–62. https://doi.org/10.1111/j.1461-0248.2010.01556.x.

Moore, Michael Thomas, and David Fresco. "Depressive Realism: A Meta-Analytic Review." *Clinical Psychology Review* 32, no. 1 (2012): 496–509. https://doi.org/10.1016/j.cpr.2012.05.004.

Norem, Julie K. *The Positive Power of Negative Thinking: Using Defensive Pessimism to Harness Anxiety and Perform at Your Peak*. New York: Basic Books, 2001.

Norem, Julie K., and Nancy Cantor. "Anticipatory and Post Hoc Cushioning Strategies: Optimism and Defensive Pessimism in 'Risky' Situations." *Cognitive Therapy and Research* 10, no. 3 (1986): 347–362. https://doi.org/10.1007/BF01173471.

———. "Defensive Pessimism." *Journal of Personality and Social Psychology* 51, no. 6 (1986): 1208–1217. https://doi.org/10.1037/0022-3514.51.6.1208.

Norem, Julie K., and Edward C. Chang. "The Positive Psychology of Negative Thinking." *Journal of Clinical Psychology* 58, no. 9 (2002): 993–1001. https://doi.org/10.1002/jclp.10094.

Peeters, Guido. "The Positive-Negative Asymmetry: On Cognitive Consistency and Positivity Bias." *European Journal of Social Psychology* 1, no. 4 (1971): 455–474. https://doi.org/10.1002/ejsp.2420010405.

Peeters, Guido, and Janusz Czapinski. "Positive-Negative Asymmetry in Evaluations: The Distinction Between Affective and Informational Negativity Effects." *European Review of Social Psychology* 1, no. 1 (1990): 33–60. https://doi.org/10.1080/14792779108401856.

Robb, Kathryn A., Alice E. Simon, and Jane Wardle. "Socioeconomic Disparities in Optimism and Pessimism." *International Journal of Behavioral Medicine* 16, no. 4 (2009): 331–338. https://doi.org/10.1007/s12529-008-9018-0.

Sharot, Tali. *The Influential Mind*. New York: Henry Holt, 2017.

———. "The Optimism Bias." *Current Biology* 21, no. 23 (2011): R941–R945. https://doi.org/10.1016/j.cub.2011.10.030.

———. *The Optimism Bias: A Tour of the Irrationally Positive Brain*. New York: Pantheon Books, 2011.

Sharot, Tali, Christoph W. Korn, and Raymond J. Dolan. "How Unrealistic Optimism Is Maintained in the Face of Reality." *Nature Neuroscience* 14, no. 11 (2011): 1475–1479. https://doi.org/10.1038/nn.2949.

Sharot, Tali, and Cass R. Sunstein. "How People Decide What They Want to Know." *Nature Human Behaviour* 4, no. 1 (2020): 14–19. https://doi.org/10.1038/s41562-019-0793-1.

Spencer, Stacie M., and Julie K. Norem. "Reflection and Distraction: Defensive Pessimism, Strategic Optimism, and Performance." *Personality & Social Psychology Bulletin* 22, no. 4 (1996): 354–365. https://doi.org/10.1177/0146167296224003.

Strunk, Daniel R., Howard Lopez, and Robert J. DeRubeis. "Depressive Symptoms Are Associated with Unrealistic Negative Predictions of Future Life Events." *Behavior Research and Therapy* 44 (2006): 861–882. https://doi.org/10.1016/j.brat.2005.07.001.

Taylor, Shelley E. *Positive Illusions: Creative Self-Deception and the Healthy Mind.* New York: Basic Books, 1989.

Taylor, Shelley E., and Jonathon D. Brown. "Illusion and Well-Being: A Social Psychological Perspective on Mental Health." *Psychological Bulletin* 103, no. 2 (1988): 193–210. https://doi.org/10.1037//0033-2909.103.2.193.

———. "Positive Illusions and Well-Being Revisited." *Psychological Bulletin* 116, no. 1 (1994): 21–27. https://doi.org/10.1037/0033-2909.116.1.21.

Tiger, Lionel. *Optimism: The Biology of Hope.* New York: Simon & Schuster, 1979.

Zell, Ethan, Jason E. Strickhouser, Constantine Sedikides, and Mark D. Alicke. "The Better-Than-Average Effect in Comparative Self-Evaluation: A Comprehensive Review and Meta-Analysis." *Psychological Bulletin* 146, no. 2 (2020). https://psycnet.apa.org/doi/10.1037/bul0000218.

Philosophical Pessimism and Afropessimism

Bailey, Joe. *Pessimism.* New York: Routledge, 1988.

Bayle, Pierre. *Historical and Critical Dictionary: Selections.* Translated by Richard Popkin. Indianapolis: Hackett, 1991.

Beiser, Frederick C. *Weltschmerz: Pessimism in German Philosophy, 1860–1900.* Oxford: Oxford University Press, 2016.

Benatar, David. *Better Never to Have Been: The Harm of Coming into Existence.* Oxford: Oxford University Press, 2006.

Caro, Hernán D. *The Best of All Possible Worlds?: Leibniz's Philosophical Optimism and Its Critics, 1710–1755.* Leiden: Brill, 2020.

Demars, Aurélien. "Le Pessimisme Jubilatoire de Cioran: Enquête sur un Paradigme Métaphysique Négatif." PhD diss., Université Jean Moulin Lyon 3, October 15, 2007. https://scd-resnum.univ-lyon3.fr/out/theses/2007_out_demars_a.pdf.

Dienstag, Joshua Foa. *Pessimism: Philosophy, Ethic, Spirit.* Princeton: Princeton University Press, 2006.

Fréron, Elie. Review of *Candide*. *Année Littéraire*, vol. 2, 204–205. Amsterdam: Michel Lambert, 1759. Reprinted in *Année Littéraire*, vol. 6. Genève: Slatkine Reprints, 1966.

Harris, George W. "Pessimism." *Ethical Theory and Moral Practice* 5, no. 3 (2002): 271–286. https://doi.org/10.1023/A:1019621509461.

Hartman, Saidiya V. *Scenes of Subjection: Terror, Slavery, and Self-Making in Nineteenth-Century America*. New York: Oxford University Press, 1997.

Kadri, Raihan. *Reimagining Life: Philosophical Pessimism and the Revolution of Surrealism*. Madison: Farleigh Dickinson University Press, 2011.

Kagan, Shelly. "Do I Make a Difference?" *Philosophy & Public Affairs* 39, no. 2 (2011): 105–141.

Leibniz, Gottfried Wilhelm. *Theodicy: Essays on the Goodness of God, the Freedom of Man, and the Origin of Evil*. Translated by E. M. Huggard. New Haven: Yale University Press, 1952.

Nadler, Steven. *The Best of all Possible Worlds: A Story of Philosophers, God, and Evil*. New York: Farrar, Straus and Giroux, 2008.

Neiman, Susan. *Evil in Modern Thought: An Alternative History of Philosophy*. Princeton: Princeton University Press, 2015.

Nietzsche, Friedrich. *Beyond Good and Evil*. Translated by R. J. Hollingdale. New York: Penguin Books, 2003.

———. *The Gay Science, with a Prelude in Rhymes and an Appendix of Songs*. Translated by Walter Kaufmann. New York: Random House, 1974.

Norlock, Kathryn J. "Perpetual Struggle." *Hypatia* 34, no. 1 (2019): 6–19. https://doi.org/10.1111/hypa.12452.

Popkin, Richard Henry. *The History of Skepticism: From Savonarola to Bayle*. New York: Oxford University Press, 2003.

Prescott, Paul. "What Pessimism Is." *Journal of Philosophical Research* 37 (2012): 337–356. https://doi.org/10.5840/jpr20123716.

Scheffler, Samuel. *Death and the Afterlife*. New York: Oxford University Press, 2014.

Schopenhauer, Arthur. *Studies in Pessimism*. New York: MacMillan, 1890.

———. *The World as Will and Representation*. Vol. 1. Mineola: Dover, 1969.

Schrader, Paul, dir. *First Reformed*. New York: A24, 2018.

Scruton, Roger. *The Uses of Pessimism and the Danger of False Hope*. New York: Oxford University Press, 2010.

Sexton, Jared. "Affirmation in the Dark: Racial Slavery and Philosophical Pessimism." *The Comparatist* 43 (2019): 90–111. https://doi.org/10.1353/com.2019.0005.

———. "Afro-Pessimism: The Unclear Word." *Rhizomes* 29 (2016). https://doi.org/10.20415/rhiz/029.e02.

Taleb, Nassim Nicholas, Rupert Read, Raphael Douady, Joseph Norman, and Yaneer Bar-Yam. "The Precautionary Principle (with Application to the Genetic Modification of Organisms)." Extreme Risk Initiative—NYU School of Engineering Working Paper Series. arXiv.org, 2014. https://doi.org/10.48550/arXiv.1410.5787.

Thacker, Eugene. *After Life*. Chicago: University of Chicago, 2010.

———. *Cosmic Pessimism*. Minneapolis: University of Minnesota Press, 2015.

Van der Lugt, Mara. *Dark Matters: Pessimism and the Problem of Suffering.* Princeton: Princeton University Press, 2021.
Voltaire. *Candide, Zadig, and Selected Stories.* Translated by Daniel Frame. New York: Signet Classics, 1961.
Vyverberg, Henry. *Historical Pessimism in the French Enlightenment.* Cambridge: Harvard University Press, 1958.
Warren, Calvin. "Black Nihilism and the Politics of Hope." *CR: The New Centennial Review* 15, no. 1 (2015): 215–248. https://doi.org/10.14321/crnewcentrevi.15.1.0215.
Wilderson, Frank, III. *Afropessimism.* New York: Liveright, 2020.
Zapffe, Peter Wessel. "The Last Messiah." In *Wisdom in the Open Air: The Norwegian Roots of Deep Ecology*, edited by Peter Reed and David Rothenberg. Minneapolis: University of Minnesota Press, 1993. 40–52.

INDEX

Abbey, Edward, 166
abolitionism, 180
Abram, David: and literacy hypothesis, 284n60; *Spell of the Sensuous*, 124–25, 127, 284n56
Abramson, Lyn, and Lauren Alloy: "Depressive Realism," 289n9; "Judgment of Contingency in Depressed and Nondepressed Students," 141–42, 146, 287n1, 288n3
acceleration, age of, 10, 132–33; problem of ethics of globalization, 87–92; problem of the human, 92–98; problem of time, 98–101. *See also* Great Acceleration
accelerationist thinking, 177
Acharya, UN Under-Secretary-General Gyan Chandra, 105
Adams, Henry, 173
Adorno, Theodor, 23, 172, 205
AfD, Germany, 73
African American satire, 183–87, 278n8

Afropessimism, 173, 177–88; and African American satire, 187; apocalyptic call for end of this world, 178, 181, 187, 188; on binary racial ontology as essential to modern civilization, 177–78, 181, 187, 208; context of emergence, 178, 278n8, 303n78; critique of by Black scholars, 181–83; on emergence of racial division through chattel slavery and trans-Atlantic slave trade, 178; on impossibility of assimilation of Black being into Humanity, 178–79, 188
agriculture: discovery of, 43, 205; ecosystem breakdown, 83; large-scale development of, 116, 120, 122, 128, 164, 205; shift to from hunting and foraging, 282n35; subsistence, 167
Ajl, Max, 83
Ali, Mona, "militarized adaptation," 76
Allen, Mike, 60, 61

Alloy, Lauren, and Lyn Abramson: "Depressive Realism," 289n9; "Judgment of Contingency in Depressed and Nondepressed Students," 141–43, 146, 287n1, 288n3
American Institute for Economic Research, 165
anarcho-primitivism, 120, 123
Anasazi. *See* Ancient Puebloans (Anasazi)
"ancestor myths," 39
anchoring, 32
Ancient Puebloans (Anasazi), xi, 225n2
Anderson, Amanda, 174
Andreotti, Vanessa de Oliveira, 119
Anglo-American Black Radical tradition, 278n8
anthropocentrism: and posthumanism, 177; and utilitarianism, 104, 272n12
antihumanism, 165, 177
anti-immigrationists, 73–74
anti-natalism, 177, 200, 203
anti-reflexivity thesis, 253n52
"Apocalypse Soon? Dire Messages Reduce Belief in Global Warming by Contradicting Just-World Beliefs" (Matthew Feinberg and Robb Willer), 64
apocalyptic thought: Afropessimism, and apocalyptic call for end of this world, 178, 181, 187, 188; apocalyptic fascism, 134; apocalyptic fictions, 17–19, 85, 135, 219; and Myth of Progress, 39, 177; and Myth of Renewal, 123, 128; and Western Christian conception of linear time, 17–18
aporia, 127
Appiah, Kwame Anthony, *Cosmopolitanism*, 271n4
Aquinas, Thomas, 153, 200
Arctic peoples, loss of traditional hunting grounds to climate change, 124
Arendt, Hannah, 93, 95, 172

Aristotle: and civic value of fear appeals, 257n78; and culture and nature, 168
Arpan, Laura, 70
Atkin, Emily, 51
Attenborough, David, 166
Augustine, Saint, *De Civitas Dei*, 175
availability heuristic, 63

Babylonian captivity, 117
Bacevich, Andrew, 175, 176
Bacon, Francis, 23; natural philosophy, 152; *Novum Organum*, 159; and philosophical assumption of scientific method, 34
Bailey, Joe, 147, 151
Bain, Paul, "Promoting Pro-Environmental Action in Climate Change Deniers" (et al.), 53–54
Bakhtin, Mikhail, and chronotope, 16
Baldwin, James, 172
Bashford, Alison, 163
Battistoni, Alyssa, 78–79
Bauman, Zygmunt: critique of Western intellectuals, 25, 26; and "ignorance of the knowers," 27, 28; *Legislators and Interpreters*, 23, 238n32
Baumeister, Roy: and effects of negative information about climate change, 60, 61; "optimal margin of illusion," 145; and self-esteem, 149
Bayle, Pierre, 171; argument with Leibniz about reason and suffering, 151–52; assertion that reason was incapable of explaining free will or evil, 151, 153; *Dictionnaire Historique et Critique*, 151, 296n67; fideism, 153, 176; on pain as more than absence of pleasure, 200; "Rosarius" article, 296n66
Beckett, Samuel, 16, 173
Bednarik, Robert, human self-domestication hypothesis, 284n60

Beiser, Frederick, 193, 300n42
Benatar, David, 168, 172; and anti-natalism, 200–202
Bendell, Jem, 41
Bendik-Keymer, Jeremy, 207
Benjamin, Walter: on Fascism, 134; and pessimist counter-tradition, 172; "Theses on the Philosophy of History," 23
Bennett, Jane, 176
Bennett, Oliver, 176
Berkes, Firket, 118
Berlant, Lauren, and "cruel optimism," 146
Berlin, Isaiah, 174
Bernauer, Thomas, 70
biases, cognitive: anchoring, 32; confirmation bias, 32; and empiricism, 32–33; hindsight bias, 32; optimism bias, 15, 32; theory of, 32–33
Biehl, Janet, 74–75
biodiversity loss, 1, 14, 35, 135
biosphere, collapse of, 44, 132
#BlackLivesMatter movement, 96, 178
Bloomberg, Michael, 78
Bookchin, Murray: and deep ecology, 266n158; post-scarcity bioregionalism, 101
border gnosis, 118
Borges, Jorge Luis, "Funes, His Memory," 33–34, 192, 241n73, 242n75
Botero, Giovanni, 166
Boulding, Kenneth, 72, 91, 92
BP oil company, US media campaign, 59
Bradford, George, 104
Bradley, Graham L., 61, 257n78
Brave New Europe, 41
Breivik, Anders Behring, 74
Bresson, Robert, 216
Brexit, 8, 121
Bridle, Jon, 135
"bright-siding," 41, 63, 146, 170, 243n104

Britain: food riots, famine, and labor riots of eighteenth century, 162–63; Poor Laws, 163–64
Broome, John, 92, 272n12
Brown, Brian, 55
Brown, Harrison, *The Challenge of Man's Future*, 166
Brown, Jonathon, 149
Brulle, Robert, 59, 71
Buckley, William F., 175
Buddhism: Buddhist pessimism, and right action and Eightfold Path, 195, 208; and desire as suffering, 199; *The Gateless Gate*, 203; role of meditation, 203
Bugden, Dylan, 69
Bush, George W., 94, 96
Byerly, Hilary, et al., 70

Cain, Susan, and "toxic positivity," 146
Callicott, J. Baird, 104, 207, 272n12
Camus, Albert, existential pessimism, 172, 195
Canadian Arctic, collapse of last fully intact ice shelf, 9
Cantor, Nancy, 148
capitalism: consumer, 199; "disaster," 78; fossil fuel (carbon), 1, 2, 23, 44–45, 75, 118, 122, 144, 164; free-market, 69, 167; global, 9–10, 108; and human suffering, 78–79; industrial, 33, 122, 205; liberal, 108, 109; progressivist techno-utopian, 103, 128; surveillance, 30; sustainable, 165; trust in, 168, 176
carbon dioxide emissions: and Great Acceleration, 9, 14, 43–44, 133, 245n114; historic increase after the UN-IPCC 1990 report, 1; and major extinction events, 2, 232n8; reductions dependent on international cooperation, 70

Carson, Anne, literacy hypothesis, 126, 284n60
Carson, Rachel, *Silent Spring*, 166
Carter, Jimmy, 94
Carver, Charles S., 68
Castel, Louis-Bertrand, coining of word optimism (*optimisme*), 151, 159
Centeno, Miguel, 46
Centre for Climate Change and Social Transformations, University of Bath, 62
Cerulo, Karen, 150
Chacoan culture: complexity of, xvi, xix–xx; growth and rapid collapse, and climate change, xi, xvii, xix–xx, 227n27, 229n37; as trading and religious center of San Juan Basin, 800 CE to 1200 CE, xiii–xiv, xvii
Chaco Canyon, New Mexico: Chacra Mesa, xvi–xvii; Chetro Ketl, xii–xv, xx; Escavada Wash, xvii; Great Houses of, xii–xv, xx; and Navajo (Diné) story of building Great Houses, xi, xiii; population estimates, 226n8; Pueblo Bonito, xv–xvi, xx; *rincones*, xvi
Chakrabarty, Dipesh, 8, 93, 113; *The Climate of History in a Planetary Age*, 97; *Provincializing Europe*, 278n8
Chang, Edward, 146, 148, 149; "Cultural Influences on Optimism and Pessimism," 290n15
Chaplin, Joyce, 163
Charbonnier, Pierre, "war ecology," 76
Charles V, Holy Roman emperor and king of Spain, 26
Chateaubriand, François-René, vicomte de, on beholding ruins, xv
Chau, Sophia, 67
Chenoweth, Erica, 79, 268n184
China, People's Republic of, largest global portion of ongoing greenhouse gas emissions, 94, 99–100
chlorofluorocarbons (CFCs), 59
chronotope, as world-producing cultural and literary narratives, 16–17, 237n11. *See also* world, as ontological concept
Cioran, Emil, 172
class divisions, US, 238n32
climate anxiety, 75
climate apartheid, 78, 106, 109
"climate bright-siding," 41, 63, 243n104
climate change: and abrupt transformation, 1, 44, 135; activist movements, 8, 69; and adaptive and mitigating strategies, 67–68; complexity of ("wicked problem"), 5, 59; denial of, 50, 66, 75, 135, 165; as existential problem, 18, 92, 112, 217; and human suffering, 102; and societal collapse, 1–4, 45–46, 116, 132, 227n29, 232n9; and societal complexity, xxii, 44, 96; and transformation and erosion of local ecological knowledge, 124. *See also* global atmospheric warming; Great Acceleration; impasse, of connected crises threatening survival of modern civilization
climate change communication, problems of, 5, 47, 49, 54–59, 87; attacks on perceived "pessimism," 60; belief in human-caused change based on cultural identification, 58; Climategate, 66–67; and complexity and existential threat of problem, 59; discourse on corrupted by corporate influence, 59; "double ethical bind" of scientists and science communicators, 54–55; failure of positive messaging about climate change to motivate, 55–56, 65, 250n32; government and corporate efforts to censor climate science, 66–67, 262n113; lack

of understanding about long-term social change, 55, 85; low level of scientific literacy, erroneous beliefs, lack of trust in science, and populist anti-elitism, 57, 253n52; and manipulation, greenwashing, scams, and hucksterism, 59; motivated reasoning and confirmation bias among audience, 56, 70, 150; partisan divide on the issue, 57–59, 65, 253nn51–53; and persuasion of independents, 77; "shifting baseline syndrome" and normalization of abnormal weather patterns, 56; strategic reframing of issue to appeal to conservative values, 69–71; tendency to underrepresent extreme possibilities, 60. *See also* fear appeals; negative messaging, and climate change

climate change models, flawed, 2, 233n13

climate change politics: and adaptive and mitigating strategies, 67–68, 78, 107; and economic and political elites, 78–79; global and national failures to respond, 6, 41; and independents, 77; partisan division, 58–59; tendency toward optimism, 235n18

climate change politics, left-wing: difficulty in identifying clear target, 84; leftist ecoterrorism, 79–86; unlikelihood of engaging in political violence, 83–84. *See also* Malm, Andreas, *How to Blow Up a Pipeline*

climate change politics, right-wing: ecofascism, 72–75; and free-market solutions, 69–70; lifeboat ethics, 71–72; reactionary responses to overpopulation and ecological crisis, 71–72; tragedy of the commons, 71

Climate Code Red (David Spratt and Philip Sutton), 41

climate ethics/justice. *See* environmental (climate) justice movement

Climate Feedback, 51, 249n24

Climategate (Climate Research Unit email controversy), 66–67

climate migration policy, 107–8

climate protests, 69, 132, 136; and increased support for activism only among those who already believe in climate change, 69

climate refugees, 8–9, 70, 107–8, 214

climate reparations, 101, 107

Clinton, Bill, 94

cliodynamics, 285n63

coal, record-breaking use of, 1, 231n4

cognitive-behavioral therapy, and reinforcement of self-deception, 142

Cohen, Daniel Aldana, 78, 79, 267n172

Colebrook, Claire, 32, 94

collapse, societal: as adaptation, xix; Chacoan culture, xi, xvii, xix–xx, 227n27, 229n37; and climate change, xvii, xx, 1–2; criticism of concept, 227n29; denial of possibility, and Myth of Progress, 46; ecological overshoot, 3, 45, 2287n29; and energy flow, xviii, 228n31; and rapid simplification, 45–47; as sociopolitical phenomenon, xviii; Tainter's theory of, xvii–xx, xxii, 9, 45–46; Western Roman Empire, xviii

The Collapse of Western Civilization (Naomi Oreskes and Eric Conway), 8

colonialism/imperialism: and American imperial ideology, 75; European conquest and genocide of indigenous peoples of Americas, 117; and European peasants, 121; and genocide, extinction, and eco-system colonialism/imperialism (cont.) breakdown, 83, 97, 101, 118–19; and integration of diverse societies into single civilizational network, 95; Jones, Henry, "Property, Territory, and Colonialism," 282n45; roots of ecological thinking, 74, 265n156; witch trials, seventeenth century, and racial colonial violence, 121, 282n45. See also Afropessimism; decolonization; postcolonial theory; Wynter, Sylvia

Columbian Exchange, 116, 152

complexity, civilizational, 3, 95; and climate change, xxii, 44, 96; and global digital networks, 96; and global energy consumption, 44; increasing technological and social, 41–44, 96; and literacy, 285n63; and Red Queen effect, 45; and reduction in problem-solving abilities, 45; as result of one-time energy surplus from fossilized carbon, 44–45; theory of diminishing returns on, xvii–xx, xxii, 9

Condorcet, Nicolas de, 159; and Myth of Progress, 39, 162; *Outlines of an Historical View of the Progress of the Human Mind*, 39

confirmation bias, 32

conflict, inevitability of, 199, 204–5

Conrad, Joseph, *Heart of Darkness*, 121–22, 283n50

consciousness, 307n44; as source of suffering, 202–4

Consejo Real de las Indias, 26. See also Las Casas-Sepúlveda debate of 1550

conservation biology: and habitat destruction, 102; and needs of impoverished communities dependent on ecosystems for survival, 102–3

conservative nostalgia, 176

conservative pessimism, 173, 175–76

constructive pessimism hypothesis, 62, 309n67

Conway, Daniel, 298n98

Conway, Erik, *The Collapse of Western Civilization* (with Naomi Oreskes), 8, 59

Cook, John, 51

Copeland, Huey, 179

Cortina, Adela, 110

COVID-19 pandemic: failure of government response to, 41; and "finite pool of worry," 260n98; and polarization, 69; psychological responses to, 143; recent skepticism toward science with regard to, 253n52; trauma and complications, 15; and weaknesses of cultural system, xxi–xxii, 9

Crist, Meehan, 310n12; "Is It OK to Have a Child?," 275n27

critical race theory, 119, 120, 177, 286n70

Cronon, William, 104

Crow (Apsáalooke) peoples: coup-sticks, planting of, 130–31, 136; forced onto reservation with treaty of May 7, 1868, 130–31; transition to a new life after end of their world, 129–32

Crusius, Patrick, 72–73, 74; "The Inconvenient Truth," 73, 264n146

cultural pessimism, 173, 176

cultural revolution, 75

cyborgs, 177

Dakota Access Pipeline, 84, 132

Dalton, Greg, 215

Daly, Herman, 160

Danaher, John, 300n40

Danowski, Déborah, 112–13, 115, 132

Dark Mountain Project, 120

Darrow, Clarence, 207

Darwin, Charles, 104, 160, 164, 166
Davis, Heather, 118
The Dawn of Everything (David Graeber and David Wengrow), 282n36
The Day After Tomorrow, 61, 258n83
death, inevitability of, 199, 207
decarbonization, obstacles to, 2, 128
de Castro, Eduardo Viveiros, 112–13, 115, 132
declinism (cultural pessimism), 176
decolonization, 117–18, 119, 120, 123, 127–28; "Keele Manifesto for Decolonizing the Curriculum," 119
Deely, John, "Umwelt," 239n36
deep ecology, 74, 266n158
deforestation, 1, 9
degrowth, xxii, 75, 122
Delany, Samuel, 34
Deleuze, Gilles, 23, 79, 277n50; *Anti-Oedipus: Capitalism and Schizophrenia* (with Félix Guattari), 277n50
Dembicki, Geoff, 59
Deneen, Patrick, 7
denial, of climate change, 50, 66, 75, 135, 165
depressive realism, theory of, 141–43, 146, 288n8, 289n9
"Depressive Realism" (Lauren Alloy and Lyn Abramson), 289n9
Derrida, Jacques, 278n8, 302n57
Descartes, René, 298n98; *Discourse on Method*, 159; *Meditations*, 152
despair, versus pessimism, 195
Desrochers, Pierre, 165–68
Devall, Bill, "A Spanner in the Woods," 266n158
Dickson-Carr, Darryl, 183
Dienstag, Joshua Foa, 166–67, 171; "forward-oriented" conception of pessimism, 171; philosophical pessimism, 172–73, 194; progress, critique of, 211
digital humanities, 177

digital literacy, 127
digital networks, global, 96
"disaster capitalism," 78
"discount problem," 98, 109
Dixon, Graham, 69
Domino, Brian, 298n99
"doomism": criticism of, 49–50, 60, 246n4, 247nn5–6, 248n8, 255n69 (*See also* Mann, Michael); Loeb on, 246n3. *See also* fear; negative messaging
"Dossier on History, Representation, and the Impossible Subject of Race" (Jared Sexton and Huey Copland, eds.), 179
Du Bois, W. E. B., 172, 278n8
Du Châtelet, Émilie, 155
Duhem, Pierre, "Duhem-Quine thesis," 36
Dunlap, Riley, 58, 253n52, 253n53
duration neglect, 202

Earth Day, 75
ecocentrism, 103–4
ecofascism, 72–75, 264n142; in America, 75–76; and American environmental thought, 74, 265n156, 266n158; definitions, 73–74; "green nationalism," 73; and historic European fascism, 74; manifestos of killers, 73, 264nn145–47
Ecofascism Revisited (Janet Biehl and Peter Staudenmeier), 74–75
ecological collapse, 1, 40, 116, 122, 166. *See also* biosphere, collapse of; climate change; ecological overshoot
ecological humanism, 9–10
"ecological Indian," racialized stereotype of, 122, 132, 283n53
ecological Marxism, 75, 101, 228n31
ecological overshoot, 2–3, 45, 92, 132, 136, 227n29

ecological thinking, racist and imperialist roots of, 74, 265n156
Ecomodernism, 101
ecopessimism, 165–69, 173, 208, 299n27; antihumanist strains, 165, 168; association with biological racism, 168; challenge to idea that culture and nature are separate domains, 168; defined, 167; and human inability to live within natural limits, 165
ecoterrorism, leftist, 79–86. *See also* Malm, Andreas, *How to Blow Up a Pipeline*
Edelman, Lee, 176, 302n57
Edwards, Paul, 235n18
"effective altruism," 101
Ehrenreich, Barbara: analysis of the Professional-Managerial Class (PMC), 238n32; *Bright-Sided*, 146, 170, 243n104
Ehrlich, Paul and Anne, *The Population Bomb*, 166
Eisenstadt, Shmuel, 227n29
Eliot, George, 174; *Middlemarch*, 202
Eliot, T. S.: "Burnt Norton," 98, 99, 288n8; *Four Quartets*, 142–43, 218–19
Ellison, Ralph, 173, 174
Emerson, Ralph Waldo: "Compensation," 141; pessimism, 172
empiricism. *See* nonadaptive cognition (scientific empiricism)
enclosure, 121, 282n45
Encyclopédie (Jean le Rond d'Alembert and Denis Diderot), 156, 157
end of the world: acceptance of, 211; and Afropessimism, 178, 181, 187, 188; fantasy version, 111; as fiction and end of a fiction, 115–16; idea of, 5; and loss of conceptual worldview, 116–17, 279n16; and meaning of "end," 114; and meaning of "end of world," 114–15; and ontological world, 114, 278n8; struggle of indigenous peoples after end of their world through imperialism, 117, 129–32
The End of the World (Déborah Danowski and Eduardo Viveiros de Castro), 112–13, 115
endowment effect, 201
Engelke, Peter, 43–44
environmental ethics, 71–72, 78, 106, 109, 137, 201, 275nn230, 308n57; ethics of globalization, 87–92, 93; virtue ethics, 102, 207, 272n12, 308n57. *See also* environmental (climate) justice movement
environmental (climate) justice movement, 70, 102–10, 207; approaches that attempt to resolve ethical issues of climate change, 102–3, 275n29; authoritative history of, 275n30; and claim that "those who have contributed the least will suffer the most," 105–10; and conflation of class, wealth, and race, 278n53; and division between human and nonhuman, 103–5; and "elite capture," 106; and fear of the poor, 109–10; and moral obligation of affluent to protect the poor from suffering, 105–9, 277n45; and narrow focus on carbon emissions, 102; and obligation to the future, 98–101, 109, 278n52
Environmental Liberation Front (ELF), 81
environmental racism, 70
environmental sabotage, and popular turn against environmentalism, 82. *See also* Malm, Andreas, *How to Blow Up a Pipeline*
Epic of Gilgamesh, 116, 166
epistemic disobedience, 118
epistemic diversity, 25

ESG, 168
ethical pessimism, 3, 5, 137, 144, 210; commitment to some future human existence, 219, 310n12; and comprehension of demands of being human, 192–93. *See also* pessimism; philosophical pessimism
ethical systems: as frameworks of values and meaning that guide behavior, 197; individualist, 87–92
ethics of globalization, 87–92
Ettinger, Philip, 213
European and American imperialism. *See* colonialism/imperialism
Evelyn, John, *Sylva*, 275n30
existential pessimism, 208
experience: as epistemic grounding, 21, 24–25, 38, 44, 56–58, 115, 127, 204; and narrative, 13, 20, 32, 217; of suffering, 200, 204; transcendental, xii, 33, 91, 93, 134, 218–19
extinction: carbon dioxide emissions and, 2, 232n8; near-term, 196; *The Sixth Extinction* (Elizabeth Kolbert), 8
Extinction Rebellion, 8

"Fact-Checking the Climate Crisis" (Will Steffen and Wolfgang Knorr), 249n24
Fanon, Frantz, 278n8; and Afropessimism, 181, 182; on decolonization, 119, 127–28; and mission of each generation, 31; "new Humanity," 30, 101; and theory of race, 179
Farnsworth, Patrick, 187
Farrer, Austin, 296n66, 297n74
Far Right Ecologism (FRE), 75
fascism, apocalyptic, 134
Fauci, Anthony, xxi
Faulseit, Ronald, xviii, 227n29, 228n30
Faye, Guillaume, 72

fear appeals, and climate change: combined with sense of efficacy, 62–63, 259n93; correlation to pro-environmental actions or intentions to act, 62, 251n35, 258n84; criticism of, 49–54, 248n8; and "negativity bias," 61; studies of effectiveness of, 60–63, 256n76, 257n78; value of, 47, 65. *See also* negative messaging, and climate change
Federation for American Immigration Reform (FAIR), 74
Federici, Silvia, *Caliban and the Witch*, 282n45
Feinberg, Matthew, 64, 69
Feldman, Sol, 62
Ferrando, Francesca, 177
Fesenfield, Lukas, 70
fictions: apocalyptic, 17–19, 85, 135, 219; end of the world, 115–16; Kermode and, 133–37; shaping of reality through socially constructed and inherited narrative and conceptual structures, 15–22, 24; and simultaneous human inhabitation of multiple layers of narrative reality, 20, 238n25. *See also* chronotope, as world-producing cultural and literary narratives
fideism, 153, 176
Fidesz, Hungary, 73
Fielding, Kelly, et al., 62
Firenze, Paul, 192
First Reformed, Paul Schrader, 213–17, 309N2
Fisher, Mark, 302n57
Foreman, Dave, 82, 266n158
fossil fuel (carbon) capitalism, 1–2, 23, 44–45, 75, 118, 122, 144, 164
Foucault, Michel, 23, 172, 278n8; "imperial boomerang," 121
Fox, Matthew, 215

Francis, Pope, 166
Franklin, Benjamin, 298n5
Franzen, Jonathan, "What If We Stopped Pretending?," 249n24
Fraser Institute, 165
Frederick of Prussia, 155–56, 157
Freedman, Andrew, criticism of Wallace-Wells, 50
free-market capitalism, 69, 167
French Revolution, 159
Fréon, Elie, review of *Candide* in *Année Littéraire*, and first recorded use of word "*pessimisme*," 158, 159
Freud, Sigmund, 172
Fridays for the Future, 8
Frye, Northrop, definition of satire, 183, 186
Fuller, Buckminster, 72
future, as final frontier, 101

Gardiner, Stephen, 92; "climate change blinders," 275n29; "Ethics and Global Climate Change," 5; and "moral corruption," 237n9; "The Role of Claims of Justice in Climate Change Policy," 275n29; three types of climate justice theories, 275n29
Gates, Bill, 78
Gendron, Peyton: ecofascist terrorist, 74; "You Wait for a Signal While Your People Wait for You," 73, 264n147
Gershman, Samuel, 144
Ghosh, Amitav: *The Great Derangement*, 8; *The Nutmeg's Curse*, 106, 279n17
Ghost Dance, 131–32, 287n80
gilets jaunes protest, France, 8
Gilio-Whitaker, Dina, 117
Gilroy, Paul: "The Black Atlantic and the Re-enchantment of Humanism," 274n20; "Fanon and Améry," 286n70

Girard, René, 80
Glissant, Édouard, and errantry, 127
global atmospheric warming: global land and ocean surface temperatures, increase since 1965, 14; inadequacy of mitigation on its own, 67–68; questionable models and optimistic narratives inhibiting action on, 2–3; ramifications of, 41, 124; research on, 2–3, 133; and tipping point into abrupt transformation, 1. *See also* climate change; climate change communication, problems of
global capitalism, 9–10, 108
global digital networks, and new levels of human social complexity, 96
global energy consumption: correlation of growth with markers of modern progress, 43; and creation of complex society, 44; one-time energy surplus from fossilized carbon, 44
globalization, ethics of, 87–92; balancing of self-interest and compassion, 87–92; and individual inefficacy, 92, 272n11; and problem of justice, 102; "Spaceship Earth," 91; and universal rights-bearing subject, 89
global land and ocean surface temperatures, increase since 1965, 14
global poverty, World Bank estimates of, 191
Global South, 88
global village, 88
Global Warming Policy Foundation, 165
Global War on Terror, 76, 178
Glover, Donald, *Atlanta*, 186
Glowacki, Donna, xviii
Godwin, William, 88; *An Enquiry Concerning Political Justice*, 160–62, 271n1, 298n5; and Malthus, 299n17; and Myth of Progress, 39, 92, 159, 165, 170

Gone, Joseph, 118, 124
Goody, Jack, 284n60
Gore, Al, 67, 94
Graeber, David, 119, 120, 282n36
Grant, Madison, 74, 168; *The Passing of the Great Race*, 97
Gray, J. Glenn, 81
Great Acceleration, 43–44, 93, 103, 205; and carbon dioxide concentrations, global atmospheric, 9, 14, 43–44, 133, 245n114
The Great Acceleration (J. R. McNeill and Peter Engelke), 43–44
"great simplification," 46, 246n123
Greenberg, Jonathan, 183
green energy transition, obstacles to, 2
green growth, 75, 76, 78, 133, 168
greenhouse-gas emissions, increase in, 41
"green nationalism," 73
Green New Deal, 101
Grundmann, Reiner, 66
Guattari, Félix, 23, 79; *Anti-Oedipus: Capitalism and Schizophrenia* (with Gilles Deleuze), 277n50

Haberl, Helmut, 255n68
Haeckel, Ernst, 166
Halberstam, Jack, 302n57
Hall, Stuart, 278n8, 302n57
Hamilton, Lawrence, 58
Hanke, Lewis, on Las Casas-Sepúlveda debate, 27
Hansen, Bue Rübner, critique of Malm's proposed political violence, 82–83, 270n205
Hansen, James, 230n1, 262n113; *Storms of My Grandchildren*, 8
Haraway, Donna J.: Chthulucene, 113; "Cyborg Manifesto," 177; and post-humanism, 176, 177
Hardin, Garrett: "Living on a Lifeboat" thought experiment, 71–72, 88, 90–92, 101, 190, 271n6; "The Tragedy of the Commons," 166
Harris, George W., "Pessimism," 301n43
Hart, Lauren, 62
Hartman, Saidiya: *Lose Your Mother*, 181; "The Position of the Unthought" (interview), 179; *Scenes of Subjection*, 179–81; "Venus in Two Acts," 181; *Wayward Lives, Beautiful Experiments*, 181
Hartmann, Eduard, 194–95
Havelock, Eric, literacy hypothesis, 284n60
Hawke, Ethan, 213
Hayhoe, Katharine, 60, 63, 69, 255n69
Hayles, N. Katharine, 176
Hazlitt, William, 160
Hebrew Bible, pessimistic currents, 169
Hegel, Georg Wilhelm Friedrich, 27, 217, 278; the Absolute, 30; *The Phenomenology of Spirit*, 278n8
Heidegger, Martin, 112, 114, 120, 168, 173, 237n11, 278
Heinonen, Kati, 149
Hersey, John, 174
Hewett, Edgar, xiii, 226n8
Hickel, Jason, 255n68
Hill, Victoria, 216
hindsight bias, 32
Hine, Dougald, 119, 120
history, lack of inherent, intelligible meaning, 39, 93, 133, 146, 157, 171–72, 199, 205–6
Holdren, John, choices for responding to climate change, 67
Holmes, Oliver Wendell, 172
Holthaus, Eric, "Stop Scaring People About Climate Change. It Doesn't Work," 51, 53–54
Homer-Dixon, Thomas, 46
Horkheimer, Max, 172

Hornsey, Mathew, 55, 58, 62
Houllebecq, Michel, 168
How Everything Can Collapse (Pablo Servigne and Raphaël Stevens), 8
Hughes, Brian, et al., 74, 265n151
human, the, thinking, 92–98; dialectical movement of social ethics from *I* to *We*, 96; global digital networks and new levels of human social complexity, 96; impermanence of universalizing representations of, 96; nationalist and identitarian politics in opposition to globalist order, 96–97; and problem of justice, 102; without the idea of progress, 97–98 "the Human," historicizing, 177. *See also* Afropessimism; Wynter, Sylvia
humanities: comparatist method, 36; relativistic and historical, 37; truth-for statements stabilizing power relations, 36
human self-domestication hypothesis, 284n60
human societies, as living systems surviving by extracting energy from environment, xviii
humans without world, 132
human universals, 36, 96, 242n88
Hume, David, 172, 176
Hunkpapa Lakota, 132
hunting and foraging: recovering of elements of culture, 120; shift to agriculture, 282n35
Hurricane Katrina, 178
Hurricane Sandy, 78

illness, as subject to forces beyond human control, 200
immigration, 72–75, 88, 107, 168
impasse, of connected crises threatening survival of modern civilization, 14–15, 218; and ethical pessimism, 3–4, 5–6, 144, 210, 211; and ethics of globalization, 87–92; and exacerbation of political conflict, 68–69, 262n123; and growth of complexity of human civilization in nineteenth and twentieth centuries, 3, 95; historical analogue, 116–17; and impossibility to allocate responsibility, 93–95; and inability of modern progressive society to respond to, 2, 85–86, 136, 144, 233n12; occurrence of greatest anthropogenic impacts on the Earth in span of one human life, 93; perceptions of, and decision to have children, 101, 275n27; and possibility of successful transition to a meaningful existence in a new world, 128–33; as prelude to new form of life, 112; reduction of, to pre-existing political narrativization, 2–3, 31–32, 46–47, 87; threat to modern cultural values of freedom, individualism, consumerism, and progress, 71
indigenous epistemologies: climate change, and transformation and erosion of local ecological knowledge, 124; local nature of embedded in specific ecological and geophysical relationships, 124–25, 128; "returning" to, 117–28, 132; shaped in primarily oral cultures, 125–26, 127, 128; transformed and obliterated by forced or adopted literacy, 125–27. *See also* decolonization; myth of renewal
industrial capitalism, 33, 122, 205
Inquisition, 117
intellectuals, and ideology, 22–32; complicity in dominating structures of society, 22–23, 24, 25; and rationalization of power, 25; role as cultural

critics and commitment to idea of progress, 25–26
"intergenerational justice," 99
internal combustion engine, 170

Jackson, Mark, 118
James, C. L. R., 181, 278n8
James, Henry, 21
James, William, 172, 198, 238n25, 278n8
Jamieson, Dale: *Reason in a Dark Time*, 5, 8, 278n52; "When Utilitarians Should Be Virtue Theorists," 92, 207, 272n12
Javeline, Debra, 67
Jeffers, Robinson, 120; "Carmel Point," 103–4
Jerome, Saint, 166
Job, Book of, 152
Jonas, Hans: heuristics of fear, 208–9; *The Imperative of Responsibility*, 204, 277n45
Jones, Henry, "Property, Territory, and Colonialism," 282n45
Jones, LeRoi, 184
"Judgment of Contingency in Depressed and Nondepressed Students" (Lauren Alloy and Lyn Abramson), 141–43, 146, 287n1, 288n3
"judgment of history," 38, 40
Jünger, Ernst, 172, 200
just war theory, 26, 28

Kaczynski, Theodore, 74, 82
Kadri, Raihan, 173
Kahan, Daniel, 57, 58
Kahneman, Daniel, 32
Kaida, Naoko and Kosuke, constructive pessimism hypothesis, 62, 309n67
Kallis, Giorgios, 100; critique of Malthus, 299n24
Kant, Immanuel, 42, 44; *Critique of Practical Reason*, xii, 226n3; and Myth of Progress, 39
Kármán line, 114
Katz, Eric, 272n12
Keats, John, and "negative capability," 218
"Keele Manifesto for Decolonizing the Curriculum," 119
Keeling, Kara, 179
Keltner, Dacher, xii
Kennedy, John F., definition of liberalism, 173–74
Kermode, Frank, 115, 116, 204; and the ambivalent critic, 28; effort to distinguish between myth and fiction, 19–21; and fictional structures through which humans make sense of world and ends, 13, 15–22, 45, 133–34, 135; and gap between reality and socially constructed human narratives, 18; on human attempts to imagine and predict apocalypse, 17–22, 114; *The Long Perspectives*, 13, 17; and philosophical assumptions behind the Shoah, 19, 20; progress as type of apocalypse, 39–40; *The Sense of an Ending*, 13–22, 219, 236n3; simultaneously contrary positions involving responsibilities of the critic, 22; and "the end," 112, 114; and time, 99
Keynes, John Maynard, 160, 164
Khanna, Parag, ecomodernist argument, 72
Kierkegaard, Søren, 131
Kimmerer, Robin Wall: *Braiding Sweetgrass*, 118; and traditional ecological knowledge, 120
King, Stephen, *The Dark Tower*, 111
Kingsnorth, Paul, 119, 120; *The Wake*, 121, 283n47
Klages, Ludwig, 168, 265n156
Klein, Naomi: "disaster capitalism," 78; *This Changes Everything*, 8

Knorr, Wolfgang, 249n24
Kofman, Sarah, 172
Kolbert, Elizabeth: *The Sixth Extinction*, 8; *Under a White Sky*, 292n30
Koteyko, Nelya, 55
Kurzweil, Ray, 39; transhumanism, 177

Lacan, Jacques, 200, 278n8
Lachapelle, Erick, 64
Lane, H. Clifford, xxi
language, through which our model of reality acquires meaning, 23–24, 239n36. *See also* experience; fictions; literacy; narratives, socially constructed and inherited; world, as ontological concept
Lantigua, David, and Las Casas-Sepúlveda debate of 1550, 27, 28
Las Casas, Bartolomeo de, debate with Juan Ginés de Sepúlveda: *Apologética historia summaria de las gentes destas Indias*, 28, 240n53; argument for transporting African slaves to the Americas, 240n55; argument that natives of New World had God-given right to self-determination, 27, 28–29; founder for modern ideas of human rights and anthropology, 28–29; *History of the Indies*, 28; *Short Account of the Destruction of the Indies*, 28
Las Casas-Sepúlveda debate of 1550: case study in social change, 29; concern with rights framed by contemporaneous norms, 27; and emergence of secular intellectual as servant of state power, 27–28; and "ignorance of the knowers," 27; on rights and power relations of European civilization, 29, 240n56; on rights and treatment of indigenous peoples in New World, 26–29. *See also* Las Casas, Bartolomeo de, debate with Juan Ginés de Sepúlveda; Sepúlveda, Juan Ginés de, debate with Bartolomeo de Las Casas; Wynter, Sylvia
Lasch, Christopher: and conservative pessimism, 175; *The True and Only Heaven*, 39–40
Latour, Bruno, 114; and modernity, 123, 124
Lear, Jonathan, *Radical Hope*, 129–33, 219
learned helplessness, theory of, 141, 288n2
LeBlanc, Steven, 122
Lee, Spike, *Bamboozled*, 180
Le Guin, Ursula, "The Ones Who Walk Away from Omelas," 189–92
Leibniz, Gottfried Wilhelm, 169; argument with Bayle about reason and suffering, 151–52, 296n66; conception of all reality as part of God's perfection, 153–54, 158, 159, 170, 193, 206; *Essays of Theodicy on the Goodness of God, the Freedom of Man and the Origin of Evil*, 151; *Histoire des Ouvrages des Savants*, 296n66; invention of word *theodicy* ("justice of God"), 151; and reforming of scholasticism, 297n74; *On the Ultimate Origination of Things*, 159
Leiserowitz, Anthony, 248n8
Lekson, Stephen: Chacoan culture, xiii; Chaco Meridian hypothesis, 229n37
Leninism, revolutionary, 79
Leopold, Aldo, 166, 197; Land Ethic, 104
Levin, Kelly, xxii
Lévi-Strauss, Claude: *La Pensée Sauvage*, 284n60; structural anthropology, 284n60; *Tristes Tropiques*, 122; wild and domesticated thought, 126, 284n60
liberal capitalism, 108, 109
liberalism, 173–74

liberal pessimism, 173–75
liberal progressivism, 175
Liere, Kent, 58
lifeboat ethics, 71–72, 78, 88, 90–92, 101, 109
Ligotti, Thomas, 168, 172; *A Conspiracy Against the Human Race*, 203
limits: of communication, 48–59, 70; of human reason, xii, 10, 24–25, 36, 46, 86, 95, 104, 135–137, 194, 198–99, 202–5; of nature, 30, 44–46, 98, 101, 107, 135, 163, 165–69, 197; of progress, 5, 6, 39, 43, 86; of resources, xxii, 2, 44, 91, 100
Limits to Growth, Club of Rome, 165, 166
Linderman, Frank, *Plenty-Coups*, 286n74
Linkola, Pentti, 72, 264n142
Lisbon earthquake: and philosophical pessimism, 157; scholarly discussion of effect on European intellectual culture, 156–57
literacy: climate, 52; development of, 35, 38, 116, 119, 121, 122, 125–26, 205; digital, 127; forced, and obliteration of indigenous ways of knowing, 125–27; literacy hypothesis, 284n60; scientific, lack of in US, 56–57; and thresholds of social and political complexity, 285n63; and transformation of human brain, 126
literary *marronage*, 127
Lomborg, Bjorn, 41, 42, 167
"longtermism," 101, 275n28
loss aversion, 201
Louis XV, 156
Lovecraft, H. P., 168
Lowe, Thomas, 61
Löwith, Karl, 39, 158, 171–72, 206
Lubarda, Balsa, Far Right Ecologism (FRE), 75
luxury emissions, 79, 107, 269n187

Ma, Yanni et al., 55
MacAskill, William, *What We Owe the Future*, 275n28
Machiavelli, Niccolò, 166
Maibach, Edward, 61
Maier, Steven F., theory of learned helplessness, 141, 288n2
Mair, John, 28
Makah, Pacific Northwest, 103
Malebranche, Nicolas, 158
Malm, Andreas, *How to Blow Up a Pipeline*, 79–86; carbon-centric and Eurocentric approach, 83; "controlled political violence," 80; diffuse target, 84; and Ende Gelände, 85; failure to recognize that violence begets violence, 81–82; "intelligent sabotage," 81; simplistic ideas of mass movements and political violence, 82–83; voluntaristic understanding of political action, 84
Malthus, Thomas Robert, 159–67, 172, 275n30, 299n17; *An Essay on the Principle of Population*, 160, 162; criticism of English poor laws, 163–64; disagreement with radical progressivism, 162–63; and limits to human progress, 163–65; pessimism, 172
Mann, Geoff, 78–79
Mann, Michael: assumption that pessimistic assessments of climate change lead to inaction, 50, 60, 67; climate bright-siding, 63; criticism of "doomism," 60, 247nn5–6, 255n70; criticism of Wallace-Wells article, 49–50, 54–55, 67, 248n8; *The New Climate War*, 60
Mann, Thomas, *Joseph and His Brothers*, 238n25
Marcus, Sara, *Political Disappointment*, 174
Marriott, David, 179
Marshall, Margaret, 149

Martel-Morin, Marjolaine, 64
Marx, Karl, 95, 160, 164, 228n31, 275n30; *Capital*, 95; *The Eighteenth Brumaire of Louis Bonaparte*, 31; and material determination of culture, 23. *See also* ecological Marxism
Marxian pessimism, 302n57
material determination of culture, 23, 24
Matlin, Margaret W., 287n1
Maupertuis, Pierre Louis, 155
Mayhew, Robert J., 162
Mbembe, Achille, 111–12, 219, 278n8
McAnany, Patricia Ann, 227n29
McCarthy, Jesse, *Who Will Pay Reparations on My Soul?*, 182–83, 187
McCright, Aaron, 253n52
McGrath, Liam, 70
McKibben, Bill, 166; *Eaarth*, 8, 73, 76
McLane, Maureen, 163
McNeill, J. R., 43–44
McQueen, Alison: apocalypse and modern political realism, 18; on climate change fear appeals, 257n78
Mémoires pour l'Histoire des Sciences et des Beaux-Arts, 151
men's rights, 73
Merchant, Carolyn, 35, 36
Merkel, Sonia, 64
Merleau-Ponty, Maurice, 278n8
methane, atmospheric, increase in, 9
Meyer, Robinson, claim that pessimism regarding climate change leads to fear and paralysis, 50–51, 66
Michaels, Walter Benn, 276n40
Middleton, Guy, *Understanding Collapse*, xix
Miéville, China, "apophatic Marxism," 310n12
Mignolo, Walter, 118
Migotti, Mark, "Schopenhauer's Pessimism in Context," 300n41, 306n24
militarism, 75, 76

Mill, John Stuart, 174
Montaigne, Michel de, 169
Moore, Sam, 75
Morris, Brandi, 62
Morton, Timothy, 176
Moser, Susanne, 8, 259n93
Mumford, Lewis, 172
Murphy's Law, 208
Musk, Elon, 39
Myth of Progress, 38–45, 169, 210, 218; and apocalyptic thought, 39, 177; conflation of material well-being with moral development and biological evolution, 42–43; at core of modern liberal arts education, 40; and cumulative human knowledge, 170; and denial of possibility of societal collapse, 46; Enlightenment progressivism roots, 4, 39; optimism in face of contradictory evidence, 40–41, 46–47; paradigmatic civilizational narrative, 3; philosophical pessimism as modern philosophical response to, 4, 5–6; philosophical pessimism as modern response to, 4, 5–6; repudiation of Christian faith and scientific rationality, 39–40; and solvability of all problems in infinity of time, 101, 170; as a structure of belief, 38–39. *See also* progress; progressivism
myth of renewal, 122–28; alienation of modern humans from nonhuman nature by industrial life, 125; and "going back," 123–24, 127; and local nature of indigeneity embedded in specific ecological and geophysical relationships, 124–25; versions of, 122–23

Naess, Arne, 203
narratives, socially constructed and inherited: and climate change, 2–3,

31–32, 46–47, 87; progressivist racialist narrative of moral and ethical evolution from primitive to modern, 3, 97–98; and shaping of reality, 15–22, 24. *See also* chronotope, as world-producing cultural and literary narratives; fictions; world, as ontological concept
nationalist decolonization, 101
nativism, American, 75
Nature Climate Change, 53, 70
Nazism, environmental and ecological aspects of, 74, 265n156
negative messaging, and climate change, 47; assumed by many climate experts to lead to fear and inaction, 50–56, 65–67, 248n8; as attention-getting, 52, 60; and availability heuristic, 63; combined with sense of efficacy, 63; and fear of adaptive rather than mitigating strategies, 67; and "finite pool of worry," 63, 260n98; greater effect upon conservatives than upon liberals, 63; less effect upon climate change activists than upon the public, 52, 63; limited or compromised evidence against, 64, 260n100; relationship to pro-environmental behavior, 52, 60, 61; and Rogers's "protection motivation theory," 60. *See also* "doomism"; fear
negativity bias, 61, 152, 201–2
négritude, 278n8
Neiman, Susan, *Evil in Modern Thought*, 156–57, 171
Neitzel, Jill, xv
neo-paganism, 73
Nerlich, Brigitte, 55
Nevi'im and *Ketuvim*, 116–17
New Dimensions 3, 190
new materialism, 177

Newsom, Gavin, xxi
NGOs, 88
Nicholson-Cole, Sophie, 51–52, 248n8
Niebuhr, Reinhold: "Christian realism," 175; liberal or conservative pessimism, 174–75; Serenity Prayer, 308n61
Nietzsche, Friedrich, 203; antipessimistic pessimism, 172, 173; on experience of reality, 24; "ignorance of the knowers," 27; perspectivism, 18, 20; radical empiricism, 278n8; and time, 171; transvaluation of all values, 30
nihilism, axiological and practical, 194
Nisbet, Erik, 58
nonadaptive cognition (scientific empiricism): and cognitive biases, 32–33; as compromised by adaptive truth-for-presuppositions and biases, 35; empirical reasoning and power to control, 34; and free will, 152; and idea of benevolent God, 152, 297n73; and ideological change, 30; and limits of reason, 36; and philosophical pessimism, 196–97, 198, 199, 207–8; and undermining of metaphysical grounds of cultural truth, 34
nonviolent civil resistance, effectiveness of, 79, 268n184
Nordgren, Anders, 148
Norem, Julie, 148, 149, 195
Norgaard, Kari, 71
Norlock, Kathryn, 207–8

Obama, Barack, 41, 94, 178
obesity, world-wide increase since 1980, 245n116
"objective knowledge," pursuit of, 24–25. *See also* experience, as epistemological grounding; nonadaptive cognition (scientific empiricism)
object-oriented ontology, 177

Ocasio-Cortez, Alexandria, 76
Offill, Jenny, 50
Ogunbode, Charles A., 62
Okoth, Kevin Ochieng, 182
Olaloku-Teriba, Annie, 181–82
Olson, David, literacy hypothesis, 284n60
O'Neill, Saffron, 51–52, 248n8
Ong, Walter, 127, 284n60
ontological binary, 27, 122, 178–86, 208, 283n47
ontological commitment, 19, 24, 29, 34, 125
ontology. *See* world, as ontological concept
optimism: compulsory ("cruel" or "toxic"), 146 (*See also* bright-siding); first appearance of word (*optimisme*), 151, 159; and Myth of Progress, 40–41, 46–47, 170; pervasiveness in modern industrial civilization, 67, 146, 170; philosophical, and belief that human reason can explain suffering and improve human condition, 170, 298n98; philosophical pessimism as modern philosophical response to, 4, 5–6; "strategic optimism," 148; techno-optimism, 300n40. *See also* optimism bias
optimism and pessimism: antithetical frames for making sense of suffering and time, 171, 195; and cause and effect problem, 149–50; as cognitive strategies, 148–49; as independent dimensions of personality, 148; and "nudging" through "choice architecture," 150; origins in Enlightenment, 151–59; in psychological discourse, 147–48; range of meanings, 147–48; with regard to climate change, 148; structures to support realistic assessment, 150. *See also* ethical pessimism; optimism bias; pessimism; philosophical pessimism
optimism bias, 3, 143–46; and cognitively "normal" people, 143; danger of among scientists, journalists, and political leaders, 145–46, 292n30; as evolutionary adaptation, 145; highly resistant to contrary evidence, 144; and ideological extremism, 145; negative consequences of, 145, 291n29; and obstruction of judgment, 15, 202, 210. *See also* Myth of Progress
Oreskes, Naomi, 8, 59; on myth of science as value-neutral, 35; "Why trust science?," 37–38
Ornish, Dean, 53–54
Ortega y Gasset, Jose, 172
Ostrom, Elinor, "Revisiting the Commons," 71
Oxford Encyclopedia of Climate Change Communication, 61

Pagden, Anthony, 97
pain and pleasure, asymmetry of, 201
paleo-ecologists, 120
Parenti, Christian: "the politics of the armed lifeboat," 79; *Tropic of Chaos*, 9
Paris Agreement, pledged emissions reductions, 9
partisan division on climate change, 57–59, 65, 253n51; based upon cultural identification, 58, 253n53; and communication problems, 58–59, 69; fostered by powerful political and financial networks, 58–59; increase in with education, 58; and trust in science, 253nn52
Patterson, Orlando, concept of "social death" or institutionalized marginality, 181, 186

Pearson, Roger, 298n93
Peele, Jordan, 186
Peirce, Charles, 238n25
perspectivism, 4, 18, 25. *See also* experience, as epistemological grounding
pessimism: childhood SES and pessimism in adulthood, 149–50; class-based variety of depressive realism, 149; versus despair, 195–96; first recorded use of word, 158, 159; intolerance for in American culture, 146, 170; and tragedy, 300n37
Pessimism Controversy, 194
pessimist thought: Buddhist pessimism, 195, 208; conservative pessimism, 175, 176; cultural pessimism, 173, 176; defensive pessimism, 148, 195, 208; ecopessimism, 165–69, 173, 208; existential pessimism, 208; liberal pessimism, 173–75; Marxian pessimism, 302n57; posthumanism, 176–77; queer pessimism, 302n57; stoic pessimism, 169, 208. *See also* Afropessimism; ethical pessimism; philosophical pessimism
Peterson, Nils, 52
Peterson, William, 299n17
philosophical optimism, belief that human reason can explain suffering and improve human condition, 170, 298n98
philosophical pessimism, 147; and acceptance of limitations in complex world, 195, 197–98; and belief in reality existing independently of human cognition and desire, 198; and collapse of theism, 300nn41–42, 301n43; and compassion, humility, and forgiveness, 209, 210; consciousness as source of suffering, 199, 202–4; critical arguments against, 193–95; and criticism of inherited theological authority, 176; as critique made by reason against reason, 171–72; and egalitarianism, 209, 210; emergence alongside and in reaction to Enlightenment progressivism, 4, 210; empirical claims grounded in, 198; and empirical insistence on facing reality, 196–97, 199, 207–8; as framework for existing in a world of suffering and uncertainty, 146; inclusion, 210; incompatibility with nihilism, 194; and inevitability of conflict, 199, 204–5; and inevitability of death, 198, 199, 207, 211; and lack of inherent, intelligible meaning of history, 171–72, 199, 205–6; and limitations of human agents, 199, 204, 211; modern philosophical response to Enlightenment optimism and the Myth of Progress, 4, 5–6; as non-systematic counter-tradition, 172–73; proto-modern and pre-modern roots, 169; and questioning of whether life is worth living, 195–96; recognition of time and effort required to change society, 7; and resilience, 208, 210; response to modern conceptions of time and suffering, 3–5, 7, 146, 170, 171, 176, 193–94; and responsibility, 207, 210; and sacrifice, 196; and suffering as existence, 199–200, 206, 207–8, 209, 211; and suicide, 195–96; and unpredictability of complex events that can never be anticipated, 199, 204; virtues of, 207–11; and Voltaire, 155–58, 159; wisdom of, 208–9. *See also* ethical pessimism; pessimism
Pine Tree Party, USA, 73
Pinker, Stephen, 41, 42; criticisms of work, 244n109; *Enlightenment Now*, 165, 167; progressivism, 165, 205

Pirenne, Henri, 121
planetary boundaries, 234n15
Plato, 5, 166
Plenty-Coups, Crow (Apsáalooke) chief, guiding of his people to a meaningful new existence beyond the end of their world, 129–32, 287n80
Plotinus, and *apophasis*, 219
polarization. *See* partisan division on climate change
political divisiveness and instability, 1, 9, 14; and COVID crisis, 69. *See also* partisan division on climate change
Pope, Alexander, *Essay on Man*, 154, 159, 297n77
Popper, Karl, 36, 174
Population Bombed! Exploding the Link Between Overpopulation and Climate Change (Pierre Desrochers and Joanna Szurmak), 165–68
positive messaging about climate change, and failure to motivate people, 55–56, 65, 250n32
positivist epistemology, 34
postcolonial theory, 97, 177, 278n8, 286n70
posthumanism, 173, 176–77; ideas of progress, 177; and philosophical pessimism, 176, 177
post-structuralist theory, 14, 176
postwar political order, breakdown of, 132
precautionary principle, 208
Prescott, Paul, 194, 195
pre-Socratics, 169
"The Professional-Managerial Class (PMC)" (Barbara Ehrenreich and John Ehrenreich), 238n32. *See also* intellectuals, and ideology
progress: versus change, 205; and cheap energy from fossil fuels, 43; and growth in global energy consumption, 43; human without the idea of, 97–98; intellectual commitment to idea of, 25–26; limits of, 5; posthumanism ideas of, 177; as type of apocalypse, 39–40; as unselfconscious myth, 45; and unstoppable climatic transformation, 6, 44. *See also* Myth of Progress; progressivism
progressivism: and belief in improving, controlling, and surpassing nature, 2; and difficulty of developing an ethical approach to ecological crisis, 101; liberal, 175; progressivist ethics, and justification of suffering in the present for sake of future, 197; racialist narrative of moral and ethical evolution from primitive to modern, 3, 97–98; and "save the planet," 92; and suffering, 101, 170–71, 193; techno-utopian capitalism, 103, 128; view of pessimism as unethical, 197
prosperity gospel, secularized, 101
Protestant Christianity, 75
Protopapadakis, Evangelos D., 264n142
Pryor, Richard, 186
Pyndick, Robert, on climate models, 233n13

queer pessimism, 302n57
Questioning Collapse (Patricia Ann McAnany and Norman Yoffee), 227n29
Quijano, Aníbal, 117
Quine, W. V., 24; "Duhem-Quine thesis," 36
Qui Parle, 179
Quiroga, Rodrigo Quian, 241n73

race: Afropessimism view of binary racial ontology as essential to

modern civilization, 177–78, 181, 187, 208; and climate justice, 70, 75, 105–6, 276n40; and ecofascism, 73; and ecopessimism, 168; history of "slow violence" against minority groups, 70; progressivist racialist narrative of moral and ethical evolution from primitive to modern, 97–98; racial concepts of difference as fundamental components of human identity and inequality, 97–98; racialized stereotype of "ecological Indian,," 122, 283n53; and roots of ecological thinking, 74, 265n156; Wynter's analysis of transition from binary between Christian and Heathen to division between Human and Savage, 27, 97, 283n47
radical environmentalism in US, 82, 270n205. See also ecofascism
radical hope, and possibility of successful transition to a meaningful existence in a new world, 128–33
Rassemblement National, France, 73
reactionary ecology, 74–75
reactionary nostalgia, 176
Reagan, Ronald, 94
reality: and epistemic diversity, 25; as experienced as a total phenomenological whole, 16–17, 237n11; language, and acquisition of meaning, 23–24, 239n36; and limits, 46; narrative, and simultaneous human inhabitation of multiple layers of, 20, 238n25; pessimist view of obligation to other beings and other realities, 198–99; philosophical pessimism and empirical insistence on facing, 196–97, 199, 207–8; "semiotic reality," 237n11; shaped by socially constructed and inherited narrative and conceptual structures, 15–22, 24. See also experience, as epistemological grounding

reason, human: Descartes and, 298n98; empirical, and power to control, 34; Enlightenment faith in, 101; Godwin and, 161; and improvement of the conditions of human existence, 25–26; Kant and, xii; and liberalism, 173–74; limits of, 10, 36, 95–96, 150, 198; motivated, 56, 70, 150; philosophical optimism and belief that human reason can explain suffering and improve human condition, 170, 298n98; philosophical (ethical) pessimism as critique made by reason against reason, 171–72, 198–99, 210, 215, 218; and suffering, 151–54, 156–59, 170–71. See also limits, of human reason

Redfield, Robert, xxi
Red Queen effect, 45
Reed, Adolph, 276n40
Reed, Ishmael, *Mumbo Jumbo*, 186
Renaissance humanism, 152
renewal, myth of, 122–28
Reser, Joseph P., 61, 257n78
Revkin, Andrew, 66
Richards, Imogen, et al., 74
Riofrancos, Thea, "security-sustainability nexus," 76
The Rise of Ecofascism (Sam Moore and Alex Roberts), 75
Roberts, Alex, 75
Roberts, David, 70–71
Robinson, Cedric, 121, 181, 278n8
Robinson, Kim Stanley, *Ministry for the Future*, 83, 270n209
Rode, Jacob, 55–56
Rogers, Ronald, "protection motivation theory," 60
Rolston III, Holmes, 272n12
Romantic pessimists, 120

Romm, Joseph, greatest myths about global warming communication, 60
Rorty, Richard, 172–73, 238n25
Rousseau, Jean-Jacques, 162, 172
ruins, xiv–xv
Russill, Chris, 54

Sageman, Marc, social identity theory of terrorism, 74, 80–81, 83
Sandler, Ronald, 207
Santayana, George, 172
Sartre, Jean-Paul, 16, 172, 278n8
satire: defined by Northrop Frye, 183; postmodern African American satire, 183–87, 278n8; and Wilderson's *Afropessimism*, 183–86
"save the planet," 92
scarcity, ethics of, 88, 101
Scheffler, Samuel, 196
Scheier, Michael F., 68
Schlesinger, Arthur, 174
Schmitt, Mark, *Spectres of Pessimism*, 302n57
Schneider, Stephen, 54
Schopenhauer, Arthur: on desire and suffering, 199; and ethics, 197; pessimism, 165, 172, 209; and suicide, 306n24; Wittgenstein and, 301n50; Zapffe and, 203
Schrader, Paul: *First Reformed*, 213–17, 309N2; *Taxi Driver*, 215–16; *Transcendental Style in Film*, 216–17, 219
Schuyler, George S., *Black No More*, 183
science: as absolutist and presentist, 37; and cognitive bias, 34–35; recognition of radical limits of scientific knowledge, 135, 287n87; socially constructed practice, 37; and trust, 37–38, 57, 253n52; and unpredictability of how ecological communities will respond to climate changes, 135
Scientific Revolution, consequences of, 34, 35–36

scientists and science communicators: and "double ethical bind" in communicating about climate change, 54–55; rejection of negative messaging, 50–56, 65–67, 248n8
"Scientists Explain What *New York Magazine* Article on 'The Uninhabitable Earth' Gets Wrong," 51
Scott, James, *Against the Grain*, 119, 282n35
Scranton, Roy: "Climate Change Is Not World War," 266n163; *Learning to Die in the Anthropocene*, 6, 8; "Memories of My Green Machine," 270n197; "Raising My Child in a Doomed World," 7–8, 275n27; *Total Mobilization*, 4; *We're Doomed. Now What?*, 6, 215
Scruton, Roger, 175
Self and Society 44, no.2 (2016), special issue devoted to depressive realism, 288n8
Seligman, Martin, theory of learned helplessness, 141, 288n2
Sepúlveda, Juan Ginés de, debate with Bartolomeo de Las Casas, 26–29, 239n46, 239n48; argument that natives were incapable of self-rule because of alleged crimes against humanity, 28; argument that Spain had the right to conquer and convert natives of New World, 26–27, 92; justification of conquest and enslavement in terms of social and moral progress, 28, 97; proponent of just war theory, 26, 28. *See also* Las Casas-Sepúlveda debate of 1550
Serenity Prayer, 208, 308n61
Servigne, Pablo, 8
Sexton, Jared, 169, 179; on response of Black scholars to Afropessimism, 181
Seyfried, Amanda, 213
Shakespeare, William, *Hamlet*, 184

Sharot, Tali, 144. *See also* optimism bias
Shepard, Paul, *Coming Home to the Pleistocene*, 119, 120
Shermer, Michael, "Why Malthus Is Still Wrong," 164–65
Shoah, 117
Shue, Henry, "Subsistence Emissions and Luxury Emissions," 107, 269n187
Sieranski, Kristen, 309n2
Sierra Club, 168
simplification: "great simplification," 46, 246n123; and societal collapse, 45–47
Singer, Pete, "Famine, Affluence, and Morality," 87–92, 190, 271n4
Singh, Nikhil, "Racial Formation in an Age of Permanent War," 277n42
Sitting Bull, killed supporting the Ghost Dance movement, 131–32, 287n80
skepticism: methodological scientific, 34, 39, 135; of optimists, 151; philosophical pessimism, 172, 176; of public, toward climate change, 56–58, 68; toward fictions through which we comprehend reality, 21; toward progressivist teleology, 7; toward science, 37–38, 57, 253n52. *See also* experience, as epistemological grounding; perspectivism
slave narrative, 180
Smil, Vaclav, 2
Smith, Adam, and Myth of Progress, 39
Smith, Nicholas, 50, 248n8
Snyder, Timothy, 76
"The Social Bases of Environmental Concern" (Kent D. Van Liere and Riley E. Dunlap), 253n53
Socrates, 136
Solnit, Rebecca, 69
Sophocles, 169; anti-natalism, 200, 202; *Oedipus Rex*, 206

"Spaceship Earth" metaphor, 72, 91, 92, 95
Spinoza, Benedict, "Omnis determinatio est negatio," 279n14
Spivak, Gayatri, 180, 278n8
Spratt, David, climatecodered.org, 41
Stanley, Samantha, 70
Staudenmaier, Peter, 74–75
Steffen, Will, 249n24
Stein, Gertrude, 173; "Composition as Explanation," 279n10
Stephan, Maria J., 79, 268n184
Stern, Paul, 53
Stevens, Raphaël, 8
Stevens, Wallace, 14, 16, 114, 238n20
Stevenson, Kathryn, 52
Stewart, Susan, *The Ruins Lesson*, xv
stochastic feedback, 127
Stoic pessimism, 169, 208
Stone, Elizabeth, 200
Stop Cop City protests, 84
"strategic optimism," 148
subsistence emissions, 107, 269n187
suffering: and capitalism, 78–79; and climate change, 102; and climate justice, 105–10; consciousness as source of, 202–4; desire as, 199; existence as, 199–200, 206, 207–8, 209, 211; import of, 137; and negativity bias, 201–2; as positive force, 200; and progressivism, 101, 170–71, 193; reason and, 151–59, 193, 296n66, 297n73; and time, and optimism and pessimism, 3–5, 7, 146, 147–50, 170, 171, 176, 193–95; as unavoidable fact of life, 199–200
Sunrise Movement, 8
Sunstein, Cass, 76
Supran, Geoffrey, 59
surrealism, 127, 173, 278n8
sustainability, 69, 108, 117, 119, 165, 166, 207
Sutton, Philip, 41
Szurmak, Joanna, 165–68

Tadiar, Neferti X. M., 179
Tainter, Joseph: "ancestor myths," 39; *The Collapse of Complex Societies*, xvii–xx, xxii, 9, 45; and predictable experience of societal simplification, 45–46; review of *Questioning Collapse*, 227n29
Táíwò, Olúfẹ́mi, 106, 108–9; and climate reparations, 107–8; "elite capture," 106
Taleb, Nassim, 208
Talmon, Jacob, 174
Tarrant, Brenton, 72, 74; "The Great Replacement," 73, 264n145
Tatum, Dillon, 174
Taxi Driver, 215–16
Taylor, Charles, "social imaginary," 237n11
technological utopianism (techno-optimism), 6, 41, 75, 103, 107, 166, 300n40
terrorism: ecofascism, 72–74; focuses attention on the violence and perpetrators rather than the grievance, 82; Global War on Terror, 76; leftist ecoterrorism, 79–86; social identity perspective theory of, 74, 80–81, 83
Thacker, Eugene, 147, 172, 176
theodicy, 151
Thiel, Peter, 78
Thomas, Greg, 182
Thomas, Hank Willis, 179
Thompson, Allen, 207
Thunberg, Greta, 73; and Fridays for the Future, 8
Tierney, John, 61
time: and apocalyptic thought, 17–18; and carbon emissions, 99–101; and conflict between addressing climate change and social justice, 100; "discount problem," 98; problem of, 98–101; and problem of justice, 102; and suffering, optimism, and pessimism, 3–5, 7, 146, 147–50, 170, 171, 176, 193–94
Todd, Zoe, 118
Tooze, Adam: *Crashed*, 9; on Malm, 79; *Shutdown*, 9
toxic pollution, 1
Traditional Ecological Knowledge, 117, 118, 121, 122–23
transcendentalism, American, 278n8
transcendental style, in film, 216–17
transhumanism, 177
trauma, cultural, 15, 68, 71, 92, 136
Trilling, Lionel, 174; *The Middle of the Journey*, 302n69
"The Trouble with Disparity" (Adolph Reed and Walter Benn Michaels), 276n40
Trump, Donald, xxi, 178
Tuck, Eve, 119, 127
Turchin, Peter, 45
Tvsersky, Amos, 32

Uexküll, Felix von, 237n11, 278n8
"uncivilization," 120, 123
"uncivilized writing," 120
United Nations Conference of the Parties, 107
United Nations Intergovernmental Panel on Climate Change (UN-IPCC), 1, 2–3, 60, 66, 107; shift of assessment reports to lower temperatures, 3, 235n17
unsustainable development, xxii, 1, 3
utilitarianism, and global climate change, 92, 104, 106–7, 272n12

Vaihinger, Hans: Kermode and, 18, 238n20; philosophy of "as if," 18
Valladolid. *See* Las Casas-Sepúlveda debate of 1550
values framing, 69–71, 102, 251n35

Van der Lugt, Mara, 147, 151–52, 171; and "fragile thinking," 209; and philosophical pessimism, 300n41
VanderMeer, Jeff, 242n75; *Annihilation*, 91–92
Van Liere, Kent D., 253n53
van Rensburg, Alexandra, 135
violence: defined, 80, 269n191; political violence, 79–85, 271n213; and terrorism, 82. *See also* Malm, Andreas, *How to Blow Up a Pipeline*
Virgil, *Aeneid*, 116
Virilio, Paul, and Nazi fortifications, xiv
virtue ethics, and climate change, 102, 207, 272n12, 308n57
Vitoria, Francisco de, 28
Voltaire (François-Marie Arouet): *Candide, or Optimism*, and lampoon of Leibniz, 154–55, 157–59, 170, 187; *Essai sur les Moeurs et l'Esprit des Nations*, 157; in Ferney, and *Candide*, 298n93; "The Lisbon Earthquake," 156; as meliorist, 158, 159, 170; and philosophical pessimism, 155–58, 159, 172
Von Hartmann, Agnes Marie Constanze, 172
Von Liebeg, Justus, 275n30
von Wright, Georg Henrik, and Myth of Progress, 38, 211

Wade, Ira, 155
Wagner, Henry R., 240n56
Wakefield, Gilbert, 162
Wallace, Alfred Russell, 166
Wallace-Wells, David, "The Uninhabitable Earth," 8, 47, 54, 77; attacks upon by journalists, climate activists, and scientists, 48–51, 54, 66; stated intention of article to scare people into action, 50
Warren, Calvin, 181

Warren, Louis, *God's Red Son*, 287n80
Warren, Rosie, 210
Watson, Richard, 104
Weather Underground, 81
Wengrow, David, 119, 120, 282n36
Western Roman Empire, collapse of, xviii
Whitehead, Alfred North, 218; definition of philosophy, 24; on experience, 24
white supremacy, 74
Whole Earth Catalog, 75
Why Civil Resistance Works (Erica Chenoweth and Maria J. Stephan), 79, 266n163
Whyte, Kyle, 118, 128
Whyte, Robert Orr, 166
wicked (or super-wicked) problems, 230n45; and complexity of climate change, 5, 59
wilderness conservation, anti-immigration trends in, 74
Wilderson, Frank, 97, 219; *Afropessimism*, 183–87; and call for end of World as we know it, 188; critique of by Black scholars, 181–83; and irresolvable contradictions within contemporary American culture, 183, 187–88; "The Position of the Unthought" (interview with Hartman), 179; work in tradition of postmodern African American satire, 183–87
Willer, Robb, 64, 69
Wilson, E. O., consilience, 30, 240n61
Winant, Howard, 98
wisdom: and anti-natalism, 200–201; and dream of Plenty-Coups, 130; and fear, 47; Leibniz and, 153; literatures of, 169; and multiple contradictory truths, 217, 218–19; and pessimism, 5, 146, 169, 192–93, 208; rarity of, 150; *sapientia* as, 242n78

wisdom literatures, 169
witch trials, seventeenth century, and racial colonial violence, 121, 282n45
Witte, Kim: extended parallel process model (EPPM), 63; and fear appeals, 60, 61
Wittgenstein, Ludwig, 39, 172, 173, 301n50
Wolfe, Cary, 176, 177
Wolsko, Christopher, 69
Woodhouse, Keith, *The Ecocentrists*, 270n205
Woolf, Virginia, "On Being Ill," 200
world, as ontological concept, 16, 23, 68, 93, 112, 113–16, 124–27, 130, 278n8
Wrangham, Richard, human self-domestication hypothesis, 284n60
Wynter, Sylvia: analysis of the Las Casas-Sepúlveda debate of 1550, 26–27, 29–31, 240n56; and Black radical tradition critique of Western civilization, 181, 278n8; and "ignorance of the knowers," 27; "new science," 101; and nonadaptive mode of cognition, 34; and ontological divisions structuring social relations, 122; and relationship between empirical science and ideological change, 30; and transition from binary between Christian and Heathen to division between Human and Savage, 27, 97, 283n47; and transition from late Christian to early modern discursive formations, 30–31; "Unsettling the Coloniality of Being/Power/Truth/Freedom," 26–31

Xue, Wen, 62

Yale Program in Climate Change Communications, 62
Yang, K. Wayne, 119, 127
Yoffee, Norman, 227n29

Zapffe, Peter Wessel, 168, 172; "The Last Messiah," 203–4
Zen Buddhism, and *kenshō* or *satori*, 219
Zerubavel, Eviatar, "social mindscape," 237n11
Zerzan, John, 119, 120
Zimmerman, Michael, 85
Zuboff, Shoshana, surveillance capitalism, 30